Kew Chromosome Conference III

KEW CHROMOSOME CONFERENCE

ORGANISING COMMITTEE

Professor K. Jones, President
Dr P.E. Brandham, Secretary/Treasurer
Miss M.A.T. Johnson
Dr A.Y. Kenton

PROGRAMME COMMITTEE

Professor M.D. Bennett
Dr P.E. Brandham
Dr G.M. Hewitt
Miss M.A.T. Johnson
Dr G.H. Jones
Professor K. Jones
Dr A.Y. Kenton
Dr J.S. Parker
Professor H. Rees

This was the third Chromosome Conference held at Kew. The first was held in 1976 and the proceedings appeared under the title Current Chromosome Research. *The second was held in 1982 and the proceedings appeared as* Kew Chromosome Conference II.

KEW CHROMOSOME CONFERENCE III

Proceedings of the Third Chromosome Conference held in the Jodrell Laboratory, Royal Botanic Gardens, Kew, England, 1–4 September 1987

Editor
P.E. BRANDHAM   *Royal Botanic Gardens, Kew*
Richmond, Surrey TW9 3DS, UK

London
Her Majesty's Stationery Office, 1988

First published in 1988

ISBN 0 11 250036 6

HMSO publications are available from:

HMSO Publications Centre
(Mail and telephone orders only)
PO Box 276, London, SW8 5DT
Telephone orders 01-622 3316
General enquiries 01-211 5656
(queuing system in operation for both numbers)

HMSO Bookshops
49 High Holborn, London, WC1V 6HB   01-211 5656 (Counter service only)
258 Broad Street, Birmingham, B1 2HE   021-643 3740
Southey House, 33 Wine Street, Bristol, BS1 2BQ   (0272) 264306
9-21 Princess Street, Manchester, M60 8AS   061-834 7201
80 Chichester Street, Belfast, BT1 4JY   (0232) 238451
71 Lothian Road, Edinburgh, EH3 9AZ   031-228 4181

HMSO's Accredited Agents
(see Yellow Pages)

and through good booksellers

Dd. 289361 10/88 C8
Printed in the United Kingdom for
Her Majesty's Stationery Office

Foreword

As in the past, because of limited available accommodation, the third Kew Chromosome Conference could not be advertised as an open meeting. Nevertheless it was attended by nearly 200 cytologists most of whom made formal contributions to the lecture or poster sessions. Their presence, dedication and enthusiasm indicates the importance attached to the examination of chromosomes and chromosome systems as an essential means of discovering their significance for the control of heredity and evolution. We may recall that it is now almost a quarter of a century since C.D. Darlington reminded his audience that the chromosomes as we see them are by no means the same things that others might imagine them to be and our continuing discoveries are themselves a vivid testimony to this.

Of course we are fully aware that in the period following his address there was a major, even catastrophic, reduction in the number of practising cytologists. In the face of the exciting fundamental discoveries at the molecular level the simple technique of microscope observation seemed to possess only superficial and residual value. The term classical cytology was used in a derogatory manner to describe an archaic subject of little contemporary relevance. Naturally with the reduction in the number of schools of cytology came the drastic loss of teachers and within a short space of time knowledge of even the rudiments of mitosis, and more particularly meiosis, was lost. Generations of students have been confused and misinformed by grossly inaccurate descriptions of these events in books written by otherwise eminent authors who should have known better. It is up to us now to point out the errors of the past and direct our students to those texts which we know to be accurate.

Cytology as we now practise it has evolved to take within its awareness the new knowledge of the chemistry of chromosomes and the new techniques which have sprung from it. Both the quantity and to a lesser extent the quality of DNA can now be measured by simple means. There is in consequence a better appreciation of the fundamentals of the C-value paradox and of the ways in which DNA changes occur during the differentiation and evolution of populations, and we are beginning to make progress in determining their significance to the organism. Electron microscopes have allowed chromosomes to be observed at times when they were unavailable to us. Their relative positions in interphase and their pairing behaviour at leptotene and pachytene give new insight with regard to development, producing quite new views on the significance of chromosome form and relative homologies. Direct observation even allows speed of replication and size of replicons to be calculated, adding yet further information on the impact of genome size.

In addition to exploring chromosomes with the aid of new techniques much remains to be discovered by detailed, thorough examination of individuals, populations and hybrids by standard methods, for it would be a mistake to believe that all was discovered during the first 50 years of cytology. Much new and exciting information is being obtained as new groups of organisms and new floras and faunas are being explored. As we do this, taking full advantage of the continuing discoveries of our molecular colleagues, we should be careful not to discard too rapidly the principles of chromosome mechanics established by illustrious predecessors. It is easy to reinterpret chromosome characteristics in the light of what we know about the undisciplined behaviour of some genes or other nucleotide sequences but we must be mindful of supramolecular controls of pairing, chiasma frequency and segregation and of the influence of sheer homologous affinities.

Cytology as we know it from observation and experiment remains the essential means of determining the ways in which chromosomes control their own destinies and those of the organisms which contain them.

Keith Jones

Preface

This is the third Chromosome Conference to be held at Kew, and each has been larger and more successful than the previous one. The proceedings of the first Conference, held in 1976, were published as "Current Chromosome Research" by Elsevier, and those of the second, "Kew Chromosome Conference II" by George Allen and Unwin in 1983. Between these first two volumes there was a great advance in our knowledge of the biology of chromosomes, and this volume clearly shows by the extent of its subject-matter that chromosome science is progressing further by leaps and bounds.

When the Conference was first publicised the organisers requested titles of papers to be submitted, and from an overwhelming response of over 100 offers these were reduced to a number that could conveniently fill four days without the necessity of running concurrent sessions, with the inevitable disruption that they cause. The result was a meeting in which current findings related to light- and electron microscopy and biochemistry of chromosomes could be discussed in a convivial environment.

The thirty-three articles in this volume represent the great majority of the papers given at the Conference by speakers from all over the world. They are grouped into four sections, the first of which is concerned primarily with structure; of chromosomes, free centromeres, the chloroplast genome and of the meiotic cell and nucleus.

The second deals with variation; natural variation of chromosome number and morphology within and between taxa, with some reports on what happens when these variants meet in the wild or are experimentally hybridised; artificial variation induced by irradiation or resulting from tissue culture; numerical variation in which one of two genomes is eliminated from a wide hybrid; and variation of DNA amount, the quantitative measurement of which can sometimes present technical problems.

The third section, on chromosome disposition, demonstrates that genomes occupy distinct domains within the interphase, mitotic and meiotic nucleus, both within species and in hybrids.

The final section addresses different aspects of chromosome pairing and crossing-over during the early stages of meiosis, the regulation of the number of chiasmata in diploids and polyploids and the structure, function and mutability of the synaptonemal complex.

The articles are presented here in a convenient and logical order, with those on similar subjects being placed together, and no hierarchy of merit is implied from that order. All are at the cutting edge of chromosome research.

P.E. Brandham

Contents

Conference Members

Abberton, M.T., Dept. of Cell and Structural Biology, Williamson Building, The University, Manchester M13 9PL

Ahmad, S.D., Dept. of Agricultural Botany, University College of Wales, Aberystwyth, Dyfed SY23 3DD

Albers, F., Botanisches Institut, Schlossgarten 3, D-4400 Munster, Federal Republic of Germany

Albini, S.M., Dept. of Genetics, University of Birmingham, P.O. Box 363, Birmingham B15 2TT

Arana, P., Depto. de Genetica, Facultad de Biologia, Universidad Complutense, 28040 Madrid, Spain

Arends, J.C., Dept. of Plant Taxonomy, Agricultural University, 37 Generaal Foulkesweg, P.O. Box 8010, 6700 ED Wageningen, The Netherlands

Armstrong, K., Cytogenetics Section, Plant Research Centre, Ottawa, Ontario, Canada

Atkinson, M.D., Plant Breeding Institute, Maris Lane, Trumpington, Cambridge CB2 2LQ

Badaracco, G., Dip. Genetica & Biologia dei Microrganismi, Universita Studi Milano, Via Festa del Perdono 7, 20122 Milano, Italy

Bailey, J.P., Botany Dept., The University, Leicester

Baratelli, L., Dept. Genetica e Biologia Microrganismi, Via Celoria 26, 20133 Milano, Italy

Barigozzi, C., Dept. Genetica e Biologia Microrganismi, Via Celoria 26, 20133 Milano, Italy

Bebeli, P., Biochemistry Dept., Rothamsted Experimental Station, Harpenden, Herts AL5 2JQ

Bennett, M.D., Jodrell Laboratory, Royal Botanic Gardens, Kew, Richmond, Surrey TW9 3DS (Address from 1.10.87)

Bennett, S.T., Botany School, South Parks Road, Oxford, OX1 3RA

Blackman, R.L., Dept. of Entomology, British Museum (Natural History), Cromwell Road, London SW7 5BD

Bojko, M., Dept. of Biological Sciences, Stanford University, Stanford, California 94305, USA

Bosshard, F., Institut de Zoologie et d'Ecologie Animale, Bât de Biologie, Faculté des Sciences, Université de Lausanne, 1015 Lausanne, Switzerland

Bougourd, S., Dept. of Biology, The University, York YO1 5DD

Boyle, P., Botany Dept., University College, Belfield, Dublin 4, Eire

Brace, J., Dept. of Biological Sciences, Manchester Polytechnic, Oxford Road, Manchester

Brandham, P.E., Jodrell Laboratory, Royal Botanic Gardens, Kew, Richmond, Surrey TW9 3DS

Breckon, G., MRC Radiobiology Unit, Chilton, Didcot, Oxfordshire OX11 ORD

Callow, R., Dept. of Cell and Structural Biology, Williamson Building, The University, Manchester, M13 9PL

Cano, M.I., Depto. de Genetica, Facultad de Biologia, Universidad Complutense, 28040 Madrid, Spain

de Caritat, A.-K., Université Louvain, Unité d'Ecologie et de Biogeographie, 4–5 Place Croix du Sud, 1348 Louvain la Neuve, Belgium

Carpenter, A.T.C., Dept. of Biology, B-022, University of California San Diego, La Jolla, California 92093, USA

Claire, H., Jodrell Laboratory, Royal Botanic Gardens, Kew, Richmond, Surrey TW9 3DS

Clark, M., Biochemistry Dept., Rothamsted Experimental Station, Harpenden, Herts AL5 2JQ

Colasante, M., Dip. di Biologia Vegetale, Universita 'La Sapienza', P. Le Aldo Moro 5, 00100 Roma, Italy

Coleman, J., Jodrell Laboratory, Royal Botanic Gardens, Kew, Richmond, Surrey TW9 3DS

Croft, J.A., 59 Rednal Road, Kings Norton, Birmingham B38 8DT

Davies, A., Dept. of Agricultural Botany, University College of Wales, Penglais, Aberystwyth, Dyfed SY23 3DD

Dickinson, H.G. Dept. of Botany, Plant Science Laboratories, University of Reading, Whiteknights, Reading RG6 2AS

Dyer, T.A., Plant Breeding Institute, Maris Lane, Trumpington, Cambridge CB2 2LQ

Eizenga, G.C., USDA-ARS, Agronomy Dept., University of Kentucky, Lexington, Kentucky 40546–0091, USA

Ellis, J.R., Botany Dept. University College, Gower Street, London WC1

Endo, T., Nara University, 1230 Horaicho, Nara, Japan

Evans, E.P., Sir William Dunn School of Pathology, University of Oxford, South Parks Road, Oxford OX1 3RE

Evans, G.M., Dept. of Agricultural Botany, University College of Wales, Penglais, Aberystwyth, Dyfed SY23 3DD

Falistocco, E., Centro Miglioramento Genetico, Piante Foraggere-CNR, Borgo XX Giugno, 74–06100 Perugia, Italy

Febles, R., Jardin Botanico 'Viera y Clavijo', Aptdo. 14, Tafira Alta, Las Palmas de Gran Canaria, Canary Islands, Spain

Feldman, M., Dept. of Plant Genetics, Weizmann Institute of Science, Rehovot 76100, Israel

Fernandez-Peralta, A.M., Depto. Genetica C-XV, Facultad de Ciencias, Universidad Autonoma de Madrid, 28049, Madrid, Spain

Ferrer, E., Depto. Biologia Celular y Genetica, Universidad Alcala de Henares, Apdo. 20, Alcala de Henares, Madrid, Spain

Ferris, C., Dept. of Cell and Structural Biology, Williamson Bldg., The University, Manchester M13 9PL

Ferwerda, M.A., Dept. of Genetics, Biological Centre, University of Groningen, Kerklaan 30, 9751 NN Haren, The Netherlands

Finch, R., 68 Holbrook Road, Cambridge CB1 4ST

Fletcher, H.L., Dept. of Biology, David Keir Bldg., Queens University, Belfast BT7 1NN

Ford, C., 156 Oxford Road, Abingdon, Oxfordshire OX14 1AF

Francis, H.A., Dept. of Agricultural Botany, University College of Wales, Aberystwyth SY23 3DD

Fredga, K., Dept. of Genetics, Box 7003, S750 07 Uppsala, Sweden

Garagna, S., Dip. di Biologia Animale, Piazza Botta 10, 27100 Pavia, Italy

Garbari, F., Istituto Botanico, Via Lucia Ghini 5, 56100 Pisa, Italy

Garcia de la Vega, C., Depto. de Biologia (Genetica), Facultad de Ciencias C–XV, Universidad Autonoma de Madrid, 28049 Madrid, Spain

Gibby, M., Dept. of Botany, British Museum (Natural History), Cromwell Road, London SW7 5BD

Giddings, G., Dept. of Agricultural Botany, University College of Wales, Penglais, Aberystwyth, Dyfed SY23 3DD

Giraldez, R., Depto. de Biologia Funcional, Universidad de Oviedo, 33071 Oviedo, Spain

Godward, M.B.E., Dept. of Biological Sciences, Queen Mary College, Mile End Road, London E1 4NS

Goicoechea, P.G., Depto. de Biologia Funcional, Universidad de Oviedo, 33071 Oviedo, Spain

Gonzalez-Aguilera, J.J., Depto. de Genetica C–XV, Facultad de Ciencias, Universidad Autonoma de Madrid, 28049 Madrid, Spain

Gosalvez, J., Depto. de Genetica C–XV, Facultad de Ciencias, Universidad Autonoma de Madrid, 28049 Madrid, Spain

Grant, C.J., Botany Dept., The University, Bristol BS8 1UG

Grant, W.F., Genetics Laboratory, P.O. Box 4000, Macdonald Campus of McGill University, Ste. Anne de Bellevue, Quebec H9X 1C0, Canada

Greilhuber, J., Institute of Botany, Rennweg 14, A–1030 Wien, Austria

Guerra, M., Depto. Biologia General, Universidad Federal Pernambuco, Recife PE, Brazil

Hagemann, S., Millergasse 15118, 1060 Wien, Austria

Hamey, Y., Dept. of Cell and Structural Biology, Williamson Building, The University, Manchester M13 9PL

Harold, B., Dept. of Applied Biology, Brunel University, Uxbridge, Middlesex UB8 3PH

Hassan, L., Dept. of Agricultural Botany, University College of Wales, Penglais, Aberystwyth SY23 3DD

Hartman, T.P.V., Dept. of Cell and Structural Biology, Williamson Building, The University, Manchester M13 9PL

Heneen, W., Dept. of Crop Genetics and Breeding, Swedish University of Agricultural Sciences, S 26800 Svalov, Sweden

Henriques-Gil, N., Colegio Universitario San Pablo C.E.U., Laboratorio de Genetica, Monteprincipe, Boadilla del Monte, 28660 Madrid, Spain

Heslop-Harrison, J.S., Plant Breeding Institute, Maris Lane, Trumpington, Cambridge CB2 2LQ

Hewitt, G.M., School of Biological Sciences, University of E. Anglia, Norwich NR4 7TJ

Holm, P.B., Dept. of Physiology, Carlsberg Laboratory, Gamle Carlsbergvej 10, DK 2500 Copenhagen Valby, Denmark

Hubner, R., Laboratoire de Cytogenetique, Rue de Bruxelles 61, FUNDP — Namur, B–5000, Belgium

Ilio de Dominicis, R., Dip. di Biologia Vegetale, Universita "La Sapienza", P. le Aldo Moro 5, 00100 Roma, Italy

James, S., Botany Dept., University of Western Australia, Nedlands, W.A. 6009, Australia

Jenkins, G., Dept. of Agricultural Botany, University College of Wales, Penglais Aberystwyth, Dyfed SY23 3DD

Johnson, D.S., Medizinische Universitat zu Lübeck, Institut fur Biologie, Ratzeburger Allee 160, D 2400 Lübeck, Federal Republic of Germany

Johnson, M.A.T., Jodrell Laboratory, Royal Botanic Gardens, Kew, Richmond, Surrey TW9 3DS

Jones, G.H., Dept. of Genetics, University of Birmingham, P.O. Box 363, Birmingham B15 2TT

Jones, K., Jodrell Laboratory, Royal Botanic Gardens, Kew, Richmond, Surrey TW9 3DS

Jones, M., 1 Castle Street, Aberystwyth, Dyfed

Jones, R.N., Dept. of Agricultural Botany, University College of Wales, Penglais, Aberystwyth, SY23 3DD

de Jong, J.H., Dept. of Genetics, Agricultural University, Generaal Foulkesweg 53, NL 6703 BM Wageningen, The Netherlands

Karp, A., Biochemistry Dept., Rothamsted Experimental Station, Harpenden, Herts. AL5 2JQ

Kaushal, P.S., Dept. of Botany, Panjab University, Chandigarh 160014, India

Kenton, A.Y., Jodrell Laboratory, Royal Botanic Gardens, Kew, Richmond, Surrey TW9 3DS

Khalfallah, N., C.N.R.S., Laboratoire G.P.D.P., 91190 Gif-sur-Yvette, France

King, I.P., Plant Breeding Institute, Maris Lane, Trumpington, Cambridge, CB2 2LQ

King, M., Museum of Arts and Sciences of the Northern Territory, P.O. Box 4646, Darwin, Australia

Lacadena, J-R., Depto. de Genetica, Facultad de Ciencias Biologicas, Universidad Complutense, 28040 Madrid, Spain

Ladizinsky, G., Faculty of Agriculture, Hebrew University, P.O. Box 12, Rehovot 76100, Israel

Laird, C.D., Dept. of Zoology NJ 15, University of Washington, Seattle, WA 98195, USA

Lange, W., Foundation for Agricultural Plant Breeding, P.O. Box 117, 6700 AC Wageningen, The Netherlands

Laurie, D., Plant Breeding Institute, Maris Lane, Trumpington, Cambridge CB2 2LQ

Leggett, J.M., Welsh Plant Breeding Station, Plas Gogerddan, Aberystwyth, Dyfed

Leitch, A.R., Plant Breeding Institute, Maris Lane, Trumpington, Cambridge CB2 2LQ

Lim., K.Y., 12 Tower House, Candover Street, London W1

Linde-Laursen, I., Risø National Laboratory, DK 4000 Roskilde, Denmark

Loidl, J., Institut fur Botanik der Universitat, Rennweg 14, A 1030 Wien, Austria

Lopez-Fernandez, C., Depto. de Biologia (Genetica), Facultad de Ciencias C–XV, Universidad Autonoma de Madrid, 28049 Madrid, Spain

Mahadevaiah, S.K., Dept. of Genetics and Biometry, Wolfson House, 4 Stephenson Way, London NW1 2HE

Malheiro, M.I., Rua dos Acores 265 1°, ESQ 4200 Porto, Portugal

Martinez, A., CEFAPRIN, Serrano 665, 1414 Buenos Aires, Argentina

Mejias Gimeno, J.A., Depto. de Botanica, Facultad de Biologia, Aptdo 1.095, 41080 Sevilla, Spain

Miller, T.E., Plant Breeding Institute, Maris Lane, Trumpington, Cambridge CB2 2LQ

Mittwoch, U., Dept. of Genetics and Biometry, Wolfson House, 4 Stephenson Way, London NW1 2HE

Moens, P.B., Dept. of Biology, York University, 4700 Keele Street, Downsview, Ontario M3J 1P3, Canada

Murray, B.G., Dept. of Botany, Auckland University, Private Bag, Auckland, New Zealand

Naranjo, C.A., Depto. de Botanica Agricola, INTA, Castelar, 1724 Buenos Aires, Argentina

Naranjo, T., Depto. de Biologia Funcional, Universidad de Oviedo, 33071 Oviedo, Spain

Narayan, R.K.J., Dept. of Agricultural Botany, University College of Wales, Aberystwyth, Dyfed SY23 3DD

Neelam, A., Dept. of Agricultural Botany, University College of Wales, Aberystwyth, Dyfed SY23 3DD

Newton, M.E., Dept. of Cell and Structural Biology, Williamson Building, The University, Manchester M13 9PL

O'Donoughue, L., Plant Breeding Institute, Maris Lane, Trumpington, Cambridge CB2 2LQ

Ooms, G., Dept. of Biochemistry, Rothamsted Experimental Station, Harpenden, Hertfordshire AL5 2JQ

Östergren, G., Dept. of Genetics, Box 7003, S750–07 Uppsala, Sweden

Oud, J.L., Dept. of Electron Microscopy and Molecular Cytology, University of Amsterdam, Plantage Muidergracht 14, NL1018 TV Amsterdam, The Netherlands

Ozhatay, N., Eczacilik Fakultesi, Farmasotik Botanik Kursusu, Universite Istanbul, Turkey

Palomino, G., Lab. de Citogenetica, Jardin Botanico, Instituto de Biologia, UNAM, C.P. 04510, A.P. 70–614 Mexico DF, Mexico

Papes, D., Dept. of Botany, Faculty of Science, University of Zagreb, Rooseveltov trg 6, 41001 Zagreb, p.p. 933, Yugoslavia

Parker, J.S., School of Biological Sciences, Queen Mary College, Mile End Road, London E1 4NS

Parkin, C.A., Regional Cytogenetics Unit, Birmingham Maternity Hospital, Edgbaston, Birmingham B15 2TG

Petitpierre, E., Lab. de Genetica, Depto. de Biologia, Facultad de Ciencias, 07071 Palma de Mallorca, Spain

Poggio, L., Dept. de Ciencias Biologicas, F.C.E.y N., UBA, 1428 Buenos Aires, Argentina

Popova, M.T., V. Kolarov Higher Agricultural Institute, Dept. of Botany, Mendeleev Str. 12, Plovdiv 4000, Bulgaria

Pijnacker, L.P., Dept. of Genetics, Biological Centre, University of Groningen, Kerklaan 30, 9751 Haren, The Netherlands

Puertas, M.J., Depto. de Genetica, Facultad de Biologia, Universidad Complutense, 28040 Madrid, Spain

Queiroz, A.T., Depto. de Botanica, Inst. Superior de Agronomia, Tapada da Ajuda, 1399 Lisboa Codex, Portugal

Rafferty, J., Dept. of Biology (Genetics), Queens University, Belfast BT7 1NN

Ramsay, G., Dept. of Agricultural Botany, University of Reading, Whiteknights, Reading RG6 2AS

Rawlins, D., John Innes Institute, Colney Lane, Norwich NR4 7UH

Redi, C.A., Dip. di Biologia Animale, Piazza Botta 10, 27100 Pavia, Italy

Rees, H., Dept. of Agricultural Botany, University College of Wales, Penglais, Aberystwyth, Dyfed SY23 3DD

Rickards, G.K., Botany Dept., Victoria University of Wellington, Private Bag, Wellington, New Zealand

Riley, R., 16 Gog Magog Way, Stapleford, Cambridge CB2 5BQ

Roca, A., Depto. de Biologia Funcional, Universidad de Oviedo, 33071 Oviedo, Spain

Ruiz Rejon, M., Depto. de Genetica, Facultad de Ciencias, Universidad de Granada, 18071 Granada, Spain

Rumpler, Y., Faculté de Médecine, Institut d'Embryologie, 11 Rue Humann, F67085 Strasbourg Cedex, France

Santos, J.L., Depto. de Genetica, Facultad de Biologia, Universidad Complutense, 28040 Madrid, Spain

Scanlon, M.J., Dept. of Agricultural Botany, University College of Wales, Penglais, Aberystwyth, Dyfed SY23 3DD

Schlarbaum, S.E., Dept. of Forestry, 240 Ellington Hall, University of Tennessee, Knoxville, TN 37916 USA

Schwarzacher-Robinson, T., Experimental Pathology Group, Los Alamos National Laboratory, NS : M888, Los Alamos, N.M.87544, USA

Searle, J.B., Dept. of Zoology, University of Oxford, South Parks Road, Oxford OX1 3PS

See, C.G., School of Biological Sciences, Queen Mary College, Mile End Road, London E1 4NS

Selldén, K., Karolinska Institutet, Dept. of Molecular Genetics, Box 60400, S 10401 Stockholm, Sweden

Sen, S., School of Biological Sciences, Queen Mary College, Mile End Road, London E1 4NS

Sentis, C., Depto. de Biologia, Universidad Autonoma de Madrid, 28049 Madrid, Spain

Setterfield, L.A., 30 Doyle Gardens, Willesden, London NW10

Shaw, D., Dept. of Population Genetics, R.S.B.S., Australian National University, Canberra, A.C.T. 2601, Australia

Shewry, P.R., Rothamsted Experimental Station, Harpenden, Hertfordshire AL5 2JQ

Smets, S., Unité d'Ecologie et de Biogeographie, Université Louvain, 4–5 Place Croix du Sud, 1348 Louvain la Neuve, Belgium

Smith, J.B., Plant Breeding Institute, Maris Lane, Trumpington, Cambridge CB2 2LQ

Snape, J., Plant Breeding Institute, Maris Lane, Trumpington, Cambridge CB2 2LQ

Soliman, M., Botany Dept., The University, Sheffield S10 2TN

Southern, D.I., Dept. of Cell and Structural Biology, Williamson Building, The University, Manchester M13 9PL

Speed, R.M., MRC Clinical and Population Cytogenetics Unit, Western General Hospital, Crewe Road, Edinburgh EH4 2XU

Spyropoulos, B., Dept. of Biology, York University, 4700 Keele Street, Downsview, Ontario M3J 1P3, Canada

Stack, S., Dept. of Biology, Colorado State University, Fort Collins, Colorado 80523, USA

Sree Ramulu, K., Research Institute ITAL, P.O. Box 48, 6700AA Wageningen, The Netherlands

Sumner, A.T., MRC Cytogenetics Unit, Western General Hospital, Crewe Road, Edinburgh EH4 2XU

Sutcliffe, M., 58 Portsea Hall, Portsea Place, London W2 2BY

Sybenga, J., Dept. of Genetics, Agricultural University, Generaal Foulkesweg 53, NL 6703 BM Wageningen, The Netherlands

Taylor, S., School of Biological Sciences, Queen Mary College, Mile End Road, London E1 4NS

Tease, C., MRC Radiobiology Unit, Chilton, Didcot, Oxfordshire OX11 0RD

Tornadore, N., Dipto. di Biologia, Sesione Geobotanica, Via Orto Botanico 15, 35100 Padova, Italy

Traut, W., Institut fur Biologie, Medizinische Universitat zu Lübeck, Ratzeburger Allee 160, D2400 Lübeck, Federal Republic of Germany

Turner, B.M., Anatomy Dept., University of Birmingham Medical School, Vincent Drive, Birmingham B15 2TJ

Valdes, B., Depto. de Botanica, Facultad de Biologia, Aptdo 1.095, 41080 Sevilla, Spain

Van de Vooren, J.G., Dept. of Plant Taxonomy, Agricultural University, 37 Generaal Foulkesweg, P.O. Box 8010, 6700 ED Wageningen, The Netherlands

Vincent, C.A., Plant Breeding Institute, Maris Lane, Trumpington, Cambridge CB2 2LQ

Vincent, J.E., Delph House, Deerpark Nurseries, Towneley Holmes, Burnley, Lancashire

Volleth, M., Dept. of Human Genetics, Schwabachanlage 10, D8520 Erlangen, Federal Republic of Germany

Volobouev, V., CNRS — UA 620, Structure et Mutagenèse Chromosomiques, Institut Curie, Section de Biologie, Pavillon Regaud, 26 Rue d'Ulm, 75231 Paris Cedex 05, France

Wahrman, J., Dept. of Genetics, Hebrew University, Jerusalem 91904, Israel

Wall, W.J., Dept. of Cytogenetics, Queen Elizabeth Hospital for Children, Hackney Road, London E2 8PS

Wallace, A.J., Dept. of Cell and Structural Biology, Williamson Building, The University, Manchester M13 9PL

Wallace, B.M.N., Dept. of Genetics, University of Birmingham, P.O. Box 363, Birmingham B15 2TT

Wang, R.R.C., USDA-ARS, Utah State University UMC 63, Logan, UT 84322, USA

Watanabe, K., Biological Institute, Faculty of General Education, Kobe University, Kobe 657, Japan

White, J.A., Dept. of Agricultural Botany, University College of Wales, Penglais, Aberystwyth, Dyfed SY23 3DD

Whitehorn, J., Dept. of Genetics, University of Birmingham, P.O. Box 363, Birmingham B15 2TT

Wolf, K., Medizinische Universitat zu Lübeck, Institut fur Biologie, Ratzeburger Allee 160, D2400 Lübeck, Federal Republic of Germany

Wilby, A., School of Biological Sciences, Queen Mary College, Mile End Road, London E1 4NS

Wischmann, B., Dept. of Physiology, Gamle Carlsbergvej 10, DK 2500 Copenhagen Valby, Denmark

Wojcik, J., Mammals Research Institute, Polish Academy of Sciences, 17–2030 Biakowieza, Poland

Xingzhi Wang, Dept. of Physiology, Gamle Carlsbergvej 10, DK 2500 Copenhagen Valby, Denmark

Xu Jie, Plant Breeding Institute, Maris Lane, Trumpington, Cambridge CB2 2LQ

Zickler, D., Laboratoire de Genetique, Bât. 400, Université de Paris Sud, 91405 Orsay Cedex, France

STRUCTURE OF MITOTIC AND MEIOTIC NUCLEI AND ASSOCIATED ORGANELLES

Centromere-like elements devoid of chromosome arms in *Megaselia scalaris* (Diptera)

K.W. Wolf, D.S. Johnson and W. Traut

Institut für Biologie der Medizinischen Universität zu Lübeck, Ratzeburger Allee 160, D-2400 Lübeck 1, Federal Republic of Germany

Compared with the numerous cell biological studies in *Drosophila* species other Diptera have received limited attention. One of these under-represented animals is the Phorid fly *Megaselia scalaris* (formerly *Aphiochaeta xanthina*). This is surprising, since it has an unusual sex determining system which was termed 'alternating sex determination' (for review see Mainx 1964).

M. scalaris has three chromosome pairs, none of which is heteromorphic. According to Mainx (1964) sex is determined by an epistatic male "realiser" located terminally on one chromosome. This factor moves to the other two non-homologous chromosomes. Although not proved rigorously, Mainx (loc. cit.) considered a translocation process to be involved in the exchange. Recently, the process was interpreted as a transposition (Green 1980).

These findings and ideas revived interest in *M. scalaris*, and we have examined the karyotype using both light and electron microscopy.

Material and methods

The animals were raised on a modified *Drosophila* medium supplemented with minced liver. For fluorescence microscopy, chromosome preparations were stained with one of two DNA-specific dyes, DAPI (Schweizer & Nagl 1976) or BAO-Feulgen (Traut & Rathjens 1973). Chromosome spreads for bright field microscopy were prepared according to Traut *et al.* (1986).

For electron microscopy, gonads of pupae were dissected in saline solution (Hayes 1953) or in Grace's insect tissue culture medium (Flow Laboratories). The tissues were embedded using conventional methods including fixation in glutaraldehyde and OsO_4, and treatment with tannic acid for better observation of microtubules. Thin sectioning, electron microscopy, and volume measurements were made according to Wolf (1987) and Wolf *et al.* (1987) respectively.

Results

Light microscopy

BAO-Feulgen stained metaphase plates from brain tissue showed three elements (Fig. 1A). By their length and the presence of a median non-fluorescent gap two metacentric pairs could be distinguished from one telocentric pair. Squashed oogonia treated with the DNA-specific dye DAPI confirmed these observations (Fig. 1B). In aceto-orcein stained preparations there were also only three elements found (Fig. 1C,D). Additional fluorescent spots or chromosomes were not seen and our findings are thus in accordance with previous observations (Tokunaga 1953).

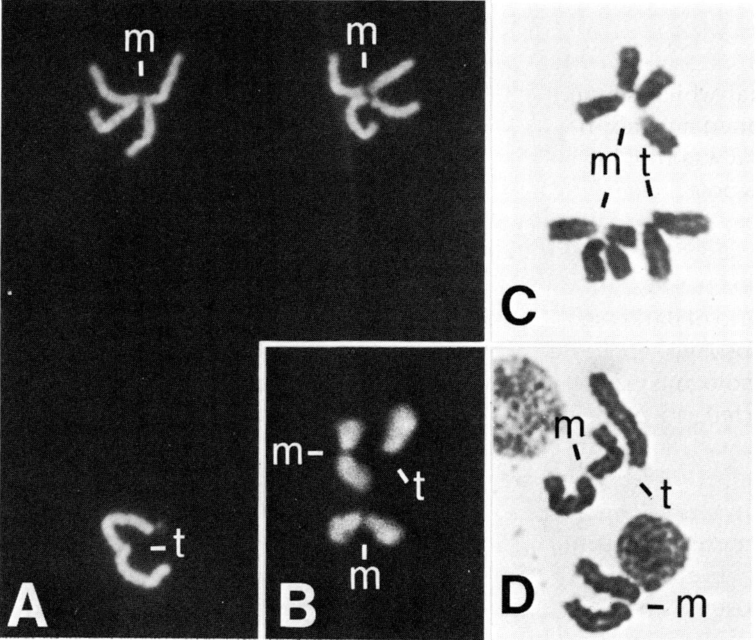

Figure 1. A BAO-Feulgen stained metaphase chromosomes from brain tissue: Three elements can be recognised. The homologues are associated close to the centromeres both in the telocentrics (t) and in the metacentrics (m), where the centromeres appear as unstained gaps. **B** DAPI stained metaphase chromosomes of an oogonium: Three chromosome pairs, two of the metacentric (m) and one of the telocentric type (t), can be seen. Note the unstained centromere region of the metacentrics. **C** Oogonium in metaphase: Both the homologous metacentrics (m) and telocentrics (t) are associated in regions close to the centromeres. **D** Spermatocyte II in metaphase: The unstained centromere region permits the distinction between the metacentric (m) and the telocentric (t) chromosomes.
A–D are light micrographs. C and D are of cells stained with aceto-orcein.

Electron microscopy

Division spindles of mitosis in somatic and germ line cells, and of meiosis were analysed in males and females (Table 1) from complete series of micrographs.

All spindles had in common a perforated spindle envelope (Fig. 2A). Sheath cells of the ovariole and spermatocytes II showed longer spindles in comparison to oogonia and spermatogonia. Spindle poles were recognised by the presence of two centrioles embedded in the cytoplasm. Cells of the type shown in Fig. 2A were interpreted as secondary spermatocytes. In these cells only one basal body was present per pole, and the chromatin volume was considerably lower than in a mitotic metaphase (data not shown). The basal body was orientated perpendicularly to the cell membrane, and flagellar outgrowth was initiated (Fig. 2A).

The darkly staining condensed chromatin of the chromosome arms had a smooth outline in prometa- and metaphase and a uniform fibrogranular texture. Material which stained considerably lighter formed the centromeres. It was positioned medially between or attached terminally to chromosome arms. This material was roughly cylindrical, the long axis being orientated parallel to the pole-to-pole axis. The centromeres consisted of an amorphous matrix interspersed with fibrillar elements. A well-defined primary constriction was apparent.

In the rod-shaped chromosomes with terminal centromeres no trace of a short arm of darkly staining chromatin was detected. Thus, the term 'telocentric chromosomes' applies here.

Homologous chromosomes of mitotic metaphase plates were laterally connected with each other in a region close to the centromeres. As is common in Diptera, chromosomes are held together by somatic pairing (see also Fig. 1 A, B, C). The centromeres of homologues, however, were not visibly connected.

A thin layer of dense material was attached to both poleward surfaces of the centromeres. Here kinetochore microtubules originated. In the present context we use the term 'standard centromere' for the centromeres attached to chromosome arms.

Surprisingly, in the spindle apparatus, depending on cell type and division stage, 1 to 8 extra elements were found (Table 1). These were identical to standard centromeres except for the lack of attached chromosome arms (Fig. 2A, B, C). We refer to these structures as 'centromere-like elements' (CLEs). Like the standard centromeres they were composed of an amorphous matrix interspersed with fibrils. A thin dense layer connected to both poleward surfaces could also be recognised (Fig. 2C). Spindle microtubules were attached to this layer. Both the height (i.e. extension along the pole-to-pole axis) and the volume of the extra centromeres were, however, smaller than those of standard centromeres (Table 2).

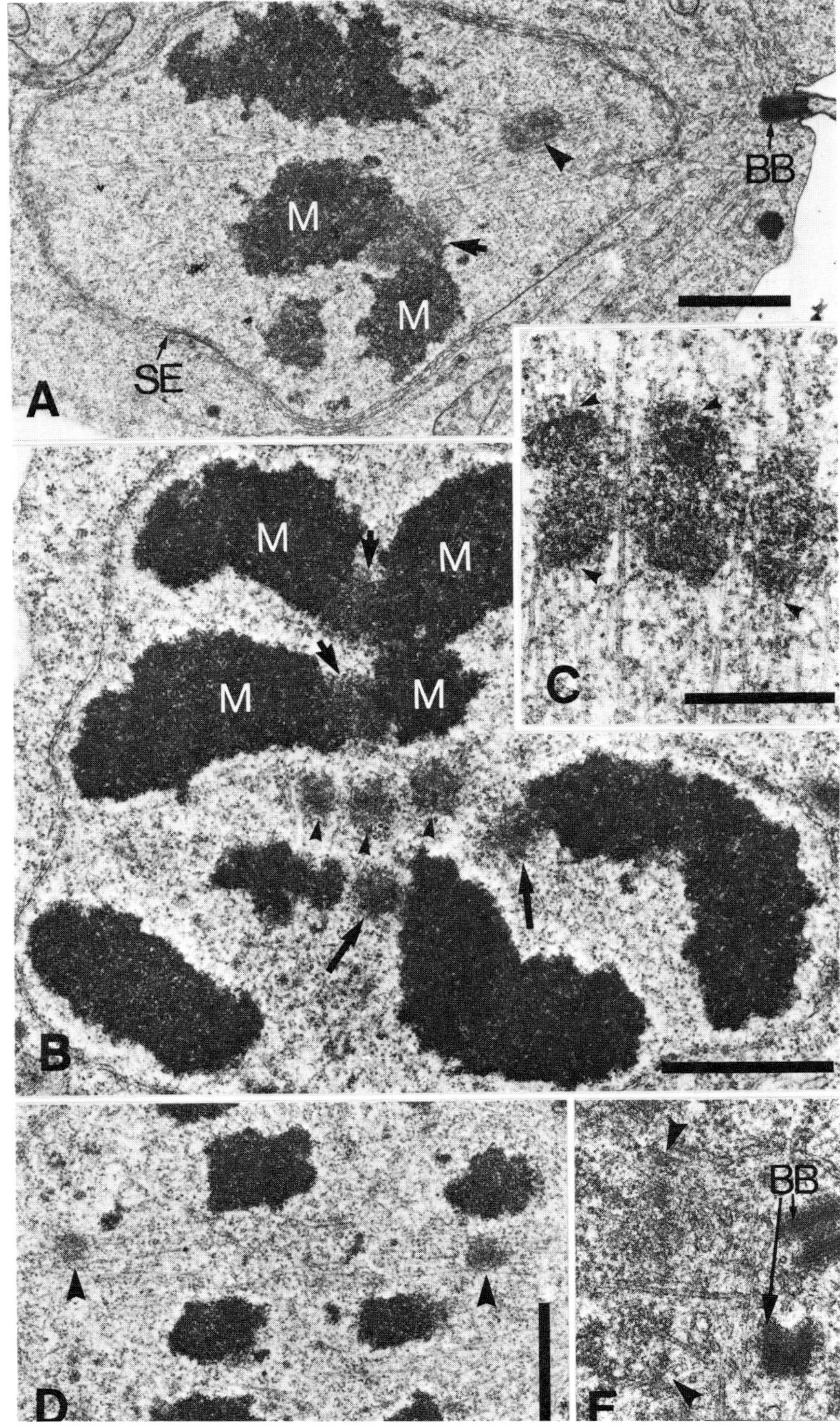

Table 1   Number of CLEs in 5 spindles recorded from serial sections.   The oogonium (1) and the sheath cell of the ovariole (2) were taken from the same animal.   The spermatogonia in meta- and anaphase (3, 4) were found in two different pupae of the same preparation.

Cell type	Stage	Number of centromere-like elements
1 oogonium	metaphase	2
2 sheath cell of the ovariole	metaphase	2
3 spermatogonium	metaphase	3
4 spermatogonium	anaphase	8
5 spermatocyte II	prometaphase	1

In spermatogonia the course of mitosis could be followed until late telophase. By the time of anaphase CLEs divided and migrated to the spindle poles similarly to regular chromosomes (Fig. 2D). In late anaphase CLEs were usually found closer to the poles than the standard centromeres. At the end of chromosome migration a polarised orientation of chromosomes was found. The centromeres could be recognised as individual bodies aligned close to the pole and perpendicular to the spindle axis. In late telophase the centromeres merged to form a perforated plate pierced by microtubules (Fig. 2E). At this stage, the irregular surface of the trailing chromosome arms indicated the onset of chromosome decondensation.

Figure 2. Electron micrographs of spermatocytes and spermatogonia.

A Prometaphase spermatocyte II: Longitudinal section through the spindle showing a metacentric chromosome (M) with standard centromere (arrow). The centromere-like element (arrowhead) is in the right half-spindle. BB=basal body. SE=spindle envelope. **B** Metaphase spermatogonium: This representative cross-section through the equatorial plate shows a metacentric chromosome pair (M) with standard centromeres (arrows). Two further centromeres (long arrows) are associated with chromosome arms, while three centromere-like elements (arrowheads) occur without direct contact to the chromatin. **C** Metaphase spermatogonium: In this longitudinal section through three centromere-like elements their connection with spindle microtubules (arrowheads) can be seen. The poleward dense plate extends over the whole centromere mass. **D** Anaphase spermatogonium: Two separating centromere-like elements (arrowheads) are visible in this longitudinal section. **E** Late telophase spermatogonium: The longitudinally sectioned spindle pole is characterised by two basal bodies (BB). Microtubules pass through a plate of centromere material close to the pole (arrowheads). Bars represent 1 μm in A, B, D; 0.5 μm in C, E.

Table 2 Comparison of height and volume between standard and centromere-like elements of four serially sectioned prometa- and metaphase spindles according to Table 1.

	Standard centromere	Centromere-like elements
Number of measurements	21	8
Mean height (nm)	660	507
SD height	102	50
Mean volume (μm^3)	0.045	0.027
SD volume	0.007	0.009

Discussion

Electron microscopy of meta- and anaphase spindles in several cell types of *Megaselia scalaris* showed varying numbers of centromere-like elements devoid of chromosome arms, in addition to the regular chromosomes.

The process of mitosis as studied in spermatogonia follows the usual pattern and involves the division of the CLEs. The long persistence of the centromeres until late telophase, when the chromosomes have assumed a polarised orientation, is somewhat exceptional.

The CLEs of *M. scalaris* have some characteristics in common with B chromosomes as listed by Rieger *et al.* (1976), such as the numerical variation between individuals and the small size relative to the A chromosomes. They can be interpreted as B chromosomes, which are reduced to the minimum required for perpetuation, the centromere.

The origin of B chromosomes is a matter of speculation (Jones & Rees 1982). The strong resemblance between the CLEs and the standard centromeres in terms of DNA content and fine-structural appearance points to a centromere origin for the extra elements in *M. scalaris*. One mechanism that could give rise to such elements is a centric fusion.

Among the five models suggested by John & Freeman (1975) for whole arm translocations one may apply for *M. scalaris*. Only one functional centromere is required for the Robertsonian metacentric. Depending on the site of chromosome breakage, a whole centromere or a fusion product of two centromere fragments may be left over with little or no material of the chromosome arms. Centric fragments usually seem to become lost after a Robertsonian translocation but one case of a stable fragment has been reported (Marks 1978). Thus, the CLEs in *M. scalaris* tissue may represent a centric fragment of a size below the resolving power of the light microscope. The numerical variation between individuals may result from segregation irregularities.

Acknowledgements

The authors wish to thank Ms. K. Baumgart for expert technical assistance and Mr. H.G. Mertl for continuously providing us with animals. The help of Dr. M. Bastmeyer (Tübingen) with DAPI staining is gratefully acknowledged.

References

Green, M.M. 1980. Transposable elements in *Drosophila* and other Diptera. *Ann. Rev. Genet.* 14, 109–120.

Hayes, R.O. 1953. Determination of a physiological saline solution for *Aedes aegypti* (L.). *J. Econ. Entomol.* 46, 624–626.

John, B. and M. Freeman 1975. Causes and consequences of Robertsonian exchange. *Chromosoma* 52, 123–136.

Jones, R.N. and H. Rees 1982. *B Chromosomes*. London, New York: Academic Press.

Mainx, F. 1964. The genetics of *Megaselia scalaris* Loew (Phoridae): A new type of sex determination in Diptera. *Am. Nat.* 98, 415–430.

Marks, G.E. 1978. The consequences of an unusual Robertsonian translocation in Celery (*Apium graveolens* var. *dulce*). *Chromosoma* 69, 211–218.

Rieger, R., A. Michaelis and M.M. Green 1976. *Glossary of Genetics and Cytogenetics*. Berlin, Heidelberg, New York: Springer Verlag.

Schweizer, D. and W. Nagl 1976. Heterochromatin diversity in *Cymbidium* and its relationship to differential DNA replication. *Exptl. Cell Res.* 98, 411–423.

Tokunaga, Ch. 1953. Life cycle and chromosome of *Aphiochaeta xanthina* Speiser from Okinawa. *Kobe College Studies* 1, 23–28.

Traut, W. and B. Rathjens 1973. Das W-Chromosom von *Ephestia kuehniella* (Lepidoptera) und die Ableitung des Geschlechtschromatins. *Chromosoma* 41, 437–446.

Traut, W., A. Weith and G. Traut 1986. Structural mutants of the W chromosome in *Ephestia* (Insecta, Lepidoptera). *Genetica* 70, 69–79.

Wolf, K.W. 1987. Cytology of Lepidoptera. I. The nuclear area in secondary oocytes of *Ephestia kuehniella* contains remnants of the first division. *Eur. J. Cell Biol.* 43, 223–229.

Wolf, K.W., K. Baumgart and W. Traut 1987. Cytology of Lepidoptera. II. Fine structure of eupyrene and apyrene primary spermatocytes in *Orgyia thyellina*. *Eur. J. Cell Biol.* 44, in press.

Analysis of metaphase chromosome structure using monoclonal antibodies and flow cytometry

B.M. Turner and A. Keohane

Anatomy Department, University of Birmingham Medical School, Vincent Drive, Birmingham B15 2TJ, UK

Progression of eukaryotic cells through the cell cycle involves a series of changes in the organisation of nuclear chromatin. While some of these changes, such as those which precede mitosis, can be studied at the light microscope level, others, such as the re-organisation which accompanies the G0–G1 transition (Walker *et al.* 1986) and the local decondensation of chromatin associated with gene transcription (Elgin 1982; Pederson *et al.* 1986), are not amenable to microscopical analysis. In view of the significance of higher-order chromatin structure for a variety of cellular functions, it is of some interest to study the mechanisms by which changes are brought about. However, work in this area is hampered by a number of experimental difficulties, particularly the problem of developing simple quantitative assays for monitoring changes in the structure and organisation of chromatin within whole nuclei and chromosomes.

We have developed an assay which uses antibodies directed against chromatin components as probes to detect changes in chromatin structure. This approach is based on the premise that changes in higher-order chromatin structure will alter the accessibility of defined chromatin antigens and thereby cause a measurable change in binding of the appropriate antibodies. The approach has obvious similarities to the use of nucleases to detect local decondensation of chromatin and to define the "nuclease hypersensitive sites" associated with actively transcribed regions of the genome (Elgin 1982). However, while nucleases, in combination with DNA probes, provide a powerful approach for studying the organisation of specific genes in chromatin, they are less suitable for the analysis of global changes in chromatin organisation, such as those associated with progression through the cell cycle. Our recent experiments suggest that antibodies may be more useful probes for studying such changes.

The feasibility of using antibodies to monitor changes in chromatin structure has been demonstrated by analysis of the binding of anti-histone antibodies to chromatin fragments. Significant changes in antibody binding result from changes in chromatin structure induced by post-translational modification of core histones (Muller *et al.* 1984) or treatment with intercalating dyes (Whitfield *et al.* 1986). We have recently extended this approach to whole nuclei and chromosomes by using the technique of flow cytometry to quantify antibody binding (Trask *et al.* 1984; Turner & Keohane 1987a). The application of this approach to the study of metaphase chromosome structure is described below.

Flow cytometric analysis of metaphase chromosomes

Flow cytometry has been used extensively for the analysis of metaphase chromosomes stained with fluorescent dyes (Gray *et al.* 1975; Young *et al.* 1981; Green *et al.* 1984). The technique involves passing the labelled chromosomes rapidly in single file across a laser beam tuned to a wavelength suitable for excitation of the chosen fluorochrome. Emitted light is detected by two photomultiplier tubes, the wavelengths detected being determined by appropriate filters. Thus, the light emitted by two different fluorochromes can be measured simultaneously for each chromosome. For the experiments described below, chromosomes have been double-labelled with the DNA binding fluorochrome propidium iodide (PI) and fluorescein (FITC)-conjugated antibodies (see Turner & Keohane 1987a). All experiments were carried out on a Becton-Dickinson FACS 440 flow cytometer equipped with a Spectra-Physics 4W argon ion laser tuned to 488 nm to excite both PI and FITC.

For most of our experiments we have used Chinese hamster ovary cells grown in tissue culture. Cells were treated with Colcemid (0.05 μg/ml) for up to two hours prior to removal of mitotic cells by shaking the flask. Metaphase chromosomes were prepared by the procedure of van den Engh *et al.* (1984) and labelled with antibodies as described by Trask *et al.* (1984) with minor modifications (Turner & Keohane, 1987a).

Antibody labelling defines two discrete chromosome populations

The results shown in Fig. 1A were obtained with chromosomes labelled with monoclonal antibody HBC-7, a mouse IgM directed against the N-terminal region of histone 2B (Turner 1982; Whitfield *et al.* 1986) and counterstained with PI. Chromosomes prepared and labelled in this way were resolved into a series of peaks on the basis of differences in DNA content. The antibody labelling procedure caused only slight loss of resolution of individual chromosomes (compare panels a, c and e). However, in contrast to the resolution achieved on the basis of DNA content, individual chromosome peaks were not distinguishable on the basis of labelling with monoclonal antibody HBC-7 (panel f). The reason for this lack of resolution becomes apparent when the results are plotted in the form of a contour plot correlating antibody binding (FITC fluorescence) with DNA content (PI fluorescence). The antibody resolves the chromosomes into two discrete populations with high and low levels of antibody binding respectively (Fig. 1B). Similar results have been obtained with a mouse monoclonal IgG antibody to double-stranded (ds) DNA (Fig. 2).

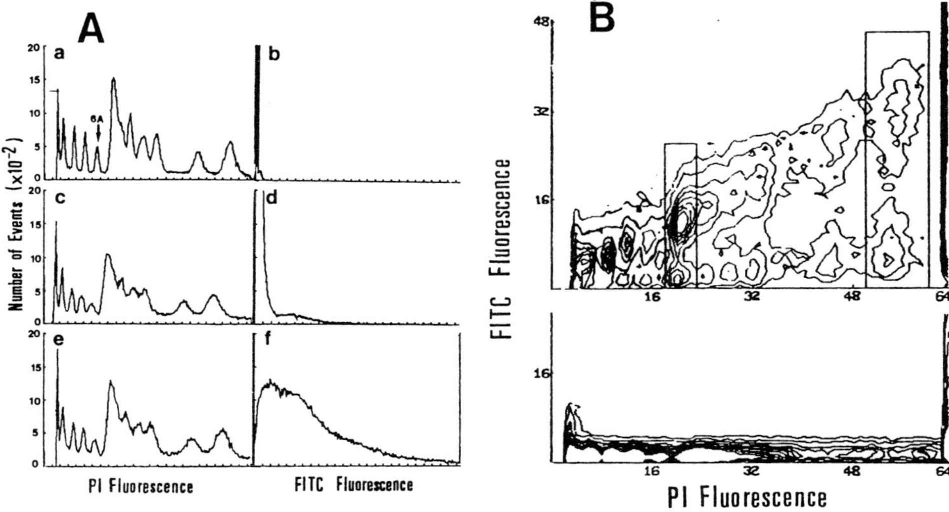

Figure 1. A The effect of antibody labelling on the fluorescence profile of CHO chromosomes. Chromosomes were either stored on ice (panels a & b), labelled with FITC-conjugated antibody to mouse immunoglobulin (panels c & d), or labelled with monoclonal antibody HBC-7 to histone 2B followed by FITC-conjugated antibody to mouse immunoglobulin (panels e & f). Chromosomes were counterstained with propidium iodide (PI) at 50 μg/ml prior to analysis. **B** Contour plots showing the correlation between antibody binding (FITC fluorescence) and DNA content (PI fluorescence) for CHO chromosomes labelled with antibody HBC-7 (panel a) and second antibody alone (panel b). 100,000 events were recorded. Contours range from 20 to 280 events. Boxes identify peaks 1 and 6.

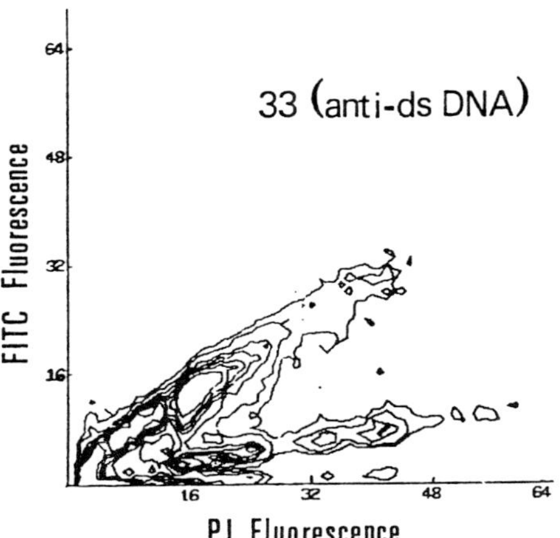

Figure 2. Contour plot showing the correlation between DNA content (PI fluorescence) and antibody binding (FITC fluorescence) for CHO chromosomes labelled with antibody 33. 50,000 events were recorded and contours range from 10 to 200 events.

We have carried out a number of experiments which effectively eliminate an artefactual origin for the two chromosome populations (Turner & Keohane 1987a). Perhaps the most telling is the demonstration that brightly- and weakly-labelled chromosomes are clearly distinguishable by fluorescence microscopy in preparations labelled with rhodamine (TRITC)-conjugated antibody and counterstained with Hoechst 33342. Thus, the two populations are not products of either the flow cytometric technique or the specific fluorochrome combination used for flow cytometric analysis.

The two separate chromosome populations have been observed consistently in different preparations, though the distribution of chromosomes between the high-labelling and low-labelling populations varies and, in occasional preparations, only a single population is seen. (The single population can be high- or low-labelling). The two populations have been found in single cell clones derived from the parent CHO cell line and in human cells (HeLa and REN 2 lymphoblastoid cells). Thus, the resolution of metaphase chromosomes into two populations on the basis of antibody binding appears to be a relatively consistent and general phenomenon. The variability, from one preparation to another, in the distribution of chromosomes between the two populations may be attributable to as yet undefined differences in cell growth or chromosome preparation methodology.

Effects of in-vitro parameters on antibody labelling

The presence of high- and low-labelling chromosome populations is insensitive to variations in buffer composition, ionic strength, pH and divalent cation concentration during preparation. While these parameters can all influence the level of antibody binding, and a specific inhibitory effect of $Cu2+$ ions has been observed (Turner & Keohane 1986b), they do not determine the presence or absence of two populations. Of the various pretreatment regimes we have tried, the ones which have given the most striking and reproducible effects have involved exposure to intercalating dyes.

Incubation of chromosomes with propidium iodide prior to labelling with HBC-7 causes a reduction in the amount of antibody bound to the high-labelling population and, at concentrations of 20 μg/ml and above, results in just a single, low-labelling population (Fig. 3). Other intercalating dyes, such as ethidium and m-AMSA, have a similar effect. Antibody 33 (to dsDNA) also shows reduced binding to chromosomes treated with intercalating dyes.

It seems likely that the inhibitory effect of intercalating dyes on antibody binding is due to a change in chromatin condensation rather than a simple blocking of antigenic sites. Evidence for this comes from experiments in which pretreatment of chromatin fragments with ethidium bromide was shown, by radio-immunoassay, to enhance the binding of antibody HBC-7, while having little effect on the binding of antibody 33 (Whitfield *et al.* 1986). This enhancement is consistent with the distortion of the nucleosome and the

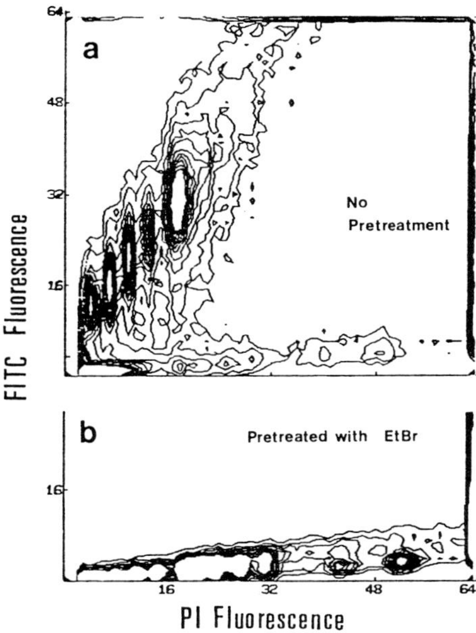

Figure 3. Contour plots showing the effect of pre-incubation with ethidium bromide (EtBr) at 50 µg/ml on binding of antibody HBC-7 to metaphase chromosomes. Each plot represents 100,000 events with contours ranging from 10 to 200 events.

modification of DNA-histone interactions which have been shown to result from treatment with ethidium (Wu *et al.* 1980). These changes enable the nucleosome to accommodate the unwinding and elongation of the DNA helix induced by intercalation. Such distortion would be expected to render the core histones more accessible, while having little effect on the accessibility of DNA. However, in nuclei and metaphase chromosomes, distortion of the nucleosome may be limited by additional constraints imposed upon the DNA. These include higher orders of folding and, more significantly, the formation of closed DNA loops by attachment to a proteinaceous matrix or scaffold. The response of chromatin in nuclei or metaphase chromosomes to intercalating dyes is therefore more likely to resemble that of closed circular DNAs in which unwinding of the helix is accommodated by changes in supercoiling. The experiments of Vinograd and co-workers have shown that such changes can push circular DNAs into a more compact conformation (Bauer & Vinograd 1968). While the complexity of chromatin prevents the direct application of the simple rules developed for naked DNA, it is reasonable to propose that treatment of nuclei or chromosomes with intercalating dyes may, under the appropriate conditions, also result in a more compact structure. Recent experiments on the effect of ethidium treatment on the size and morphology of interphase nuclei in permeable CHO cells support this possibility (B.M. Turner unpublished).

Structural changes which may explain the two chromosome populations

The two populations of metaphase chromosomes resolved by labelling with antibodies to histone 2B or DNA must differ in such a way that the accessibility of both these antigens is altered. This points to a general shift in chromosome structure rather than local changes involving specific DNA-histone interactions. Results presented elsewhere (Turner & Keohane 1987a) suggest that the two populations represent an *in-vivo* change in chromosome structure associated with transit through mitosis.

In attempting to interpret the antibody binding data in structural terms, it is important to bear in mind that the ability of antibodies to penetrate the metaphase chromosome is inevitably limited. Most antibody binding would be expected to occur at or close to the surface. The superficial nature of antibody labelling is often apparent on microscopical examination. These observations suggest two general ways in which binding of antibodies can be influenced. The first is to change the degree of condensation of the chromatin so as to alter the depth to which the antibody can penetrate. The second is to contract the entire chromosome in such a way as to reduce its surface area, without necessarily altering the level of chromatin condensation. The latter could be accomplished by some form of "quaternary" coiling, as originally outlined by DuPraw (1966).

Alterations in chromosome structure which could influence antibody binding are shown diagrammatically in Fig. 4. The basic model is that formulated by Rattner & Lin (1985) and incorporates both chromatin loops attached to a

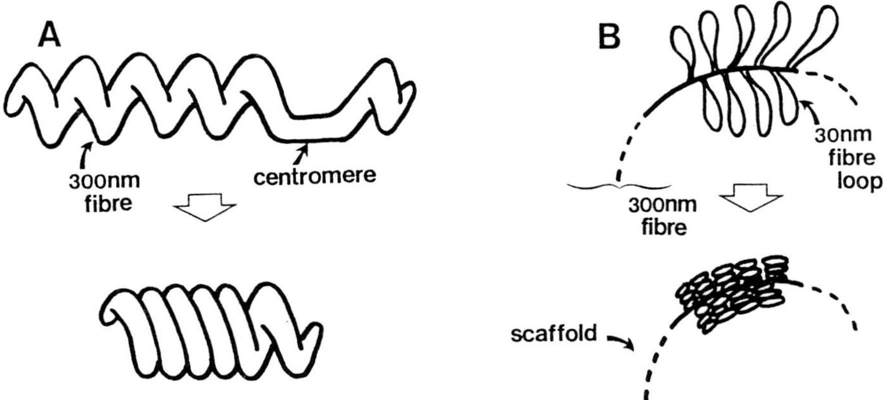

Figure 4. Two types of structural change which could contribute to changes in the accessibility of chromosomal antigens. **A** Contraction of the chromosome is brought about by coiling of a 300 nm fibre. The consequent reduction in surface area causes a reduction in antibody binding. Only a single chromatid is represented in the diagram. **B** The 300 nm fibre consists of loops of chromatin, in the form of a 30 nm fibre, attached to a proteinaceous scaffold. Folding or coiling of these loops results in condensation of chromatin within the 300 nm fibre and a reduction in antibody penetration.

proteinaceous scaffold (Paulson & Laemmli 1977; Pienta & Coffey 1984) and coiling of a 200–300 nm fibre. Evidence for such a chromatin fibre has accumulated over a number of years, largely from studies of whole-mount metaphase chromosomes by scanning and transmission electron microscopy (DuPraw 1966; Harrison *et al.* 1981; Rattner & Lin 1985).

Contraction of the chromosome by coiling, shown in Fig. 4A, will ultimately have the effect of reducing the surface area accessible to antibody. At a particular stage in contraction adjacent gyres will be sufficiently close to exclude antibody and thereby reduce the level of labelling. In order to account for the presence of two discrete populations, it is necessary to propose that the contraction proceeds at the same rate throughout the chromosome. Differential contraction within different regions of the chromosome would result in a progressive reduction in labelling rather than a rapid transition.

An alternative explanation for the antibody labelling results lies in changes in the organisation of the chromatin loops which make up the 300 nm fibre. Microscopical evidence suggests that the chromatin in these loops is in the form of 30 nm fibres (Labhart & Koller 1982; Rattner & Lin 1985) presumably with the solenoidal structure proposed by Klug and co-workers (Finch & Klug 1976). Changes in the organisation of the 30 nm fibre, either by folding or supercoiling, would be expected to alter antibody accessibility and, providing the change progressed rapidly through the chromosome, could give rise to the two populations detected by antibody labelling. This is shown diagrammatically in Fig. 4B.

If the inhibitory effect of intercalating dyes is indeed mediated by alterations in coiling of the 30 nm fibre (Fig. 4B), then the inhibitory effect should also be seen in interphase cells, in which the chromatin is organised into loops but does not form the 300 nm gyres seen in metaphase chromosomes. Experiments to test this prediction have shown that binding of antibody HBC-7 to interphase nuclei in permeabilised CHO cells is inhibited by ethidium over the same concentration range that is effective against metaphase chromosomes (B.M. Turner & A. Keohane, in preparation).

Mechanisms for structural change

Our experiments have shown that antibodies provide sensitive probes for detecting changes in the higher order structure of chromatin in both metaphase chromosomes and interphase nuclei. The mechanism by which these changes are brought about remains uncertain and there are clearly many possibilities which deserve investigation. The effect of intercalating dyes on antibody binding raises the possibility that chromatin compaction in both metaphase and interphase cells may be mediated by changes in DNA supercoiling. Of course, the changes induced by intercalating dyes *in vitro* need not necessarily resemble those which occur *in vivo*, though the end result in terms of antibody binding may be the same. It will therefore be of interest to investigate possible *in vivo* mechanisms and particularly the role of enzymes which can change DNA supercoiling, namely the topoisomerases (Wang 1985). Topoisomerase II is a major component of the metaphase chromosome scaffold (Earnshaw *et al.* 1985; Gasser *et al.* 1986) and may well play both an enzymatic and structural role in chromosome organisation. Recent experiments suggest that the combined use of specific inhibitors of topoisomerase and antibody probes will prove valuable for investigation of this possible mechanism for structural change.

Acknowledgements

We thank Anne Milner and Roger Bird for help with the FACS, Judy Fantes for the REN 2 cell line and Adrienne Morgan for antibody 33. Our work is supported by a grant from the Cancer Research Campaign.

References

Bauer, W. and J. Vinograd 1968. The interaction of closed circular DNA with intercalative dyes I. The superhelix density of SV40 in the presence and absence of dye. *J. Mol. Biol.* 33, 141–171.

DuPraw, E.J. 1966. Evidence for a "folded-fibre" organization in human chromosomes. *Nature* 209, 577–581.

Earnshaw, W.C., B. Halligan, C.A. Cooke, M.S. Heck and L.F. Liu 1985. Topoisomerase II is a structural component of mitotic chromosome scaffolds. *J. Cell Biol.* 100, 1706–1715.

Elgin, S.C.R. 1981. DNAase I-hypersensitive sites of chromatin. *Cell* 27: 413–415.

Finch, J.T. and A. Klug 1976. Solenoidal model for superstructure of chromatin. *Proc. Natl. Acad. Sci. USA* 73, 1897–1901.

Gasser, S.M., T. Laroche, J. Falquet, E. Boy de la Tour and U.K. Laemmii 1986. Metaphase chromosome structure. Involvement of topoisomerase II. *J. Mol. Biol.* 188, 613–629.

Gray, J.W., A.V. Carrano, L.L. Steinmetz, M.A. Van Dilla, D.H. Moore, B.H. Mayall, and M.L. Mendelsohn 1975. Chromosome measurement and sorting by flow systems. *Proc. Natl. Acad. Sci. USA* 72, 1231–1234.

Green, D.K., J.A. Fantes, K.E. Buckton, J.K. Elder, P. Malloy, A. Carothers and H.J. Evans 1984. Karyotyping and identification of human chromosome polymorphisms by single chromosome flow cytometry. *Hum. Genet.* 66, 143–146.

Harrison, C.J., M. Britch, T.D. Allen and R. Harris 1981. Scanning electron microscopy of the G-banded human karyotype. *Exp. Cell Res.* 134, 141–153.

Labhart, P. and T. Koller 1982. Involvement of higher-order chromatin structures in metaphase chromosome organization. *Cell* 30, 115–121.

Lewis, C.D., J.S. Lebkowski, A.K. Daly and U.K. Laemmli 1984. Interphase nuclear matrix and metaphase scaffolding structures. *J. Cell Sci. Suppl.* 1, 103–122.

Muller, S., A. Mazen, A. Martinage and M.H.V. Van Regenmortel 1984. Use of histone antibodies for studying chromatin topography and the phosphorylation of chromatin subunits. *EMBO J.* 3, 2431–2426.

Paulson, J.R. and U.K. Laemmli 1977. The structure of histone-depleted metaphase chromosomes. *Cell* 12, 817–828.

Pederson, D.S., F. Thoma and R.T. Simpson 1986. Core particle, fibre and transcriptionally active chromatin structure. *Ann. Rev. Cell Biol.* 2, 117–147.

Pienta, K.J. and D.S. Coffey 1984. A structural analysis of the role of the nuclear matrix and DNA loops in the organization of the nucleus and chromosome. *J. Cell Sci. Suppl.* 1, 123–135.

Rattner, J.B. and C.C. Lin 1985. Radial loops and helical coils coexist in metaphase chromosomes. *Cell* 42, 291–296.

Trask, B., G. van den Engh, J. Gray, M. Vanderlaan and B. Turner 1984. Immunofluorescent detection of histone 2B on metaphase chromosomes using flow cytometry. *Chromosoma* 90, 295–302.

Turner, B.M. 1982. Immunofluorescent staining of human metaphase chromosomes with monoclonal antibody to histone H2B. *Chromosoma* 87, 345–357.

Turner, B.M. and A. Keohane 1987a. Antibody labelling and flow cytometric analysis of metaphase chromosomes reveals two discrete structural forms. *Chromosoma* 95, 263–270.

Turner, B.M. and A. Keohane 1987b. Cu2+-dependent changes in metaphase chromosome structure assayed by antibody labelling and flow cytometry. *Biochem. Soc. Trans.* 14, 1166–1167.

Walker, L., G. Guy, G, Brown, M. Rowe, A.E. Milner and J. Gordon 1986. Control of human B-lymphocyte replication I. Characterization of novel activation states that precede the entry of Go B cells into cycle. *Immunology* 58, 583–589.

Wang, J.C. 1985. DNA topoisomerases. *Ann. Rev. Biochem.* 54, 665–697.

Whitfield, W.G.F., G. Fellows and B.M. Turner 1986. Characterization of monoclonal antibodies to histone 2B. Localization of epitopes and analysis of binding to chromatin. *Eur. J. Biochem.* 157, 513–521.

Wu, H.-M., N. Dattagupta, M. Hogan and D.M. Crothers 1980. Unfolding of nucleosomes by ethidium binding. *Biochemistry* 19, 626–634.

Young, B.D., M.A. Ferguson-Smith, R. Sillar and E. Boyd 1981. High-resolution analysis of human peripheral lymphocyte chromosomes by flow cytometry. *Proc. Natl. Acad. Sci. USA* 78, 7727–7731.

The chloroplast genome — its function, structure and evolution in vascular plants

T.A. Dyer

Plant Breeding Institute, Maris Lane, Trumpington, Cambridge CB2 2LQ, UK

Chloroplast DNA (cpDNA) is ideally suited for study by the techniques of molecular biology. Consequently it has been extensively examined in many different types of plant using these techniques and this work culminated recently in the complete sequencing by two Japanese groups of the cpDNA of tobacco (Shinozaki *et al.* 1986) and *Marchantia* (Ohyama *et al.* 1986).

As a result of these and numerous other studies of cpDNA we now have a large amount of information concerning its structure. These data can be used to determine: a) how it has evolved, b) how the plants of which it is part have evolved, and c) what it encodes.

Differences in the cpDNA of higher plants are particularly suitable for use as markers in evolutionary studies because the cpDNA is simply inherited from one generation to the next. This is because the cpDNA is uniparentally (usually maternally) transmitted in most plants and in higher plants recombination between the DNA of different chloroplasts is either very infrequent or does not occur at all. Also, mixed populations of plastids sort out rapidly to homogeneity during normal mitotic development of plants. Thus once a change is fixed in a chloroplast population, it does not become redistributed at each sexual cycle as do changes in the nuclear DNA and because of this changes are accumulated sequentially.

Coding properties

Most chloroplast proteins are in fact coded for by the nucleus and synthesised in the cytoplasm. Of the thousand or so proteins which probably occur in the chloroplast, only about 80 are encoded by the cpDNA and with very few exceptions the distribution of the genes in the chloroplast and nucleus is the same in all the plants studied.

It is clear now that the principal role of cpDNA is to facilitate the synthesis of thylakoid proteins and to code for many of them. To achieve this it codes for many components of the machinery in chloroplasts for the transcription and translation of the chloroplast genome and for about half of the thylakoid proteins involved in energy transduction during photosynthesis (Table 1). In contrast, nearly all stromal proteins and those of the bounding membranes are nuclear encoded.

Table 1.   Identified genes in the chloroplast genome of vascular plants.

Function	Complex	Gene designations	Number (total)
Energy transduction in thylakoids	Photosystem I	psa (frx)	5 (8?)
	Photosystem II	psb	6 (9)
	Cytochrome b-f	pet	3 (5)
	ATP synthase	atp	6 (9)
	NADH dehydrogenase	ndh	6 (26?)
Translation of cp-encoded proteins	Ribosomal proteins	rps, rpl	19 (55?)
	tRNAs	trn	30-32
	rRNA	rrn	4
	Factor	inf	1 (?)
Transcription of cpDNA	RNA polymerase	rpo	3 (4)
Various	Stroma	rbc	2 (1000?)
	Outer membrane	mbp	
Unknown	-	ORF	28

In each of the four main protein complexes in the thylakoids involved in photosynthesis, chloroplast and nuclear encoded polypeptides associate to form the functional units and the same is true for the ribosomes and probably for the RNA polymerase as well. Thus, in higher plants at least, there are no chloroplast protein complexes that are coded for by the cpDNA only.

Not all chloroplast genes have been identified yet. Numerous open reading frames (ORFs) have been found in sequencing cpDNA. Where the coding properties of such sequences are found to be relatively conserved in different types of plant and they are transcribed and associated with appropriate regulatory sequences they can reasonably be expected to be functional. In order to identify these ORFs much can be deduced from the derived amino acid sequences of their putative products. It is worth noting that the functions of many of the chloroplast genes have been deduced only from their similarity to known genes of other organisms and consequently have yet to be rigorously characterised.

Perhaps the most unexpected finding is that six of the chloroplast genes resemble those in human mitochondrial DNA for components of the NADH ubiquinone oxidoreductase complex of the respiratory chain. This suggests that chloroplasts may have a respiratory activity as was first suggested by Bennoun from his work with the alga *Chlamydomonas reinhardii* (Bennoun 1982).

Origins of the chloroplast genome

In many features the genome of chloroplasts resembles that of prokaryotic organisms more closely than that of the nucleus. This indicates beyond reasonable doubt that chloroplasts were derived from this group of organisms, probably by endosymbiosis. For example, the control sequences in most instances are very similar, so similar in fact that the chloroplast sequences can function quite satisfactory in bacteria or bacterial extracts. Also, as in prokaryotes, several genes are co-transcribed (nuclear genes are not) and the clustering of co-transcribed genes is strikingly similar. Another prokaryotic feature is the presence of two pairs of overlapping genes (psbDC and atpBE). Furthermore, the degree of homology between higher plant chloroplast and cyanobacterial genes is particularly high, reaching over 75 per cent in some instances (psbA). However, because the size of the chloroplast DNA is approximately 1/30th the size of the genome size of a prokaryote and most of the genes for chloroplast proteins are in the nucleus, there must have been a massive transfer of genetic material from the endosymbiont at some stage during its evolution into an organelle.

A fascinating observation is that some of the sequences homologous to those found in chloroplasts are found in mitochondria and in the nucleus (see Timmis & Steele-Scott 1984). These data suggest that the genomes of the nucleus, chloroplast and mitochondria are not completely isolated from one another. The high degree of similarity of the sequences at the different sites also indicates that the transfer of DNA occurs quite frequently in evolutionary time and certainly must have occurred much more recently than during the bulk transfer of genetic information to the nucleus on establishment of the endosymbiont as an organelle.

In view of the pronounced prokaryotic nature of cpDNA, it was rather surprising to find that, in contrast to bacteria, several chloroplast genes contain introns. Seven tRNA genes and nine protein-coding sequences contain introns. The most bizarre instance of a split gene is for the ribosomal protein S12 where the gene has three exons, two of which are separated by a relatively small intron but the third is located far apart on a different DNA strand, so that in chloroplasts there must be a mechanism to splice together segments from separate RNA molecules. The origin of the introns is of particular interest in relation to considerations of cpDNA evolution. The introns do not appear to be derived from other chloroplast sequences and consequently may be of extraplastidic origin. In this regard it is of particular interest that multiple repeated elements have recently been reported to occur in subclover cpDNA which is what one might expect if there has been transposon integration (Palmer *et al.* 1987a).

Clearly it is of paramount importance to know whether DNA can be inserted into chloroplasts when considering how to transform the chloroplast genome by the introduction of new sequences.

Basic structure of cpDNA in vascular plants and its evolution

All chloroplast DNAs of vascular plants have the same basic structure and those which differ from this have obviously been derived from the basic form. This consists of a circular molecule containing two segments which are inverted repeats of one another and which are separated by large and small single copy regions (Fig. 1). In ancestors of vascular plants, the mosses and liverworts, the inverted repeat consists of little more than a duplicated ribosomal RNA cistron.

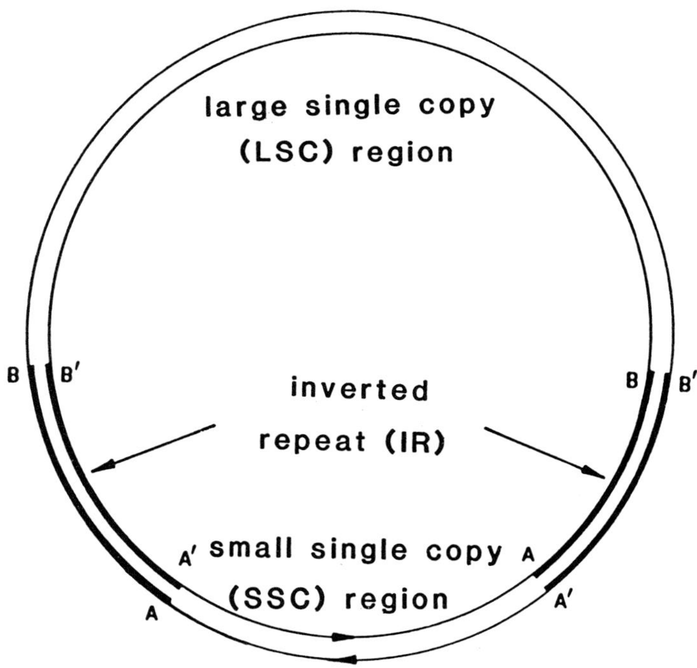

Figure 1. Basic pattern of chloroplast DNA morphology in higher plants.

It is somewhat larger in the ferns (Palmer & Stein 1986) and, in geranium chloroplast DNA, it is triple-sized encompassing two thirds of the genome! (Palmer *et al.* 1987b). Exceptions to this basic form are found in certain legumes such as the pea and broad bean in which one segment of the inverted repeat has been lost. Deletion of part of the inverted repeat has been accompanied in these plants by extensive gene rearrangements suggesting that the IR may help stabilise the genome.

In higher plants so far studied the cpDNA molecule itself is found to be reasonably conserved in size. It is smallest (about 120,000 base pairs) in legumes in which one of the segments of the inverted repeat has been deleted; most are between 135,000 and 160,000 base pairs, and the largest found so far is that of geranium (217,000 base pairs), because in this so much of the genome is duplicated. The structure of the cpDNA in vascular plants is obviously much more uniform than in the algae and there is a major discontinuity in cpDNA structure between the vascular plants and those algae which have been studied so far. Thus there is little doubt that all vascular plants had a common origin. The results give little indication of which protists might have been the progenitors of the vascular plants but as yet only a small selection of the algae have been examined due to difficulties in extracting the cpDNA in a form suitable for further analysis.

Examination of rearranged genomes demonstrates that the primary building blocks of cpDNA are the transcriptional units which in many cases consist of several genes. Apparently these may be shuffled about quite extensively without ill effects. However, even some of the transcriptional units found in all higher plants are not identical in the algae so that different transcriptional units have become established in these organisms.

In comparisons of closely related plants it is easiest to detect differences in their cpDNA using restriction enzymes. Either target sites for these enzymes may be gained or lost due to single base changes or the fragments produced may be changed in size due to the insertion or deletion of lengths of DNA. The more extensive rearrangements which have occurred in some evolutionary lineages have been detected primarily by homologous sequence and gene mapping. Some similar rearrangements have occurred in widely differing plants. For instance the inversion of a 50 kb segment of the large single copy region has occurred in the legumes and *Oenothera* and a similar but smaller inversion is found in cereals (Howe 1985; Quigley & Weil 1985) and in lettuce (Jansen & Palmer 1981). This indicates the possibility that certain regions may be more predisposed to change than others. Certainly point mutations occur much more frequently in non-transcribed spacer regions between genes than in the genes themselves and in their control sequences (Koller *et al.* 1982). Furthermore, changes tend to be less frequent in the inverted repeat than in genes in single copy regions and in third positions of codons which would not alter coding properties (Zurawski & Clegg 1987).

In general, the conclusions concerning plant evolution derived from comparisons of cpDNA agree well with those derived by more traditional taxonomic methods. However, they may supplement these particularly when extreme situations are considered where techniques of comparative morphology are perhaps less reliable. For example, in hexaploid wheat comparisons of cpDNA have been useful in attempts to resolve the origin of the B genome, as the donor of this was also the donor of the chloroplasts (Bowman *et al.* 1983). At the other extreme the exciting possibility exists that this type of comparison will help determine the origin of the angiosperms and of all vascular plants.

Some recent reviews that describe in greater detail than here various aspects of the chloroplast genome organisation and expression are as follows: comparative organisation; Bohnert *et al.* (1982), Dyer (1984), Palmer (1985), Zurawski & Clegg (1987); expression; Dyer (1985), Gray (1986), Weil (1987).

References

Bennoun, P. 1982. Evidence for a respiratory chain in the chloroplast. *Proc. Natl. Acad. Sci. USA* 79, 4352–4356.

Bohnert, H.J., E.J. Crouse and J.M. Schmitt 1982. Organization and expression of plastid genomes. In *Encyclopedia of Plant Physiology (New Series) Vol. 14B, Nucleic Acids and Proteins in Plants*, B. Parthier and D. Boulter, eds, 475–530. Springer-Verlag.

Bowman, C.M., G. Bonnard and T.A. Dyer 1983. Chloroplast DNA variation between species of *Triticum* and *Aegilops*. Location of the variation on the chloroplast genome and its relevance to the inheritance and classification of the cytoplasm. *Theor. Appl. Genet.* 65, 247–262.

Dyer, T.A. 1984. The chloroplast genome: its nature and role in development. In *Chloroplast Biogenesis*, N.R. Baker and J. Barber, eds, 23–69. Elsevier Science Publishers BV.

Dyer, T.A. 1985. The chloroplast genome and its products. *Oxford Surveys Plant Mol. and Cell Biol.* 2, 147–177.

Gray, J.C. 1986. Wonders of chloroplast DNA. *Nature* 322, 501–502.

Herrmann, R.G., P. Westhoff, J. Alt, P. Winter, J. Tittgen, C. Bisariz, B.B. Sears, N. Nelson, E. Hurt, G. Hauska, A. Viebrock and W. Sebald 1983. Identification and characterization of genes for polypeptides of the thylakoid membrane. In *Structure and Function of Plant Genomes*, O. Ciferri and L. Dure, eds, 143–153. Plenum Press.

Howe, C.J. 1985. The endpoints of an inversion in wheat chloroplast DNA are associated with short repeated sequences containing homology to *att*-lambda. *Curr. Genet.* 10, 139–145.

Jansen, R.K. and J.D. Palmer 1987. Chloroplast DNA from lettuce and *Barnadesia* (Asteraceae): structure, gene localization and characterization of a large inversion. *Curr. Genet.* 11, 553–564

Koller, B., H. Delius and T.A. Dyer 1982. The organization of the chloroplast DNA in wheat and maize in the region containing the LS gene. *Eur. J. Biochem.* 122, 17–23.

Ohyama, K., H. Fukuzawa, T. Kohchi, H. Shirai, T. Sano, S. Sano, K. Umesono, Y. Shiki, M. Takeuchi, Z. Chang, S. Aota, H. Inokuchi and H. Ozeki 1986. Chloroplast gene organization deduced from complete sequence of liverwort *Marchantia polymorpha* chloroplast DNA. *Nature* 322, 572–1574.

Palmer, J.D. 1985. Comparative organization of chloroplast genomes. *Ann. Rev. Genet.* 19, 325–354.

Palmer, J.D. and D.B. Stein 1986. Conservation of chloroplast genome structure among vascular plants. *Curr. Genet.* 10, 823–833.

Palmer, J.D. and W.F. Thompson 1982. Chloroplast DNA rearrangements are more frequent when a large inverted repeat sequence is lost. *Cell* 29, 537–550.

Palmer, J.D., B. Osorio, J. Aldrich and W.F. Thompson 1987a. Chloroplast DNA evolution among legumes; loss of a large inverted repeat occurred prior to other sequence rearrangements. *Curr. Genet.* 11, 275–286.

Palmer, J.D., J.M. Nugent and L.A. Herbon 1987b. Unusual structure of geranium chloroplast DNA: a triple-sized inverted repeat, extensive gene duplications, multiple inversions, and two repeated families. *Proc. Natl. Acad. Sci. USA* 84, 769–773.

Quigley, F. and J.H. Weil 1985. Organization and sequence of five tRNA genes and of an inidentified reading frame in the wheat chloroplast genome: evidence for gene rearrangements during the evolution of chloroplast genomes. *Curr. Genet.* 9, 495–503.

Shinozaki, K. and 22 other authors 1986. The complete nucleotide sequence of the tobacco chloroplast genome: its gene organization and expression. *EMBO J.* 5, 2043–2049.

Timmis, J.N. and N. Steele-Scott 1984. Promiscuous DNA: sequence homologies between DNA of separate organelles. *TIBS* 9, 271–273.

Weil, J.H. 1987. Organization and expression of the chloroplast genome. *Plant Sci.* 49, 149–157.

Zurawski, G. and M.T. Clegg 1987. Evolution of higher-plant chloroplast DNA-encoded genes: implications for structure - function and phylogenetic studies. *Ann. Rev. Plant Phys.* 38, 391–418.

Interaction between the nucleus and cytoskeleton during the pairing stages of male meiosis in flowering plants

J. Sheldon*, C. Willson† and H.G. Dickinson†

†Department of Botany, School of Plant Sciences, University of Reading, Whiteknights, Reading, RG6 2AS

*Department of Plant Science, School of Botany, University of Oxford, South Parks Road, Oxford, OX1 3RA

Despite years of active investigation, chromosome pairing during meiosis remains poorly understood. Some spectacular progress has certainly been made with regard to the biochemical events that accompany the alignment and synapsis of homologues (Stern & Hotta 1984) and data are now available on the mechanics of the complex three-dimensional re-organisation involved (von Wettstein 1984). Nevertheless, recent studies of this system have produced more paradoxes than solutions. For example, there is evidence that the nuclear envelope may play a significant part in the control of pairing (von Wettstein 1984), but the fact remains that isolated chromosome fragments will pair quite satisfactorily. Equally, whilst mechanistic investigations point to the occurrence of important events during meiotic prophase, a number of biochemical and structural studies suggest that there is some "pre-synaptic alignment" during the pre-meiotic period. Indeed, as the relationship between individual haploid genomes both at mitosis and interphase becomes better understood (Heslop-Harrison & Bennett, 1983), pairing starts to emerge as a process which must involve the alignment of organised haploid sets of chromosomes, rather than of random stretches of chromatin scattered within the nucleus.

Historically, colchicine has proved a most powerful tool in the investigation of pairing processes. As early as 1938, Walker in *Tradescantia* and Dermen in *Rhoeo* were able to show that colchicine treatment would effectively prevent the formation of chiasmata. In a similar investigation of *Allium*, Levan (1939) reported that the drug focussed its effect on synapsis. The precise timing of the colchicine effect has however yet to be resolved fully; work with *Triticum* hybrids (Dover & Riley 1973) provided conclusive evidence that cells were sensitive during the pre-meiotic mitosis, and insensitive once meiosis had begun. Bennett *et al.* (1973) were able to localise the timing of drug sensitivity more exactly, reporting it to occur over a period from anaphase of the pre-meiotic mitosis to some point in pre-meiotic G_1. This early effect of colchicine indicated that it acted on pre-synaptic alignment of homologues, an inference supported by the early work of Dermen (1938) in *Rhoeo*, where drug-treated cells were discovered to become tetraploid, having undergone complete bivalent formation. These observations, combined with those of Brown & Stack (1968) on pre-meiotic pairing in *Rhoeo*, suggest that synapsis is a lengthy process and has its origins either during the pre-meiotic mitosis, or in the interphase that follows it. Certainly it would seem that biochemical and physiological events are put in train long before any structural manifestation of synapsis may be observed. Despite the fact that pollen mother cells of the Gramineae seem to be insensitive to colchicine once meiosis has commenced, other monocotyledons are clearly sensitive. Shepard *et al.* (1974) provided persuasive evidence that the microsporocytes of *Lilium* respond to treatment with 1×10^{-3}M colchicine when applied as late as zygotene. As in most of the recent studies, the effect of the drug was measured by its influence on pairing *per se*, rather than on the formation of chiasmata. Further, Shepard *et al.* (1974) were able to show that while homologues condensed and formed individual lateral elements, these did not combine to form the characteristic synaptonemal complexes.

Despite this somewhat confusing evidence, there seems little doubt that pairing is controlled by a colchicine-sensitive process which can extend from the pre-meiotic mitosis until the beginning of meiosis itself. Since the major effect of the drug is known to be the inhibition of tubulin polymerisation, it is perhaps natural to look to either the pre-meiotic or meiotic spindles for some explanation of the effect. However, Dover & Riley (1973) used chloral hydrate to show that spindle inhibition had little or no effect on pairing. The involvement of tubulin in pairing is nevertheless also indicated by studies into the action of the Ph1 gene in the Gramineae (Riley & Chapman 1958; Driscoll & Darvey 1970). The dosage of this gene seems to regulate the stringency of pairing, and will determine whether pairing is homologous or homeologous. Most recently, authors have reported this gene to be active in the regulation of tubulin metabolism (M. Feldman & L. Avivi, this volume).

The advent of cellular and immuno-localisation techniques has permitted the examination of the tubulin cytoskeleton during this critical period of colchicine sensitivity. Preliminary results on *Lilium* (Dickinson & Sheldon 1984) indicate a progressive change from a cortical microtubular cytoskeleton to a relatively complex system of microtubules running over the surface of the nucleus. Whilst this system resembles the mitotic prophase cytoskeleton in many ways, it is not preceded by a pre-prophase band (PPB), and there is preliminary evidence that some of the microtubules may be associated with chromosomal termini (Sheldon & Dickinson 1986). Since pairing can be affected by the application of colchicine at these late stages, in *Lilium* at any rate (Shepard *et al.*, 1974), the possibility exists that these early prophase cytoskeletons are in some way involved in the synaptic process. Even a cursory consideration of the three-dimensional re-organisation of the chromatin accompanying pairing would indicate that microtubules, which essentially can only either shorten or lengthen or provide some degree of rigidity, are unlikely to be the agents directly responsible for the pairing of individual chromosome termini. Indeed, this clearly cannot be the case since pairing will occur without any contact with the nuclear envelope. However, in the vast preponderance of species examined so far, pairing is accompanied by the groups of chromosomal termini into a relatively restricted area of the nuclear envelope, forming a structure sometimes referred to as the synezetic knot or bouquet (Thomas & Kaltsikes 1976). Indeed, these authors (Thomas & Kaltsikes 1977) also reported that pairing in a pentaploid hybrid formed from a hexaploid Triticale and *Triticum turgidum* was sensitive to colchicine only after the pre-meiotic S phase, and that this sensitivity continued to early zygotene. It thus remains possible that this aggregation of the chromosomal termini is mediated by a colchicine-sensitive assembly, perhaps the cytoskeleton which invests the nucleus at prophase.

The possibility must also be considered that there are colchicine-sensitive structures other than microtubules present during the immediate pre-meiotic period. A case in point is the fibrillar material reported by Bennett *et al.* (1979) as occurring during pre-meiotic interphase, and connecting chromatin to the inner face of the nuclear membrane as well as to other sections of chromatin. Interestingly, these fibrils appear at the time when the cell is senstive to low temperatures (Bayliss & Riley 1972) and colchicine treatment (Dover & Riley 1973). Disappointingly, these structures have not been the subject of further investigation.

In order to explore the possibility that the microtubular cytoskeleton investing the nuclear envelope during early pre-meiotic prophase is involved in some way in the gathering-together of the chromosome termini to form the synezetic knot or bouquet, we have examined the disposition of the cytoskeleton in pollen mother cells in a range of species, including members of the Solanaceae, a family which has the advantage of possessing a range of meiotic stages in each anther, and of having clearly-defined synezetic associations.

The cytoskeleton in pre-meiosis

The extreme fragility and tight packing of the cells of the archesporial mass makes cytoskeletal investigations difficult. The few preparations made successfully with *Lilium* cells in pre-meiotic mitosis and early pre-meiotic interphase suggest that their microtubular organisation resembles very closely that of somatic cells. Certainly the cytoskeletons characteristic of late pre-meiosis in a number of species, including *Lilium*, *Lycopersicon* and *Rhoeo*, feature small numbers of highly distributed sub-cortical microtubules (Fig. 2). They are not grouped in wefts, and are difficult to detect under the electron microscope. These microtubules are lost at about the time when the pollen mother cells become distinguishable from the tapetal layer. Although this is the juncture at which the PPB would be formed in mitotic cells, no such structure is observed

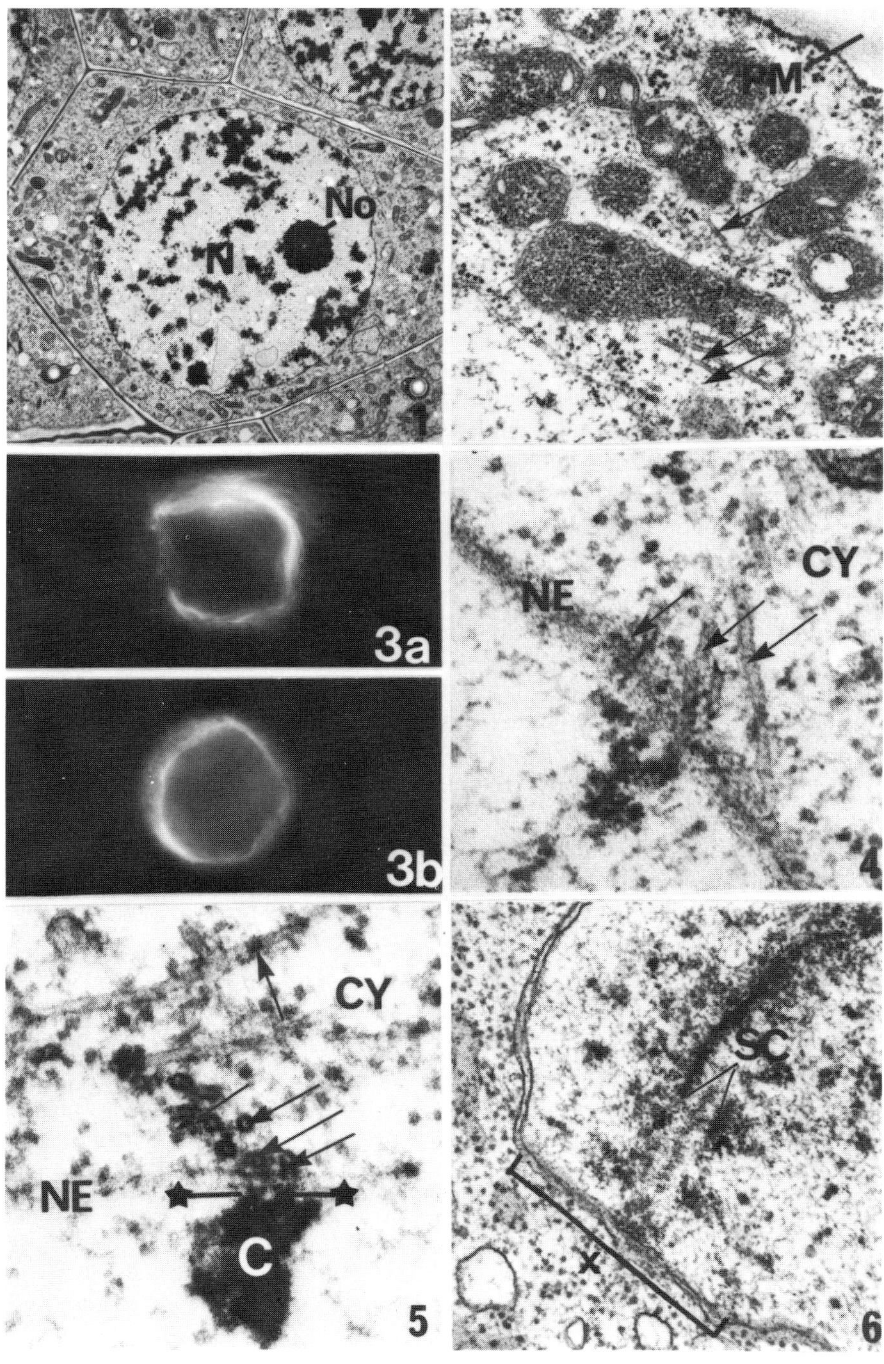
No
N
3a
3b
PM
NE
CY
4
CY
NE
C
5
SC
NE
CY
x
6

in these cells. As the callosic special wall of the meiocyte begins to thicken, the first sign of the exonuclear cytoskeleton characteristic of meiotic prophase becomes visible.

While there appear to be small differences between the species that we have examined, and indeed between our observations and those of others (Hogan 1987), there is universal agreement that a PPB is not formed prior to meiosis. New information on the origin of the PPB probably provides a clue as to why this is so. Very recent results (K. Roberts, personal communication) indicate that the band also contains a very sensitive form of actin, not normally preserved in fixed material. It has been suggested that the band is formed by the actin system's acting to draw together the helically-organised tubulin cytoskeleton, characteristic of most somatic meristematic cells, to form a band encircling the nucleus. If this is the case, the absence of a helically-organised cytoskeleton from young meiocytes — perhaps because they are synthesising the 1-3 linked glucan callose rather than cellulosic microfibrils — perhaps explains why a tubulin PPB is not detected. Since the active component of the PPB would seem to be a form of actin, there is, of course, the possibility that it is present in meiocytes but has yet to be detected.

Microtubular changes during meiotic prophase

In all species examined so far, a conspicuous microtubular cytoskeleton is formed over the surface of the nuclear envelope early in meiotic prophase (Figs 3a and b). In the Solanaceae and other groups of plants with small nuclei, the

Figures 1–6. Fig. 1. Young meiocyte of *Lycopersicon esculentum*. The chromatin is starting to condense in the nucleus (N) and the nucleolus (NO) is clearly visible. Neither lateral elements nor synaptonemal complexes are visible at this early leptotene stage. EM × 3,450. **Fig. 2.** Detail of the cytoplasm of an early leptotene meiocyte of *Lycopersicon esculentum*, such as that depicted in Fig. 1. Small numbers of cytoplasmic microtubules (arrows) are running fairly close to the plasma membrane (PM). EM × 38,000. **Fig. 3a.** Very early prophase cytoskeleton of *Lilium henryii*. Note that it assumes a characteristic 'lemon shape' and that while most of the microtubules are aggregated around the nucleus, some still extend into the cytoplasm. Immunofluorescence micrograph using antitubulin × 780. **Fig. 3b.** As Figure 3a. Immunofluorescence micrograph using antitubulin × 820. **Fig. 4.** Late leptotene nuclear envelope of a meiocyte of *Lycopersicon esculentum*. An array of microtubules (arrows) can be seen appressed to the nuclear envelope (NE) subjacent to which chromatin (C) may be seen. Other microtubules are running in the cytoplasm (CY). EM × 94,500. **Fig. 5.** As Fig. 4, but in *Lilium henryii*. Again populations of microtubules (arrow) may be seen both appressed to the nuclear envelope (NE) and running within the cytoplasm (CY). In this micrograph there is some suggestion of a physical connection (*) between the chromatin (C) and the nuclear envelope. EM × 88,000. **Fig. 6.** Pachytene nucleus of *Lycopersicon esculentum* depicting the termination of a synaptonemal complex (SC) at the nuclear envelope. While details at the point of attachment are not very clear, the chromosome terminus obviously stabilises a comparatively large area of the nuclear envelope (X). EM × 39,250.
The techniques used in the preparation of material for Figs. 1–6 are described in Sheldon & Dickinson (1986).

organisation of this cytoskeleton is difficult to discern using immuno-fluorescence methods. In the Liliaceae large wefts of microtubules may be observed extending from apparent 'centres' lying close to the nuclear envelope (Sheldon & Dickinson, 1986). Whether these centres are the sites of tubulin polymerisation is far from clear, and we have yet to see them in meiocytes from the Solanaceae. In all species, the electron microscope reveals this cytoskeleton to be composed of microtubular wefts, comprising between three and ten tubules, running very close to the outer surface of the nuclear envelope (Figs 4 and 5). As has been reported earlier (Dickinson & Sheldon, 1984) there is some evidence of connections between elements of the cytoskeleton and the subjacent chromatin (Fig. 5). This occurs over a very restricted period towards the end of leptotene, when individual lateral elements are visible within the nucleus. Once pairing has taken place and synaptonemal complexes may be seen extending from plaques affixed to the inner face of the nuclear envelope (Fig. 6), this association is no longer observed. It must be emphasised that it has yet to be established that each chromosome is associated with an individual microtubular weft, and in some taxa (e.g. *Lycopersicon*) it has proved difficult to find any regular association between microtubules and chromosomal material. The relationship between the microtubules apparently adpressed to the outer face of the nuclear envelope and the chromatin associated with the inner face beneath them is also unclear, and no specialised structure can be observed to join the two. Further, with our present knowledge of this system, it has yet to be established that this association is between microtubules and chromosome termini; it remains possible that any part of the chromatin may be involved in this interaction.

Whether or not this cytoskeleton is intimately associated with chromosome termini at any point, it persists throughout the remainder of meiotic prophase. Earlier investigations using immuno-fluorescence techniques (Sheldon & Dickinson 1986) suggested that microtubules de-polymerised during the diplotene stage, only to re-form as the meiotic spindle during metaphase I. This observation has always been contrary to the electron microscopic evidence, which shows the presence of microtubules during diplotene and diakinesis. In more recent experiments, during which it has proved possible to stablise the highly labile diplotene nuclear envelope, the cytoskeleton has proved to be detectable using immuno-fluorescence methods. However, these technical difficulties have prevented us from discovering details of the transformation of the tubulin 'exoskeleton' investing the nucleus, into the organisation characteristic of the spindle.

The post-meiotic cytoskeleton, which takes the form of radial arrays of microtubules generated from organising centres of the nuclear envelope, is described at length elsewhere (Dickinson & Sheldon 1984; Sheldon & Dickinson 1986), but only for plants which follow a sequential meiotic division. In species which carry out simultaneous meiotic division — such that four meiotic products exist in the same cytoplasm — these radial arrays commence formation before the new callosic cell wall is laid down between the microspores (Hogan 1987). In all plants examined, this radial system of microtubles remains until the commencement of intine synthesis, when the more familiar cortical cytoskeleton of microtubules re-appears.

The microtubular cytoskeleton involved in meiotic pairing?

Considerable circumstantial evidence undoubtedly exists pointing to an involvement of the protein tubulin in meiotic pairing. This evidence comes not only from the immunofluorescence and electron microscopic studies, but also from experiments with colchicine and studies of the Ph gene. Most data point to an involvement of the prophase cytoskeleton in this process but, running contrary to this line of evidence is information from the Gramineae (e.g. Dover & Riley 1973) where the colchicine sensitivity appears far earlier. It is possible that there are events at these early stages, perhaps involving the fibrils reported by Bennett *et al.* (1979), which are peculiar to the Gramineae. Alternatively, it may be that in these plants the prophase cytoskeleton is unusually resistant to the effects of the drug. Clearly, there is some worrying variability in response to treatment, for Thomas & Kaltsikes (1977) reported that a Triticale-derived pentaploid hybrid was sensitive to colchicine early in the zygotene stage. Until the sensitivity of these plants to colchicine is finally resolved, these data should perhaps not be used to counter or support any particular hypothesis.

If the prophase cytoskeleton is in some way involved in pairing, it cannot constitute the sole mechanism by which homologous termini are brought together. As has been discussed earlier, this would be stereologically impossible in view of the known function of tubulin polymers. We should therefore look for a process which would involve the attachment of microtubular wefts to parts of the chromatin, and their contraction or elongation. Since the cytoskeleton is seen only in this form for a very short period towards the end of leptotene, when only the lateral elements of synaptonemal complex are visible, it is possible that the cytoskeleton is involved in the chromosome movement that takes place solely at that stage — the formation of the synezetic knot. Such a process would be entirely consistent with the properties of microtubules, in that the skeleton would simply act as a massive unipolar spindle aggregating the chromosome points of attachment into a relatively small area of the nuclear envelope. As yet, we have no evidence that the points of attachment are telomeric. However, by far the greater proportion of chromatin in contact with the nuclear envelope during meiotic prophase is known to be telomeric (von Wettstein 1984), so it would be surprising were it not so. Perhaps the most difficult aspect of this model to explain is the apparent attachment of microtubules to the chromatin through the nuclear envelope. However, telomeric material is known to differ from the main body of the chromatin and were it to include a microtubular binding molecule, there is no valid reason why it should not penetrate the envelope and make contact with microtubules on the outer surface. The mechanism by which the wefts of microtubules could locate chromosome termini is far from clear, but presents a conceptional problem no greater than does the linkage of the meiotic spindle to the kinetochores.

Despite these intriguing circumstantial data, an involvement of the microtubular cytoskeleton in chromosomal synapsis remains far from proven. Only when the three-dimensional organisation of these tubulin assemblies is fully understood, and details are available concerning the apparent connection to microtubules and chromosomes, will it be possible to test accurately the validity of this hypothesis.

Acknowledgement

The authors wish to thank Simon Brookes for photographic assistance.

References

Bayliss, M.W. and R. Riley 1972. Evidence of premeiotic control of chromosome pairing in *Triticum aestivum*. *Genetical Research* 20, 201–212.

Bennett, M.D., K. Rao, J.B. Smith, and M.W. Bayliss 1973. Cell development in the anther, the ovule and the young seed of *Triticum aestivum* L. var. Chinese Spring. *Phil. Trans. Roy. Soc. Lond. B.* 266, 39–81.

Bennett, M.D., J.B. Smith, S. Simpson and B. Wells 1979. Intranuclear fibrillar material in cereal pollen mother cells. *Chromosoma* 71, 289–332.

Brown, W.V. and S.M. Stack 1968. Somatic pairing as a regular preliminary to meiosis. *Bull. Torrey Bot. Club* 95, 369–378.

Dermen, H. 1938. A cytological analysis of polyploidy induced by colchicine and by extremes of temperature. *J. Hered.* 29, 211–229.

Dickinson, H.G. and J.M. Sheldon 1984. A radial system of microtubules extending between the nuclear envelope and the plasma membrane during early male haplophase in flowering plants. *Planta* 161, 86–90.

Dover, C.A. and R. Riley 1973. The effect of spindle inhibitors applied before meiosis on meiotic chromosome pairing. *J. Cell Sci.* 12, 143–161.

Driscoll, C.J. and N.L. Darvey 1970. Chromosome pairing: effect of colchicine on an isochromosome. *Science* 169, 290–291.

Heslop-Harrison, J. and M.D. Bennett 1983. Prediction and spatial order in haploid chromosome components. *Proc. Roy. Soc. Ser. B.* 218, 211–213.

Hogan, C.J. 1987. Microtubule patterns during meiosis in two higher plant species. *Protoplasma* 138, 126–136.

Levan, A. 1939. The effect of colchicine on meiosis in *Allium*. *Hereditas (Lond.)* 25, 9–26.

Riley, R. and V. Chapman 1958. Genetical control of the cytologically diploid behaviour of hexaploid wheat. *Nature* 182, 713–715.

Sheldon, J.M. and H.G. Dickinson 1986. Pollen wall formation in *Lilium*; the effect of chaotropic agents and the organisation of the microtubular cytoskeleton during pattern development. *Planta* 168, 11–23.

Shepard, J.E., R. Boothroyd and H. Stern 1974. The effect of colchicine on synapsis and chiasmata formation in microsporocytes in *Lilium*. *Chromosoma* 44, 423–437.

Stern, H. and Y. Hotta 1984. Chromosome organisation and the regulation of meiotic prophase. In: *"Controlling Events in Meiosis"*, C.W. Evans and H.G. Dickinson, eds, *Symp. Soc. Exper. Biol.* 38, 161–177.

Thomas, J.B. and P.J. Kaltsikes 1976. A bouquet-like attachment plate for telomeres in leptotene of rye revealed by heterochromatic staining. *Heredity* 36, 155–162.

Thomas, J.B. and P.J. Kaltsikes 1977. The effect of colchicine on chromosome pairing. *Can. J. Genet. Gytol.* 19, 231–249.

von Wettstein, D. 1984. The synaptonemal complex and genetic segregation. In: *"Controlling Events in Meiosis"*, C.W. Evans and H.G. Dickinson, eds, *Symp. Soc. Exper. Biol.* 35, 195–231.

Walker, R.I. 1938. The effect of colchicine on microspore mother cells and microspores of *Tradescantia paludosa. Amer. J. Bot.* 25, 280–285.

STRUCTURAL AND NUMERICAL VARIATION OF CHROMOSOMES IN NATURE AND CULTURE

Critical reassessment of DNA content variation in plants

J. Greilhuber

Institute of Botany, University of Vienna, Rennweg 14, A–1030 Wien, Austria

The phenomenon of genome size variation in plants, in particular at the infraspecific level, environmentally induced, and in the course of ontogeny, is a topic of current interest in evolutionary and developmental karyology (Nagl 1979; Bachmann *et al.* 1985; Walbot & Cullis 1985). The present contribution aims at a critical discussion of some reports on unorthodox genome size variation, and is not directly concerned with "normal" DNA content variation, which is well known to occur between closely related species or even at still lower taxonomic levels. Here, topics of concern are DNA contents in conifers, which are apparently outstanding examples of DNA content instability (see for instance Berlyn *et al.* 1987), and developmentally mobile genomic DNA contents, which are reported, among others, in *Sambucus racemosa* as "floral DNA", and in *Hedera helix* as a phenomenon connected with the physiological switch from the juvenile to the adult growth phase (Kessler & Reches 1977; Nagl 1979; Schäffner & Nagl 1979).

It will be shown that methodological problems associated with cytophotometric DNA content determination in plants have been grossly underestimated in the past, and that a significant part of the reports on fluctuating genomic DNA contents can be attributed to unrecognised stoichiometric errors induced by plant tannins.

Material and Methods

Details are given in Greilhuber (1986, 1988) and König *et al.* (1987). Staining for nuclear DNA was performed with the Feulgen technique, after fixation with either methanol-acetic acid (3:1) or neutral formaldehyde, and hydrolysis with 5N HCl at 20° C. Measurements were performed by scanning densitometry on a Leitz MPV II cytophotometer interfaced to a PDP 11/23 computer steered by program CELANB, and to a minor extent by two-wavelength cytophotometry on the same instrument, as indicated in the Tables. In all instances *Allium cepa* root-tips were processed as an internal standard, in which 1C was taken as 16.75 pg (see Bennett & Smith 1976). Calibrated values are given as $\bar{x} \pm SD$, when every nucleus of the probe was calibrated against one randomly selected onion nucleus, or as $\bar{q}_w \pm s_{q_w}$, when relative values and their standard deviation ($\bar{q} \pm s_q$)

were calculated from each slide, and weighted means and their standard deviation ($\bar{q}_w \pm s_{q_w}$) were then formed (see Krug 1981). Both methods of calculation are equivalent and give practically the same standard deviations.

"Self-tanning" — a new and important source of stoichiometric error in cytophotometric determination of nuclear DNA content in plants

DNA content determinations in plants are usually performed cytophotometrically on nuclei subjected to the Feulgen reaction (Feulgen & Rossenbeck 1924), which is commonly regarded as one of the oldest and most specific reactions in histochemistry. The accuracy of the method for reproducibility of data is usually estimated to be as good as 5 to 10% (Bennett & Smith 1976), but observant plant cytologists occasionally have been aware that this is possibly not true for all materials. In particular, DNA contents in conifers seemed to have been especially controversial. Selected examples from the literature are given in Table 1, and it can be seen that very divergent estimates have been made in the same

Table 1. Some published examples of discordant Feulgen-DNA values in two conifer genera, Picea and Pinus.

Species	DNA value	Author
Picea lambertiana	34.7 pg[1]	Dhillon 1980
	87.7 pg (2C)[2]	Rake et al. 1980
Picea glauca	40.34 pg (2C)[2]	Teoh & Rees 1976
	19.3 pg (2C)[2]	Rake et al. 1980
Pinus caribaea		
var. caribaea	11.53 pg (2C)[2]	Berlyn et al. 1987
var. hondurensis	21.15 pg (2C)[2]	" "
var. bahamensis	25.23 pg (2C)[2]	" "
	37.5 – 87.5 (3C–7C)[1]	" "

[1] per interphase nucleus from root tips of dormant embryos

[2] from seedling root tips

species. Moreover, dormant embryos are sometimes reported to have more DNA, and more variable DNA content, than is expected from a normal embryo, which should have its nuclear DNA contents mainly at 2C, 4C, and perhaps in between. For instance, in dormant embryos of *Pinus coulteri, P. rigida*, and *P. caribaea*, nuclear DNA contents ranging from 2C up to 7C were reported, and in all three species the extra DNA went away shortly after germination (Dhillon *et al.* 1978; Patel & Berlyn 1982; Berlyn *et al.* 1987), thus constituting evident cases of DNA variation in development. Genetic instability in the sense of fluctuating DNA contents is also reported from tissue culture of conifers, often in connection with auxin application (Berlyn *et al.* 1986, 1987).

A solution to this unorthodox DNA content variation appeared when I found that in conifers meristematic root tip cells accumulate considerable amounts of non-hydrolysable tannins in their vacuoles. These extravasate after the commonly-used alcoholic fixations, because they remain soluble, tan the meristem, and strongly interfere with the Feulgen reaction (Greilhuber 1986). Nuclear staining is reduced to 25% compared with non-tanned nuclei, and the absorbance spectrum of the nuclei is distorted. Clearly, in the presence of tannins the Feulgen reaction is useless for quantitative purposes, if no precaution is taken to overcome the "self-tanning error". Fortunately, fixation with neutral formaldehyde paralyses this stoichiometric error by polymerising these tannins within the vacuoles, so that they cannot interfere with Feulgen-DNA staining later. Thus, formaldehyde fixation produces DNA values which are more correct. A comparison of Feulgen-DNA content determinations after methanol-acetic acid and formaldehyde fixation shows that this type of stoichiometric error is of course not only a problem in conifers, but in all plants known to contain tannins (in the wider sense) in certain tissues, e.g. ferns, *Rosaceae, Geraniaceae, Juncaceae* and *Cyperaceae* (Table 2). This also holds true for plants (e.g. *Quercus* spp.) containing tannins of the hydrolysable type, such as gallotannins. In the absence of interfering compounds no divergence of DNA estimates is evident after different types of fixation (Table 3).

Table 2. Nuclear Feulgen-DNA measurements in species in which fixation-dependent differences of genome size estimates were found. All differences are highly significant, $p < 0.0001$ (from Greilhuber 1986, 1988).

Species	Feulgen-DNA 1C (pg)					Ratio MAA/F
	Formaldehyde (F)			Methanol-acetic acid (MAA)		
	$\overline{q}_w \pm s_{q_w}$	$\underline{n}$		$\overline{q}_w \pm s_{q_w}$	$\underline{n}$	
Gymnocarpium robertianum[2]	5.89 ±0.48	(50)		2.90 ±0.89	(20)	0.492
Ginkgo biloba[2]	9.88 ±0.87	(50)		7.69 ±1.30	(50)	0.778
Picea pungens f. glauca[2]	18.17 ±0.78	(50)[4]		4.45 ±0.52	(50)	0.245
Larix decidua[2]	11.46 ±0.79	(100)		2.83 ±0.56	(50)	0.247
Pinus mugo[2]	20.16 ±1.57	(50)[4]		5.94 ±0.62	(30)[4]	0.295
Pinus cembra[2]	24.16 ±2.37	(50)[4]		7.01 ±0.71	(30)[4]	0.290
Taxus baccata[2]	11.05 ±0.52	(50)		6.58 ±2.39	(100)	0.596
Quercus petraea[1]	0.897 ±0.100	(50)		0.525 ±0.065	(50)	0.585
Pelargonium radula[2]	8.09 ±0.72	(50)		4.81 ±0.86	(50)	0.595
Cyperus haspan[2]	0.445 ±0.038	(50)		0.133 ±0.03	(50)	0.299
Juncus effusus[2]	0.308 ±0.031	(50)		0.092 ±0.018	(50)	0.299
Luzula forsteri[2]	0.704 ±0.089	(50)		0.234 ±0.065	(50)	0.332

[1] Root tips from germinating seeds.

[2] Root tips from potted plants.

[3] Meristems from stem buds.

[4] Values obtained by two-wavelength cytophotometry.

Table 3. Nuclear Feulgen-DNA measurements in species in which no or only marginally significant fixation-correlated differences in genome size estimates were found (from Greilhuber 1988).

Species	Feulgen-DNA 1C (pg)				Ratio MAA/F	ANOVA (p)
	Formaldehyde (F)		Methanol-acetic acid (MAA)			
	$\overline{q}_w \pm s_{q_w}$	n	$\overline{q}_w \pm s_{q_w}$	n		
Collinsia heterophylla[1]	2.06 ±0.25	(50)	2.02 ±0.20	(50)	0.981	0.3792
Brassica napus[1]	1.14 ±0.12	(50)	1.14 ±0.14	(50)	1.000	1.0000
Artemisia dubia[2]	8.84 ±0.51	(50)	8.86 ±0.87	(50)	1.002	0.8888
Artemisia argentea[2]	5.13 ±0.36	(50	5.08 ±0.53	(50)	0.990	0.5823
Rhoeo spathacea[2]	7.25 ±0.45	(50)[3]	7.24 ±0.37	(50)[3]	0.999	0.9036
Puschkinia scilloides[2]	6.90 ±0.53	(50)	6.87 ±0.32	(50)[4]	0.996	0.7326
Scilla greilhuberi[2]	13.30 ±1.07	(50)	12.90 ±0.86	(50)[4]	0.970	0.0420
Scilla persica[2]	20.92 ±1.56	(50)	21.02 ±0.79	(50)[4]	1.005	0.6868
Scilla siberica[2]	31.66 ±2.40	(50)	32.28 ±1.67	(148)[4]	1.020	0.0451
Scilla bulgarica[2]	5.15 ±0.54	(50)	5.14 ±0.38	(97)[4]	0.998	0.8965

[1] Root tips from germinating seeds.

[2] Root tips from potted plants.

[3] Values obtained by two-wavelength cytophotometry.

[4] Values obtained by two-wavelength cytophotometry, previously published in Greilhuber (1977), Deumling & Greilhuber (1982), and Greilhuber & Speta (1985).

That tannins and related compounds are the cause of distorted Feulgen staining can be shown further by artificially applying phenolics such as (+)-catechin, quercetin and tannic acid to methanol-acetic acid-fixed root tip meristems of *Allium cepa*, a species having no tannins in the root tips. These compounds distort Feulgen staining similarly to natural plant tannins (Table 4).

Table 4. Examples of reduction in Feulgen-staining after external application of polyphenol solutions to root tips of <u>Allium cepa</u> fixed with methanol-acetic acid.

	Feulgen-staining in % of control	
	$\overline{q}_w \pm s_{q_w}$	<u>n</u>
(+)-Catechin[1]	24.27 ± 2.51	30
(+)-Catechin[2]	22.77 ± 2.82	30
Quercetin[3]	55.84 ± 2.77	30
Tannic acid[4]	43.32 ± 3.42	30

[1] saturated solution in distilled water

[2] 40 mg/ml in methanol

[3] saturated solution in methanol

[4] 100 mg/ml in distilled water

A general conclusion which can be derived from these results is that all data on DNA amounts in conifers, where non-additive fixatives were used (such as Carnoy no. 2), differ strongly from the true values. The "self-tanning error" consistently results in DNA content estimates which are too low and vary erratically, erroneously being interpreted up to now as showing genuine DNA variation (Berlyn *et al.* 1987). The few authors who preferred to use formaldehyde as a fixative were unable to confirm some previous reports, for instance on geographical DNA content variation in certain conifer species (Teoh & Rees 1976). Unfortunately, Carnoy no. 2 (ethanol-chloroform-glacial acetic acid, 6:3:1) has been the recommended and most frequently used fixative in quantitative conifer cytology (Dhillon *et al.* 1983; Berlyn *et al.* 1986, 1987).

"Floral DNA" in *Sambucus racemosa* — an artefact

"Floral DNA" (Wardell & Skoog 1973) is a term which was used by Nagl *et al.* (1979) to denote a transitory increase in DNA content of nuclei in young flower buds of several plants, namely *Scilla decidua, Rhoeo discolor* and *Sambucus racemosa*. This increase over the basic genome size, as measured in vegetative buds or root tips, applies also to mitotic nuclei, which means that the "amplified" extra DNA is somehow integrated into the chromosome, probably for several or many mitotic steps. In a similar way, as originally suggested by Wardell & Skoog (1973), Nagl *et al.* (1979) and Nagl (1979) interpreted their data on DNA increase as a physiological process connected with floral induction. According to these authors, in *Sambucus racemosa* vegetative buds have a 2C value of 21.1 pg, while floral buds have 29.5 pg, which is about 40% more than in vegetative cells. This was reported to occur in early spring, when flowers are differentiated, but are usually still in the pre-meiotic stage. DNA extracted from flower buds differed significantly from vegetative bud DNA. Tm, percent GC, and amount of repetitive DNA were all higher in floral buds. However, no evidence for floral DNA was found in a related species, *Sambucus nigra*, by cytophotometrical and biochemical means.

Since the genus is known to accumulate tannins of the non-hydrolysable type (catechin derivatives) in certain cells, it occurred to me that stoichiometric errors as described above might provide a better explanation for "floral DNA" than genuine DNA variation. Vegetative and flower buds of *Sambucus racemosa* and *S. nigra* were therefore fixed in methanol-acetic acid and in neutral formaldehyde, using *Allium cepa* root tips as standard, and compared by Feulgen cytophotometry.

In *S. racemosa*, mitotic nuclei from vegetative meristems, and ovary tissue and anthers from young flower buds consistently yielded the same values, namely 11.42 pg (1C-level) in the mean. After methanol-acetic acid fixation consistently lower and more variable values were obtained. This was clearly due to "self-tanning", which was strongest in the organ having the largest amount of tannins, i.e. the anthers, whose epidermal cells at this stage of development contain large vacuoles filled with polyphenols. Since Nagl *et al.* (1979) obviously used ethanol-acetic acid (3:1) as fixative, it is clear that their data are obsolete. Moreover, my measurements of formaldehyde-fixed meristems refute the idea of "floral DNA" in this plant (Table 5).

In *Sambucus nigra*, genome size was estimated from formaldehyde-fixed vegetative buds to be 12.28 pg (1C). After methanol-acetic acid fixation divergent values were obtained. Due to the presence of tannin cells there was much staining reduction around these cells in slides from vegetative buds and young leaves, and the means were lower than after formaldehyde fixation. However,

Table 5. <u>Sambucus racemosa</u>. Feulgen-DNA values in picograms (1C) of mitotic nuclei from vegetative and young flower buds, fixed with formaldehyde and methanol-acetic acid. Values obtained after formaldehyde fixation are not different from each other (p = 0.5081). All other values differ from each other significantly (p < 0.0001).

Tissues	Formaldehyde		Methanol-acetic acid	
	$\overline{q}_w \pm s_{q_w}$	$\underline{n}$	$\overline{q}_w \pm s_{q_w}$	$\underline{n}$
Vegetative buds[1]	11.50 ± 0.59	50	8.86 ± 1.48	50
Ovary[2]	11.37 ± 0.60	50	9.52 ± 0.78	50
Anthers[2]	11.40 ± 0.56	50	5.00 ± 0.85	50

[1] fixed in March and August

[2] fixed in late March and early April

embryonal flower buds, which did not contain tannins at that stage of development, gave a value the same as that found after formaldehyde fixation (Table 6). It remains unexplained why Nagl *et al.* (1979) did not also find

Table 6. <u>Sambucus nigra</u>. Feulgen-DNA values in picograms (1C) of mitotic nuclei from vegetative and embryonal flower buds, fixed with formaldehyde and methanol-acetic acid. Vegetative buds fixed with formaldehyde and embryonal flower buds fixed with methanol-acetic acid are not different from each other (p = 0.2786), all other values differ significantly (p < 0.0001).

Fixative	Tissues	Feulgen-DNA	
		$\overline{q}_w \pm s_{q_w}$	$\underline{n}$
Formaldehyde	Vegetative buds[1]	12.28 ± 1.02	100
Methanol-acetic acid	Vegetative buds[2]	7.86 ± 2.80	30
	young leaves below inflorescence[3]	9.99 ± 2.44	80
	embryonal flower buds[3]	12.46 ± 0.81	50

[1] fixed in March and August

[2] fixed in April

[3] fixed in March

evidence for "floral DNA" in *Sambucus nigra*, in which the fixation-dependent differences between vegetative and floral buds are expressed better than in *S. racemosa*. Moreover, it should be noted that in Nagl *et al.* (1979) the higher values, which presumably stem from less strongly tanned nuclei, are larger than the present values by 1.29-fold in *S. racemosa*, and by 1.24-fold in *S. nigra*.

Negative evidence for differential DNA replication associated with juvenile-adult phase change in the ivy, *Hedera helix*

Hedera helix is another plant reported to undergo significant ontogenetic changes in cellular and nuclear DNA content, proportion of repetitive DNA, base content and chromatin structure. In this case they are reported to occur in connection with the physiologically important and morphologically conspicuous shift from the juvenile to the adult growth form. Kessler & Reches (1977) stated that there are 8.0 and 6.7 pg DNA per cell in juvenile and adult leaves respectively. Kinetic genome size estimates were taken to mean that nuclei are approximately 2C in juvenile and 4C in adult leaf tissue. Re-association data indicated that the DNA increase was caused by differential replication of certain sequences rather than by simple endo-reduplication. Schäffner & Nagl (1979) also found a significant difference in Feulgen DNA content of nuclei in juvenile and adult bud meristems and leaf tissue. Although the authors did not comment on this point, they found, unlike Kessler & Reches (1977), more DNA in adult than in juvenile tissues, namely 6.2 and 3.6 pg (2C + 71% and 2C, respectively). Much more heterochromatin was found in juvenile than in adult tissues. However, more repetitive DNA was stated to occur in the adult phase. This was interpreted by Schäffner & Nagl (1979) to mean "that the change in nuclear DNA content between juvenile and adult branches of the ivy is due to a polyploidization step in which the heterochromatin is not participating", i.e. a certain form of differential DNA replication. Nagl (1979) speculated "that under-replication to the near-tetraploid state of the meristematic nuclei causes changes in the growth parameters so that branches of the adult phase develop, which display altered leaf form, altered geotropism and many other changes". It seems that Schäffner & Nagl (1979) and Nagl (1979) interpreted their data as evidence that in adult meristems a relatively stable doubling of euchromatic DNA occurs in the genome, and that this somehow remains integrated in the adult chromosome complement, because it is obviously passed on from mitosis to mitosis and is diminuated later, but prior to the emergence of a new juvenile ivy plant.

In my laboratory a Feulgen-cytophotometric re-investigation of this topic was undertaken for two reasons (König *et al.* 1987). First, this seemed appropriate with respect to the importance of these data for the understanding of regulatory mechanisms in plant development. Second, the subject is important for practical quantitative karyosystematics, because one needs to know whether there is a certain genome size characteristic at least for a given individual, if not for a given species, or whether there is an array of values, each pertaining to a certain developmental stage.

The data are summarised in Table 7. Genome size from mitotic nuclei was measured from vegetative shoot meristems, young leaves, and root tips obtained from cuttings. Nuclear DNA contents were also measured from mitotically quiescent shoot meristems fixed during February, March and late September,

Table 7. DNA content in picograms (1C) of mitotic nuclei (M) and interphase nuclei (G_1) in the juvenile and adult phase of Hedera helix (from König et al. 1987).

Tissues	Nuclei	Juvenile phase $\bar{x}\pm SD$	n	Adult phase $\bar{x}\pm SD$	n	p
Shoot meristem (summer)	M	1.46±0.23	128	1.46±0.17	129	0.963
Young leaves	M	1.49±0.16	102	1.46±0.14	87	0.073
Root tips	M	1.49±0.24	88	1.52±0.17	89	0.386
Shoot meristem (spring and fall)	G_1	1.54±0.22	75	1.56±0.14	83	0.371
Old leaves (upper side)	G_1	1.34±0.29	80	1.36±0.37	75	0.805
Old leaves (lower side)	G_1	1.20±0.23	38	1.19±0.16	28	0.735

Note: n, number of nuclei; p, probability (t-test)

and from leaf parenchyma. No evidence for phase-correlated differential genome size or nuclear DNA contents was found. Negative evidence was also obtained for phase-correlated differences in heterochromatin content (see König *et al.* 1987 for details). Resting shoot meristems showed DNA contents slightly enhanced (by 4.7%) compared with mitotic nuclei, but this is a small difference demanding instrumental rather than biological explanations (e.g. glare, residual distributional error). Nuclei from old leaves showed moderately, but significantly reduced values, but there was no difference between phases. This deviation from the expected 2C values in aged tissues requires further investigation.

It is therefore obvious that "differential replication" in the ivy, in the sense as described by Schäffner & Nagl (1979) and Nagl (1979), is not real. It must be an artefact due to some unrecognised technical error, although in this case plant phenolics do not seem to have played a major role (see König *et al.* 1987).

Conclusions

From the present data it would appear that much evidence in favour of ontogenetically fluctuating genomic and nuclear DNA contents in plants has been accepted too credulously. It is shown that the use of alcoholic fixatives in plant cytophotometry is the most likely way to create errors, because these fixatives are solvents for tannins, which react with chromatin during fixation and consequently strongly disturb quantitative DNA staining. This has been the case with conifers especially. Neutral formaldehyde is superior as a fixative, because it is apparently able to abolish tannin reactivity. This is evident in the case of non-hydrolysable tannins, which are stably polymerised within the vacuole. It remains to be seen, of course, whether this method of fixation is a panacea against all possible and perhaps still unrecognised interfering plant metabolites.

However, not in every case discussed here are the reports on ontogenetically variable genomic DNA amounts solely due to errors of the "self-tanning" type. For instance, in *Sambucus racemosa* "floral DNA" is considered not to be real, because no DNA content variation is evident between vegetative and floral buds after formaldehyde fixation, while condensed tannins create such variation after alcoholic fixation. On the other hand, published DNA contents in *Sambucus racemosa* and *S. nigra* (Nagl *et al.* 1979) diverge from the present measurements in such a way that additional sources of methodologically-produced variation must be assumed. The same is true of "differential replication" associated with the juvenile-adult phase change in *Hedera helix* (Schäffner & Nagl 1979), which is shown to be not real, but where no obvious explanation in terms of "self-tanning" by secondary plant metabolites is available.

Acknowledgements

My thanks are due to Dipl.-Ing. Christiane König for her expert EDP work. The study was supported by the Austrian Research Fund (FWF), grant P5363.

References

Bachmann, K., K.L. Chambers and H.J. Price 1985. Genome size and natural selection: observations and experiments in plants. In *The Evolution of Genome Size*, T. Cavalier-Smith, ed., 267–276. Chichester, New York: John Wiley & Sons.

Bennett, M.D. and J.B. Smith 1976. Nuclear DNA amounts in angiosperms. *Phil. Trans. R. Soc. London B*, 274, 227–274.

Berlyn, G.P., R.C. Beck, and M.H. Renfroe 1986a. Tissue culture and the propagation and genetic improvement of conifers: problems and possibilities. *Tree Physiol.* 1, 227–240.

Berlyn, G.P., M.K. Berlyn and R.C. Beck 1986b. A comparison of internal standards for plant cytophotometry. *Stain Technol.* 61, 297–302.

Berlyn, G.P., A.O. Anoruo, R.C. Beck and J. Cheng 1987. DNA content polymorphism and tissue culture regeneration in Caribbean pine. *Can. J. Bot.* 65, 954–961.

Dhillon, S.S. 1980. Nuclear volume, chromosome size and DNA content relationships in three species of *Pinus*. *Cytologia* 45, 555–560.

Dhillon, S.S., G.P. Berlyn and J.P. Miksche 1978. Nuclear DNA content in populations of *Pinus rigida*. *Amer. J. Bot.* 65, 192–196.

Dhillon, S.S., J.P. Miksche and R.A. Cecich 1983. Microspectrophotometric applications in plant science research. In *New Frontiers in Food Microstructure*, D.B. Bechtel, ed., 27–69. St. Paul, Minnesota: The American Association of Cereal Chemists, Inc.

Feulgen, R. and H. Rossenbeck 1924. Mikroskopisch-chemischer Nachweis einer Nucleinsäure vom Typus der Thymonucleinsäure und die darauf beruhende elektive Färbung von Zellkernen in mikroskopischen Präparaten. *Z. physiol. Chemie* 135, 203–248.

Greilhuber, J. 1986. Severely distorted Feulgen-DNA amounts in *Pinus (Coniferophytina)* after non-additive fixations as a result of meristematic self-tanning with vacuole contents. *Can. J. Genet. Cytol.* 28, 409–415.

Greilhuber, J. 1988. "Self-tanning" — a new and important source of stoichiometric error in cytophotometric determination of nuclear DNA content in plants. *Pl. Syst. Evol.* (in press).

Kessler, B. and S. Reches 1977. Structural and functional changes of chromosomal DNA during aging and phase change in plants. *Chromosomes Today* 6, 237–246.

König, C., I. Ebert and J. Greilhuber 1987. A DNA cytophotemetric and chromosome banding study in *Hedera helix (Araliaceae)*, with reference to differential DNA replication associated with juvenile-adult phase change. *Genome* 29, 498–503.

Krug, H. 1980. *Histo– und Zytophotometrie*. VEB Gustav Fischer Verlag Jena. Nagl, W. 1979. Differential DNA replication in plants: a critical review. *Z. Pflanzenphysiol.* 95, 283–314.

Nagl, W., B. Frisch and E. Frölich 1979. Extra-DNA during floral induction? *Pl. Syst. Evol., Suppl.* 2, 111–118.

Patel, K.R. and G.P. Berlyn 1982. Genetic instability of multiple buds of *Pinus coulteri* regenerated from tissue culture. *Can. J. Forest Res.* 12, 93–101.

Rake, A.V., J.P. Miksche, R.B. Hall and K.M. Hansen 1980. DNA reassociation kinetics of four conifers. *Can. J. Genet. Cytol.* 22, 69–79.

Schäffner, K.-H. and W. Nagl 1979. Differential DNA replication involved in transition from juvenile to adult phase in *Hedera helix (Araliaceae)*. *Pl. Syst. Evol., Suppl.* 2, 105–110.

Teoh, S.B. and H. Rees. 1976. Nuclear DNA amounts in populations of *Picea* and *Pinus* species. *Heredity* 36, 123–137.

Walbot, V. and C. Cullis. 1985. Rapid genomic change in higher plants. *Ann. Rev. Plant Physiol.* 36, 367–396.

Wardell, W.L. and F. Skoog. 1973. Flower formation in excised tobacco stem segments. III. DNA content in stem tissue of vegetative and flowering tobacco plants. *Pl. Physiol.* 52, 215–220.

The inter-relationship of G-banding, C-banding pattern and nucleolus organiser structure in Anuran amphibians

M. King

Natural Sciences Division, Museum of Arts and Sciences of the Northern Territory, P.O. Box 4646, Darwin N.T. 5794, Australia

Approximately 830 species of the 3,521 extant forms of Anuran amphibians have now been examined chromosomally to some degree (Frost 1985; King in press). While many have been studied with relatively inadequate techniques, several hundred have been analysed with one or more of the banding procedures, and a very few have been investigated with particular molecular probes. There is a high degree of conservation in terms of basic chromosome number and form, with many families being inseparable in this respect. Particular families have members with substantially reorganised genomes, e.g. the Eleutherodactylidae and Artholeptidae, (Bogart 1981; Bogart & Tandy 1981), in which most species vary significantly in their chromosome number and morphology and have undoubtedly established many fusion and inversion differences. Nevertheless, conservation of chromosome number and form within most families of Anurans appears to be the rule (King in press).

Unfortunately, this is a superficial impression and it is now known that species sharing the same chromosome number and morphology may differ markedly in terms of the internal distribution of heterochromatin, as has been demonstrated by C-banding and/or fluorescence analyses (Schmid 1978a and b. 1980a, b and c, 1983; King 1980; Anderson 1986). Indeed, in the Hylid genera *Litoria* and *Cyclorana*, no two species have identical karyotypes in terms of their C-band pattern, and many differ by substantial whole-arm blocks of heterochromatin.

The Anuran amphibians also characteristically possess very large nucleolus organiser regions (NORs) on at least one pair of chromosomes. These often show high levels of heteromorphism between homologous NORs (King 1980; Schmid 1982). Schmid (1982) claimed that 67% of the 260 specimens analysed with silver staining, fluorescence and C-banding showed fixed hetero-morphisms in NOR size, i.e. occurring in all cells within a specimen. Bickham & Rogers (1985) also found that 38% of 43 species of turtles were fixed for similar size differences.

In the Hylid genera *Litoria* and *Cyclorana* a complex relationship in the size and form of NORs was determined (King 1980; King *et al.* 1979) and these will be dealt with here.

NOR structure in *Litoria* and *Cyclorana*

There are five distinct types of secondary constriction in *Litoria* and *Cyclorana* species which can be defined on the basis of their mitotic behaviour, C-band pattern and NOR activity (King 1980). Of these, type 1 and 2 (Fig. 1) were termed 'despiralised' constrictions, with massive heteromorphisms in size between

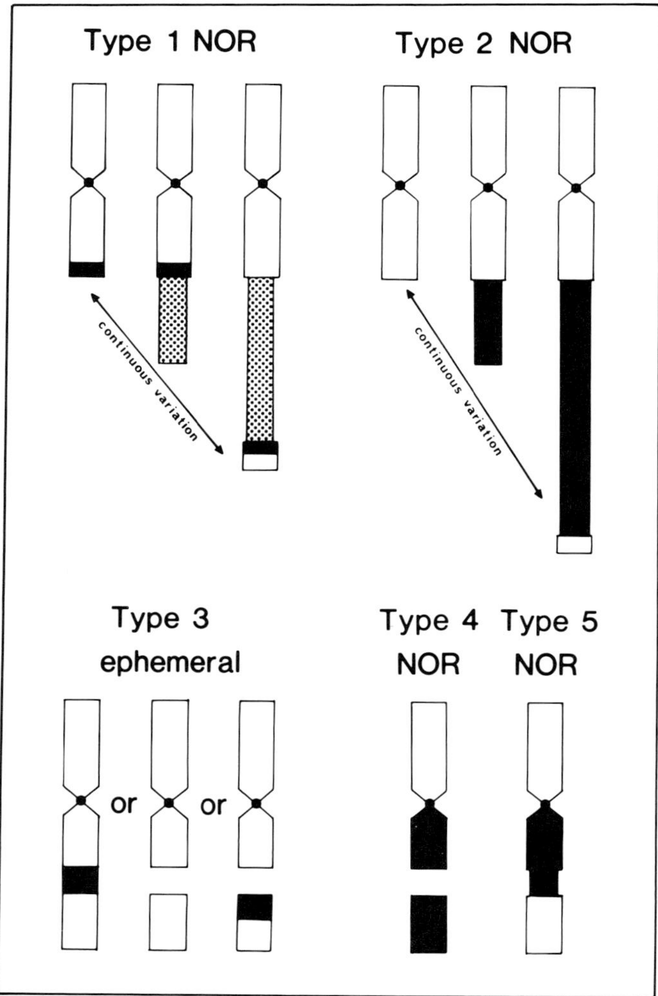

Figure 1. Idiogram showing the relationship between C-banding pattern and the mitotic form of secondary constrictions in *Litoria* and *Cyclorana* species. Note that the type 1 and 2 NORs may show a continuous gradation in size although here shown as three morphs. Some species with type 1 and 2 constrictions have a terminal satellite, and their C-band pattern is illustrated in that element on the right of each type. Type 1, 2, 4, and 5 secondary constrictions are NORs, whereas type 3s are not.

homologues within cells, and between cells within animals, thought to be the result of a packing effect of the DNA within the NOR region, i.e. with some NORs being highly condensed and others less so. Type 1 and 2 NORs are also differentiated by the distribution of heterochromatin (Fig. 1). Type 1 constrictions have C-positive grey heterochromatin within the constriction, and a narrow band of dark C-positive heterochromatin on the periphery of the constriction (either proximal or distal to it). Type 2 constrictions have dark C-positive heterochromatin within the entire constriction but lack the peripheral C-bands. When C-banding does not work effectively, this heterochromatin may be grey rather than black.

Type 3 constrictions are ephemeral in expression. They are often associated with C-bands and do not appear to be NORs, at least in terms of silver staining (King, 1980).

Type 4 and 5 NOR constrictions are more conservative in form, generally lacking the substantial heteromorphisms of types 1 and 2. They are distinguished by their C-band pattern. Type 4 NORs have large C-bands or C-blocks on either the proximal, distal or both margins of the secondary constriction, which itself appears as an achromatic gap (Fig 1). Type 5 NORs have a single large peripheral C-positive block associated with the constriction. Thus, types 1, 2 and 4, 5 are distinguished on the distribution of heterochromatin, whereas types 1, 2 and are distinguished from types 4 and 5 on the basis of the despiralised effect expressed by the former. Significantly, types 1 and 2 are terminal or subterminal, whereas types 4 and 5 are generally interstitial or subterminal.

Two basic questions have been addressed in the present study:
1. What are the type 3 ephemeral constrictions and can they be related to any chromosome structure?
2. Are heteromorphisms in type 1 and 2 'despiralised' constrictions a packing effect, or does this variation reflect polymorphism of ribosomal DNA amount between homologues within cells and between cells within animals?

In an attempt to answer these questions two approaches have been made: firstly to C- and G-band a number of species with different constriction types in an attempt to relate the position of type 3 constrictions to additional structural characteristics; secondly, an 18S + 26S ribosomal probe was hybridised to the chromosomes of species possessing the different constriction types, to determine whether the 'despiralised' effect was due to DNA packing or a difference in DNA amount.

Techniques

The G-banding technique used was essentially that of Sites *et al.* (1979) with a reduction in the time of exposure to trypsin. The C-banding technique is that of King (1980).

The G-banding technique, when applied to amphibians, requires high resolution premetaphase cells derived either from short-term leucocyte cultures, or from other culture techniques. Other forms of chromosome preparation, e.g. from air-dried intestinal or spleen cells, fail to produce G-banding with the trypsin technique. Moreover, it will become apparent that both C- and G-banding methods must be used in tandem. This stems from the complex inter-relationship between the distribution of heterochromatin and the G-banding pattern.

The *in situ* hybridisation technique used in this study is detailed in King *et al.* (1986). Both 18S + 16S *Xenopus laevis* and *Drosophila* probes were used.

G-banding pattern and secondary constriction structure

The G-banding technique is not universally successful when applied to amphibian chromosomes, and a number of workers have argued that the chromosomes of these organisms do not have the structural attributes which will permit G-banding. Thus, Schmid (1978a and b, 1982) and Schempp & Schmid (1981) reasoned that G-banding was absent because:

1. There is a high degree of chromosomal condensation in these organisms.
2. DNA content per unit chromosome length is higher in amphibians than in vertebrate species which G-band.
3. The greater degree of spiralisation of amphibian chromosomes results in the distance between the individual light and dark bands being too small to be resolved by a light microscope.

While some of these propositions may have some merit (see below), this cannot be said for the fourth argument, by Holmquist *et al.* (1982), who made a distinction between higher and lower vertebrates, suggesting that lower forms do not have the capacity to G-band with the trypsin technique because they possess an ancient pattern of genome organisation based on a uniform series of replicon clusters. The differentiated G-band pattern of higher vertebrates is recently derived. The argument was refuted by Stock (1984), who indicated the extensive literature showing G-banding in lower vertebrates.

All of the above arguments were proposed despite the publication of a clearly G-banded chromosome (albeit one), of *Xenopus muelleri* in a comparative study of amphibians, snakes, birds, mammals and turtles (Stock & Mengden, 1975). More recently, Wiley (1982) published a series of G-banded Hylid cells, although resolution was poor, Sekiya & Nekagawa (1983) produced an incomplete G-banded cell of *Xenopus laevis*, Stock (1984) showed G-banding in *Xenopus muelleri*, and King (1985) compared C and G-banding in *Litoria rothi*. Undoubtedly, the paucity of G-banded cells indicates problems with this technique in amphibians. Nevertheless, their presence in the above species negates all statements to the contrary.

A major barrier to the realisation of effective G-banding arises from the uniformity of G-banded euchromatin in certain amphibians. That is, relatively similar dark chromomeric bands are separated by small light interbands in some Anurans, whereas the achromatic interbands are often very large in organisms such as mammals (Wurster-Hill & Gray, 1973). The result is that more condensed G-banded cells of anurans may show large, uniformly dark euchromatic blocks like those in crocodiles (King *et al.* 1986) and tortoises (Dowler & Bickham 1983). Consequently, in frog species in which hetero-chromatic areas are very small, whether they be paracentromeric, interstitial or telomeric, the classic C-positive, G-negative bands that we are led to expect are not readily visible. Thus, species with heterochromatically depauperate karyomorphs may appear to be unbanded when the G-banding technique is used, e.g. many of the amphibians examined by Schmid. In species which have a high degree of heterochromatinisation, the G-band picture is very different.

In *Litoria* and *Cyclorana*, both heterochromatin addition and euchromatin transformation (*sensu* King 1980) appear to have been common evolutionary processes. While most species differ from each other by the amount and distribution of heterochromatin, they also share a generally uniform 2n = 26 biarmed karyotype in which the particular arm ratios are most similar. An exceptional species, *L. infrafrenata*, has 2n = 24 biarmed elements, but the remainder have the 2n = 26 format.

In the G- and C-band analysis undertaken in the present study, the following species were examined: *Litoria coplandi, L. lesueuri, L. peroni, L. raniformis, L. rothi, L. rubella, Cyclorana novaehollandiae, C. australis* and *C. platycephalus*. All of these species have a single major NOR (*L. raniformis* has two), and all have ephemeral type three constrictions. Observations are summarised in Table 1.

The G-band pattern of *Litoria* and *Cyclorana* species is directly related to the C-band pattern. Three distinct relationships can be seen in the heterochromatic component:

1. C-positive/G-negative bands: Generally, large heterochromatic blocks appear as light or achromatic interbands when G-banded. Paracentromeric C-bands generally have this appearance, but this is not obligatory (Fig. 2).

2. C-positive/G-positive bands: Large heterochromatic C-blocks and certain paracentromeric C-bands may be darkly stained after G-banding. These may be slightly darker than the euchromatic bands (Fig. 2).

3. Enhanced bands: When large C-blocks are present in a chromosome arm and show a G-negative pattern, the euchromatin adjacent to that block may become intensively stained, producing a very dark G-band. This 'enhanced' band may be the same size or smaller than the adjacent C-block (Fig. 2). Many of those blocks which are clearly additional heterochromatin appear to have associated enhanced bands.

It is clear that G-bands in *Litoria* and *Cyclorana* are plastic in form. Their complexity is directly related to the C-band pattern. Not only do C and G-bands at the same chromosomal site have a variable inter-relationship (C+/G+, C+/G-) but, most significantly, C-bands have the capacity to modify the adjacent euchromatic regions into enhanced bands. Moreover, certain euchromatic

Table 1. The correlated postion of enhanced bands (+ or -) with that of type 3 secondary constrictions. Short arm (SA); long arm (LA). Proximal (P); interstitial (I); distal (D) to centromere.

Species	|	Chromosome number											
	1	2	3	4	5	6	7	8	9	10	11	12	13
Litoria coplandi	-	LAD +	SAI +	LAP +	SAI +	-	-	-	LAI +	-	-	-	-
L. lesueuri	-	-	-	-	-	-	-	SAI +	-	-	-	-	-
L. peroni	-	LAI +	SAI +	-	-	-	-	SAI +	-	-	SAI +	-	-
L. raniformis	-	-	-	-	-	SAI +	-	SAI +	-	-	LAI +	-	-
L. rothi	-	-	-	-	SAI +	-	LAD +	S&LAI +	-	-	-	-	-
L. rubella	-	-	-	-	LAP +	SAI +	SAI +	LAI +	-	-	LAI +	-	-
Cyclorana australis	LAD +	LDI +	LAD +	-	LAD +	-	LAD +	-	-	-	LAI +	-	-
C. novaehollandiae	LAI +	LAD +	-	LAP +	-	-	-	-	-	-	-	-	-
C. platycephalus	-	-	LAD +	-	-	-	-	-	-	-	-	-	-

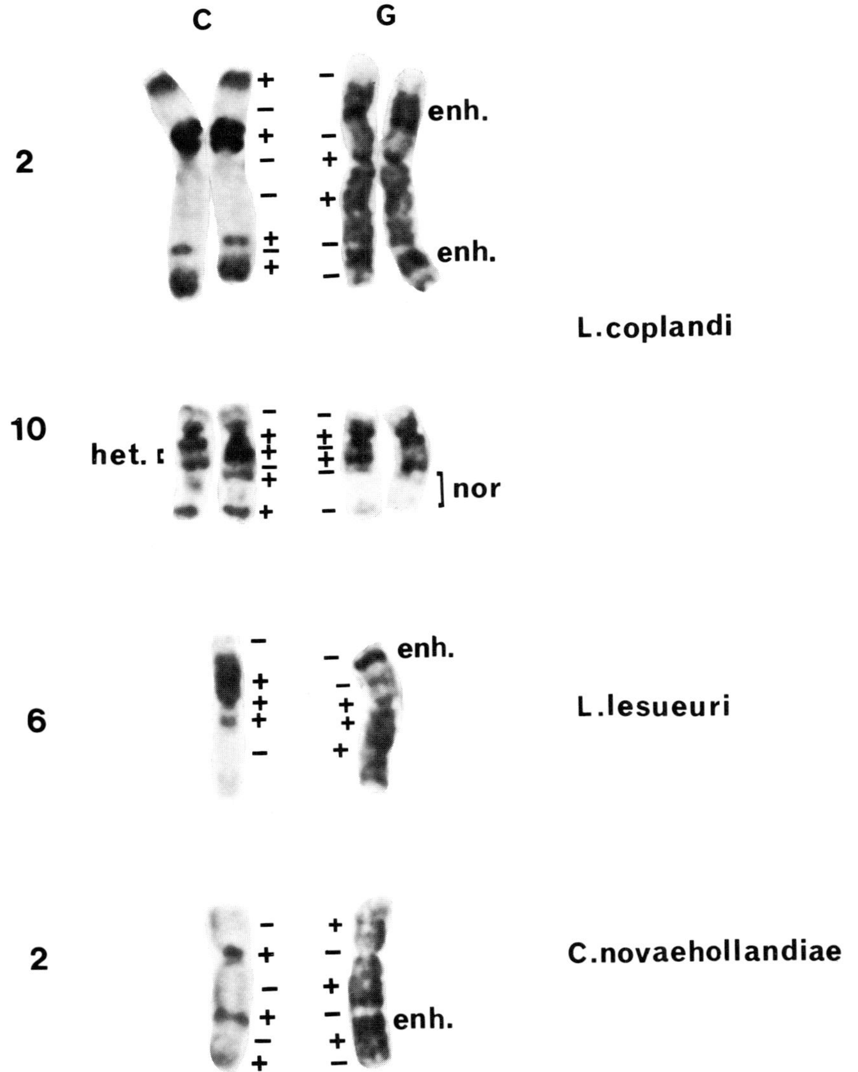

Figure 2. Representative G- and C-banded chromosomes showing the relationship between C- and G-band pattern. Note that C+/G+, C+/G- and C+/peripheral enhanced bands (enh.) are present. Note also that C-/G-, C-/G+ interrelationships also exist. The size of the enhanced band is not directly related to that of the adjacent C+ block.

areas may also have either a positive or negative reaction (Fig. 2). Thus, species with a relatively undifferentiated karyotype, in terms of the distribution of heterochromatin, also generally have a 'simple' G-banded format.

Concomitantly, C-band changes fixed during the course of evolution produce a more complex and unpredictable modification of the G-banded complement than might be expected. Clearly, the phylogenetic value of G-bands as conserved markers is open to question when such plasticity can be demonstrated.

The position of enhanced bands correlates with that of type 3 ephemeral constrictions, and is quite clearly shown in the case of *Litoria lesueuri* (Fig. 3). This is undoubtedly not an obligatory relationship, for these secondary constrictions are known to be ephemeral in their appearance. Moreover, enhanced bands provide a structural basis for type 3 constrictions in *Litoria* and *Cyclorana*, yet in other amphibians these constrictions may be correlated with the position of C-bands or particular euchromatic regions (Schmid 1982 and Fig. 1). While the possibility exists that enhanced bands are the site of significant loci such as the 5S ribosomal RNA cistrons this remains to be investigated. Nevertheless, one concludes that type 3 ephemeral constrictions may not be functional regions analogous to NORs, but preparative artifacts induced by underlying structures, a conclusion strengthened by their variable and erratic appearance.

The *in situ* hybridisation of 18S + 26S ribosomal cistrons

In a joint study with Nelida Contreras (Australian National University, Canberra) and Rodney Honeycutt (Harvard University, Cambridge, Massachusetts) an analysis was made on a series of *Litoria* and *Cyclorana* species with the *in situ* hybridisation technique. An 18S + 26S probe derived from *Xenopus laevis* and a second probe derived from *Drosophila* were hybridised onto the mitotic chromosomes of eight *Litoria* and one *Cyclorana* species. Essentially the same results were obtained with either probe and these data will be provided in detail elsewhere (King, Conteras and Honeycutt in manuscript).

The species utilised (see Table 2) possess the four major types of nucleolus organising constriction and several also possess type 3 constrictions:

Label was never associated with any of the type 3 constrictions, but the major NOR constriction was labelled in every case. The type 1 and 2 despiralised constrictions showed several significant characteristics as follows.

The substantial heterogeneity in size between homologues within and between cells was directly reflected in the grain count, this extraordinary variation being due to absolute differences in the amount of ribsomal DNA present at the NOR (Fig. 4).

Figure 3. a. The mitotic chromosomes of *Litoria lesueuri* (2n = 26). Type 3 constrictions are present on pair 8. A type 4 NOR is present in pair 10. **b.** The C-banded karyotype of *Litoria lesueuri*. Note the major C-blocks on pairs 6 and 8, and associated with the NOR on pair 10. **c.** The G-banded complement of *Litoria lesueuri*. Note the association of enhanced bands with major C-blocks on pairs 6 and 8. Relate the position of the enhanced band on pair 8 to the type 3 constriction on the same element in 3a.

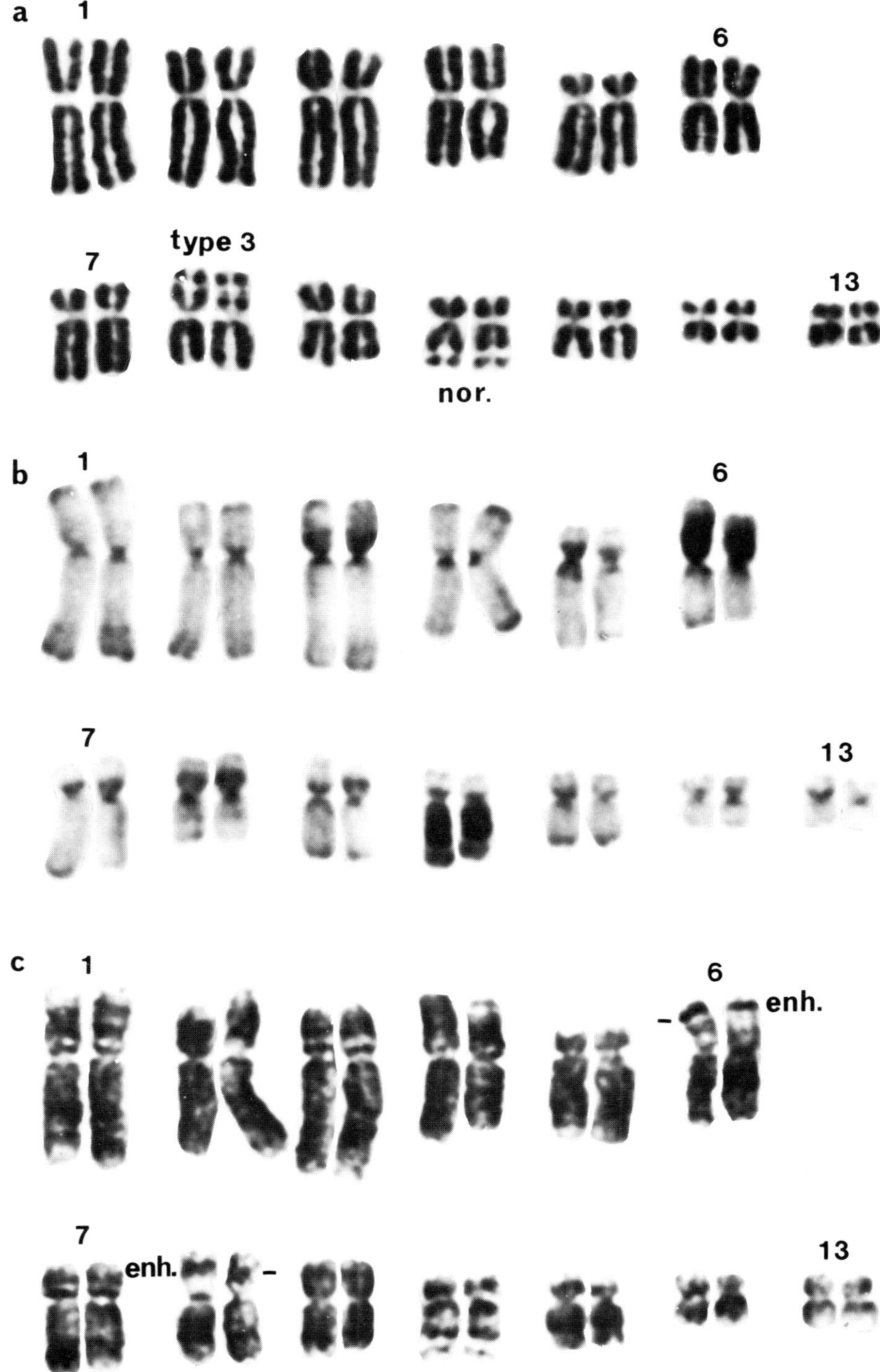

Table 2. Species analysed in 18S + 26S <u>in-situ</u> hybridization experiments and their constriction types

	NOR type	Presence of type 3 constriction
Litoria caerulea	1	−
L. coplandi	4	+
L. lesueuri	4	+
L. moorei	2	−
L. nannotis	5	−
L. peroni	1	+
L. raniformis	2 (2x)	+
L. rothi	1	+
Cyclorana novaehollandiae	4	+

This mosaicism of form was due to two processes: First, the deletion of large blocks of ribosomal DNA. Either 0, 1, or 2 homologues had labelled sites present at the NOR in a cell. The absence of a labelled site was correlated with a corresponding reduction in constriction size. Homologues were therefore polymorphic for NOR deletion. Second, the quantity of ribosomal DNA varied substantially and often erratically between homologues. This polymorphic variation appeared to be almost continuous, and ranged from minute NORs to a three-fold amplification in size of the most common NOR size. Since both types of polymorphism (deletion and amplification), were superimposed on the form of the NOR, these were effectively mosaic structures. The differences between homologues within cells can be seen quite clearly in a cell from *L. moorei* (Fig. 4a) where an amplification/deletion heterozygosity of the NOR is shown. A substantial amplification/partial deletion heterozygosity is also illustrated in a cell from *L. lesueuri* (Fig. 4b). In Fig. 4c and d, two cells on the same slide from an individual of *L. nannotis* clearly show mosaicism in NOR size: 4c is heterozygous, 4d is homozygous.

Fixed heterozygosity for differences in the amount of ribosomal DNA which characterise a single specimen have been detected on numerous occasions in amphibians using the *in situ* hybridisation technique (Miller & Brown 1969; Macgregor & Kezer 1973; Kezer & Macgregor 1973; Hennen *et al.* 1975; Macgregor *et al.* 1977; Vitelli *et al.* 1982). However, these heterozygosities were fixed in each particular specimen. Studies on *Triturus* have also shown that additional minor sites may be present within individuals (Nardi *et al.* 1977; Batistoni *et al.* 1978; Morgan *et al.* 1980). However, the highly polymorphic amplification/deletion mosaicism present in the type 1, 2 constrictions of *Litoria* has not been reported previously. Most significantly, this form of variation, whilst being most pronounced in type 1 and 2 constrictions, is also present in animals with type 5 constrictions (*L. nannotis*) and at a lower incidence in animals with type 4 constrictions (*L. lesueuri, L. coplandi* and *C. novaehollandiae*) (see Fig. 4b). Interstitial or subterminal type 4 constrictions are characterised by possessing large blocks of heterochromatin on the periphery of the NOR. Such

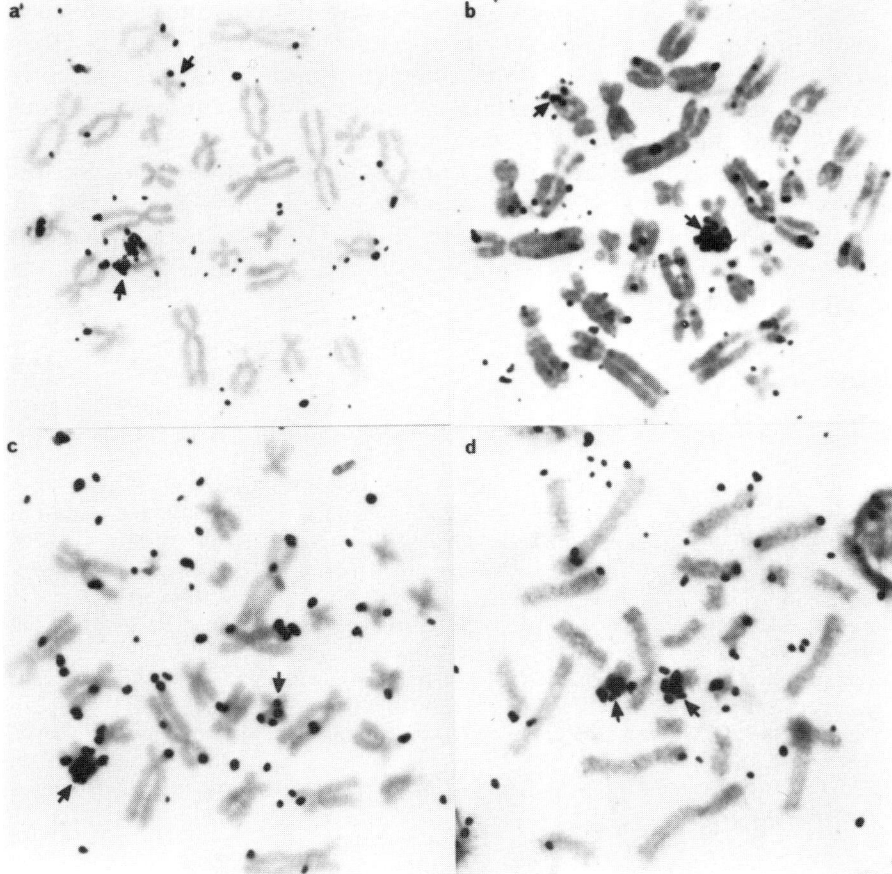

Figure 4. Autoradiographs showing the position of 18S + 26S ribosomal DNA in the genomes of three species of *Litoria*. **a.** A cell from *Litoria moorei* showing an amplification/deletion heterozygosity of the NOR. Note the absence of any label on one of homologues and the contrast to its partner (arrows). **b.** A cell from *Litoria lesueuri* showing an amplification/partial deletion heterozygosity (arrows). **c.** and **d.** Adjacent cells from an autoradiograph of *Litoria nannotis* showing heterozygosity (c) and homozygosity (d) in NOR size.

structures may well reduce the incidence of breakage within the NOR by mechanically stabilising that site (Hsu, 1974) or by preventing recombination in that chromosome arm. Moreover, the more terminal position of type 1 and 2 NORs and their lack of supporting heterochromatin exposes them to the combined effects of association, breakage, recombination, unequal crossing-over, and sister chromatid exchange. All of these processes may be implicated in establishing the variation observed in these unusually plastic NORs.

In summary, the initial questions posed on the nature of the type 3 constrictions and the type 1 and 2 'despiralised' constrictions have been answered in part;

1. Type 3 constrictions are often associated with the position of enhanced G-bands in *Litoria* and *Cyclorana*. Their ephemeral behaviour suggests that they may be preparative artifacts induced by these structures. While they are clearly not the site of 18S + 26S ribosomal cistrons, the possibility cannot be ruled out that they mark the position of 5S loci or other significant sites.

2. The 'despiralised' type 1 and 2 constrictions of King (1980) are not the product of DNA packing, but reflect absolute differences in the quantity of ribosomal DNA present between homologues and cells. Both amplification and deletion polymorphisms appear to be responsible for this variation, resulting in a complex mosaicism of NOR form.

References

Anderson, K. 1986. *A Cytotaxonomic analysis of the Holarctic Treefrogs in the genus Hyla.* PhD. Thesis, New York University.

Bastistoni, R., F. Andronico, I Nardi, and G. Barsacchi-Pilone 1978. Chromosome location of the ribosomal genes in *Triturus vulgaris meridionalis* (Amphibia Urodela). III. Inheritance of chromosomal sites for 18S + 28S ribosomal RNA. *Chromosoma* 65, 231–240.

Bickham, J.W. and D.S. Rogers. 1985. Structure and variation of the nucleolus organizer regions in Turtles. *Genetica*, 67, 171–184.

Bogart, J.P. 1981. Chromosome studies in *Sminthillus* from Cuba and *Eleutherodactylus* from Cuba and Puerto Rico (Anura: Leptodactylidae). *Life Sci. Contrib.*, 129, 1–22.

Bogart, J.P. and M. Tandy 1981. Chromosome lineages in African Ranoid frogs. *Monitore Zool. Ital.*, supp. XV No. 5, 555–91.

Dowler, R.C. and J.W. Bickham 1983. Chromosomal relationship of the tortoises family Testudinae. *Genetica*, 58, 189–197.

Frost, D.R. (Ed.). 1985. *Amphibian Species of the World.* Allen Press Inc. and Assoc. Sys. Coll., Lawrence, Kansas, U.S.A.

Hennen, S., S. Mizuno and H.C. Macgregor 1975. *In situ* hybridization of ribosomal DNA labelled with [125]Iodine to metaphase and lampbrush chromosomes from newts. *Chromosoma* 50, 349–369.

Holmquist, G., M. Gray, T. Porter, and J. Jordan 1982. Characterization of Giemsa dark and light-band DNA. *Cell*, 31, 121–129.

Hsu, T.C. 1974. A possible function of constitutive heterochromatin: the bodyguard hypotheses. *Proc. XII. Int. Congress of genet. Symp. 14: Chromosome structure*, pp. 136–150.

Kezer, J. and H.C. Macgregor 1973. The nucleolar organizer of *Plethodon cinereus cinereus* (Green). II. The lampbrush nucleolar organizer. *Chromosoma* 42, 427–444.

King, M. 1980. C-banding studies on Australia Hylid frogs: Secondary constriction structure and the concept of euchromatin transformation. *Chromosoma* 80, 191–217.

King, M. 1985. Chromosome markers and their use in phylogeny and systematics. In *Biology of Australian Frogs and Reptiles*, eds G. Grigg, R. Shine, H. Ehmann, Surrey, Beatty and Sons, Sydney. pp. 165–175.

King, M. 1986. Chromosomal repatterning in crocodiles: C, G and N-banding and the *in-situ* hybridization of 18S and 26S rRNA cistrons. *Genetica*, 70, 191–201.

King, M. *The Amphibia "Animal Cytogenetics" 4. Chordata 2*. Gerbruder Borntrager. Stuttgart. (in Press).

King, M., M.J. Tyler, M. Davies, and D. King 1979. Karyotypic studies on *Cyclorana* and associated genera of Australian frogs. *Australian Jour. Zool.*, 27, 699–708.

Macgregor, H.C. and J. Kezer 1973. The nucleolar organizer of *Plethodon cinereus cinereus* (Green). I. Location of the nucleolar organizer by in-situ nucleic acid hybridization. *Chromosoma*, 42, 415–426.

Macgregor, H.C., M. Vlad and L. Barnett 1977. An investigation of some problems concerning nucleolus organizer in salamanders. *Chromosoma* 59, 283–299.

Miller, L. and D.D. Brown 1969. Variation in the activity of nucleolar organizers and their ribosomal gene content. *Chromosoma* 28, 430–444.

Morgan, G.T., H.C. Macgregor and A. Colman 1980. Multiple ribosomal gene sites revealed by *in-situ* hybridization of *Xenopus* rDNA to *Triturus* lampbrush Chromosomes. *Chromosoma* 80, 309–330.

Nardi, I., G. Barsacchi-Pilone, R. Batistoni and F. Andronico 1977. Chromosome location of the ribosomal RNA genes in *Triturus vulgaris meridionalis* (Amphibia, Urodela) II. Intraspecific variability in number and postion of the chromosome loci for 18S + 28S ribosomal RNA. *Chromosoma* 64, 67–84.

Schempp, W. and M. Schmid 1981. Chromosome banding in Amphibia VI. BrdU-replication patterns in Anura and demonstration of XX/XY chromosomes in *Rana esculenta. Chromosoma* 83, 697–710.

Schmid, M. 1978a. Chromosome banding in Amphibia. II. Constitutive heterochromatin and nuceolus organizer regions in Ranidae, Microphylidae and Rhacophoridae. *Chromosoma* 68, 131–148.

Schmid, M. 1978b. Chromosome banding in Amphibia I. Constitutive heterochromatin and nucleolus organizer regions in *Bufo* and *Hyla. Chromosoma* 66, 361–388.

Schmid, M. 1980a. Chromosome evolution in Amphibia. In *Cytogenetics of Vertebrates*, ed. H. Muller. Birkhauser Verlag Basel, Boston, Stuttgart pp. 4–27.

Schmid, M. 1980b. Chromosome banding in Amphibia. V. Highly differentiated ZW/ZZ sex chromosomes and exceptional genome size in *Pyxicephalus adspersus* (Anura, Ranidae). *Chromosoma* 80, 69–96.

Schmid, M. 1980c. Chromosome banding in Amphibia. IV. Differentiation of GC- and AT-rich chromosome regions in Anura. *Chromosoma* 77, 83–103.

Schmid, M. 1982. Chromosome banding in Amphibia. VII. Analysis of the structure and variability of NORs in Anura. *Chromosoma* 87, 327–344.

Schmid, M. 1983. Evolution of sex chromosomes and heterogametic systems in Amphibia. *Differentiation* 23, 513–522.

Sekiya, K. and H. Nakagawa. 1983. Cytogenetics of *Xenopus laevis* chromosomes. *Experientia* 39, 786–787.

Sites, J.W., J.W. Bickham, M.W. Haiduk and J.W. Iverson 1979. Banded karyotypes of six taxon of Kinosternid turtles. *Copeia*: 692–698.

Stock, A.D. 1984. The occurrence of G-bands in the mitotic chromosomes of the amphibian *Xenopus muelleri. Genetica* 64, 225–228.

Stock, A.D. and G.A. Mangden 1975. Chromosome banding pattern conservations in Birds and nonhomology of chromosome banding patterns between Birds, Turtles, Snakes and Amphibians. *Chromosoma* 50, 69–77.

Vitelli, L., R. Batistoni, F. Andronico, I. Nardi and G. Barsacchi-Pilone 1982. Chromosomal localization of 18S + 28S and 5S ribosomal RNA genes in evolutionarily diverse Anuran Amphibians. *Chromosoma* 84, 475–491.

Wiley, J.E. 1982. Chromosome banding patterns in treefrogs (Hylidae) of the eastern United States. *Herpetologica*, 38, 507–520.

Wurster-Hill, D.H. and C.W. Gray 1973. Giemsa banding patterns in the chromosomes of twelve species of cats (Felidae). *Cytogenet. Cell Genet*, 12, 377–397.

Synaptonemal complexes in Robertsonian translocation heterozygotes in lemurs

C. Ratomponirina, B. Brun and Y. Rumpler

Faculté de Médecine, Institut d'Embryologie, 11 rue Humann, 67085 Strasbourg Cédex, France

Robertsonian translocations have played an important role in the chromosomal evolution of the lemurs, particularly in the genus *Lemur*, where the use of banding techniques has shown the existence of several species and subspecies differing by such rearrangements (Rumpler & Dutrillaux 1976; Hamilton & Buettner-Janusch 1977). The karyotypes of the six species and seven subspecies of the genus *Lemur* differ by 32 chromosomal rearrangements of which 29 are Robertsonian translocations. In order to find relationships between Robertsonian translocations and speciation in lemurs, crossbreeding was performed between species and subspecies of lemurs with differing karyotypes, and their fertility was tested. Hybridisation attempts have shown the ease of crossbreeding different species and subspecies of *Lemur* (Moses *et al.* 1979) and it has been shown that some of these categories of hybrids are fertile (Ratomponirina *et al.* 1982).

Material and Methods

Crossbreeding was performed between *L. fulvus* (2n=60), *L.f. collaris* (2n=52), *L.f. albocollaris* (2n=48), *L. macaco* (2n=44) and *L. coronatus* (2n=46). For the different hybrids, we studied:
a) fertility, in trying to breed them with *L. fulvus*;
b) spermatogenesis on testicular biopsies, which were performed under general anaesthesia, and fixed in Bouin fixative for histological studies;
c) meiosis with the optical microscope, following the technique of Evans *et al.* (1964), and with the electron microscope, using the technique of microspreading and silver staining (Moses 1977; Dresser & Moses 1980). The best metaphase spreads for study with the electron microscope were first selected with the light microscope. A Philips 60 KV electron microscope was used to photograph the grids using a 9 mm Kodak film.

Results

Three different categories of hybrids were obtained.

1) Fertile hybrids

These are the hybrids between *L. fulvus* and *L.f. albocollaris* (2n=54) and the complex hybrids: "Isidore" (2n=54), "Volmus" (2n=55), "Cesar" (2n=47). These hybrids reproduce in captivity. As shown in Table 1, these hybrids present respectively 4, 6, 5 and 3 trivalents. Spermatogenesis is not significantly different from that of the pure species and produces numerous spermatozoa (Ratomponirina *et al.* 1982) [Table 2].

Table 1. Relation between the number of trivalents, multivalents and spermatogenesis in different lemur hybrids.

Name and kind of hybrid	2n	Number of trivalents	Number of chromosomes included in multivalent	Spermatogenesis
Loft (LFF x LFA)	54	4	0	
Isidore ([LFF x LFA] x LFF)	54	6	0	like in pure
Volmus (Isidore x LFF)	55	5	0	animals
Cesar (Volmus x LFF)	47	3	0	
Felix (LFF x LMA)	52	8	0	
Aviateur (LFF x LMA)	52	8	0	
Potin (LFF x LMA)	52	8	0	variable but
Espion (LFF x LMA)	52	8	0	always reduced
Gros (LFF x LMA)	52	8	0	
Romulus (LFF x LMA)	52	8	0	
Christophe (LFA x LMA)	46	2	14	major
Bassam (LCO x LMA)	45	0	6, 11	perturbations

Table 2. Relation between abnormal meiotic configurations and spermatogenesis in different lemur hybrids.

Hybrids	Number of cells studied	Pseudo-chains	Hypercondensed and unpaired sites	XY-autosome associations	Spermatids/germ cells relations
Loft	15	0	0	0	0.52
Isidore	18	0	0	0	0.64
Volmus	7	0	0	0	0.41
Cesar	11	0	1	0	0.47
Felix	20	1	3	2	0.39
Aviateur	15	0	1	1	0.47
Potin	13	1	2	0	0.29
Espion	13	2	4	2	0.16
Gros	34	6	11	1	0.25
Romulus	9	0	6	0	0
Christophe	10	0	-	0	0.08
Bassam	11	0	9	0	0

During meiosis, the association between one metacentric and two acrocentric chromosomes gives rise to a trivalent. The trivalents show a delay in pairing, the short arms of the acrocentric chromosomes remaining free at early pachytene. This pairing is most often completed later, and at late pachytene synapsis of most trivalents is complete, including 2 acrocentric chromosomes at the level of non-homologous heterochromatic arms. The pairing of trivalents from the same cell is not synchronised, some showing completed synapsis [Fig. 1b] while others still show considerable free portions [Fig. 1a].

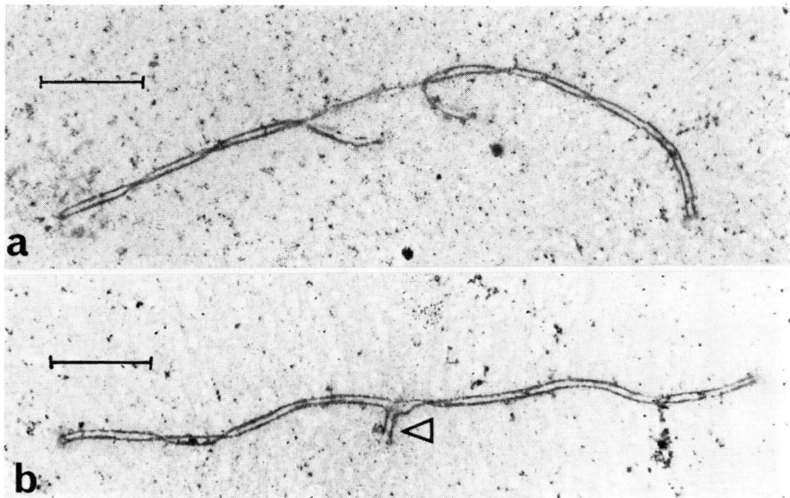

Figure 1a, 1b. Trivalents in a lemur hybrid LFA × LFF (bar = 5 μm) **a**: the trivalent is composed of a metacentric axis and two acrocentric axes, of which the long arms form synaptonemal complexes with the metacentric arms. Synapsis is incomplete in the proximal part of the metacentric chromosome, and those of the acrocentric chromosomes remain unpaired. The short arms are divergent and are evidently the last to pair. **b**: pairing has progressed; the non-homologous short arms of the acrocentric chromosomes (arrow) have synapsed to form a heterosynapsis with a normal-appearing synaptonemal complex.

2) Regularly sterile hybrids

These are the hybrids *L. macaco* × *L.f. albocollaris* (2n=46) and *L. macaco* × *L.f. coronatus* (2n=45).

Gametogenesis is strongly disturbed. When the first stages are present, including spermatocytes I, the majority of germ cells degenerate after the pachytene stage, although spermatogenesis appears complete in a few seminiferous tubes. Nevertheless, the ratio of spermatozoa is clearly lower than in fertile hybrids (Table 2).

The meiotic preparations of the hybrid *L. macaco* × *L.f. albocollaris* show, in addition to the bivalents, 2 trivalents and 1 multivalent involving 14 chromosomes [Fig. 2a]. Meiotic analysis of hybrid *L. macaco* × *L.f. coronatus* shows that 2 multivalents are produced: a long chain involving 11 chromosomes and a ring involving 6 chromosomes [Fig. 2c]. The synapsis of the 2 trivalents of the first hybrid is delayed. The synapsis of the long chains has never been seen

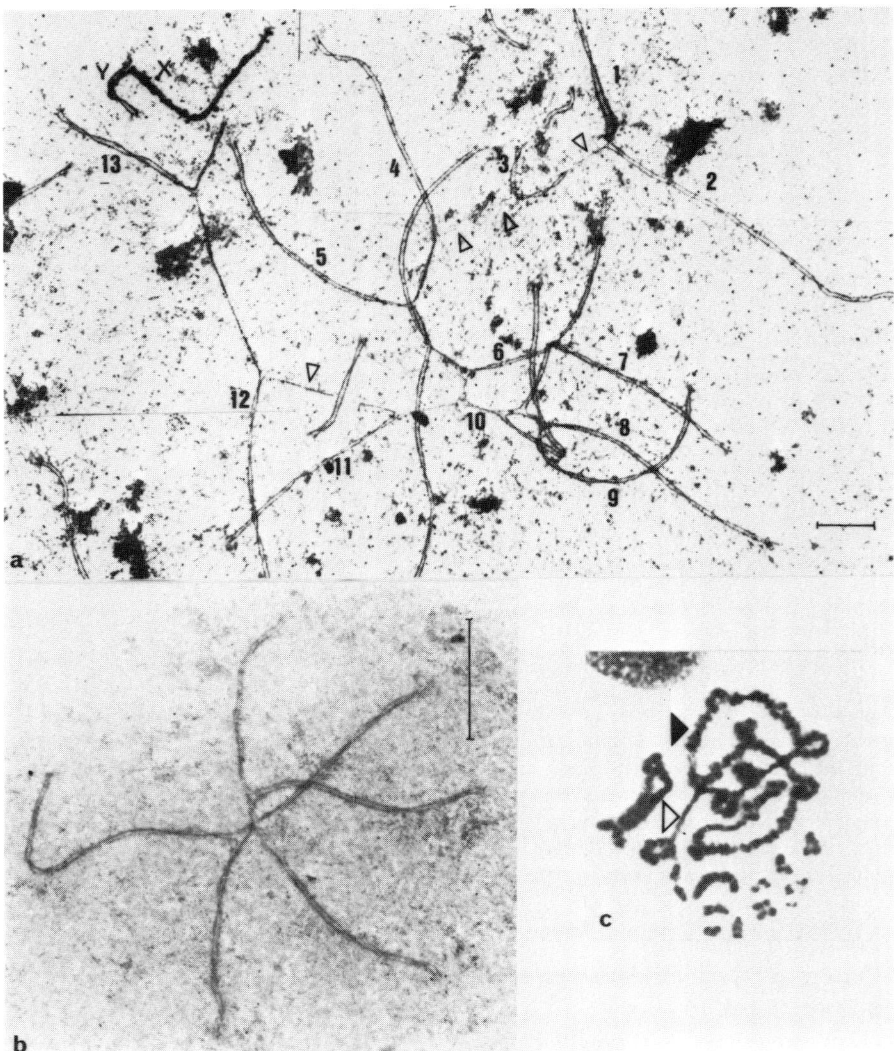

Figure 2a, 2b, 2c. Multivalents (bar = 5 μm): **a**: multivalent synaptonemal complex (SC) of a chain in a hybrid (LFA × LMA) where 14 chromosomes are involved; the synapsed arms are arbitrarily numbered from 1 to 13. Arrows show the parts which are visible with difficulty because of their stretching. **b**: multivalent synaptonemal complex of a ring in a hybrid LMA × LCO where 6 arms are involved and almost completely synapsed. **c**: micrograph of a ring (full arrow) and a chain appearing at diakinensis stage (open arrow) in a hybrid LMA × LCO to be complete in the 21 cells studied, when the ring often seems completely synapsed [Fig. 2b]. Even in late pachytene, large segments of the chromosomes involved in the chain remain asynapsed. In some spermatocytes, parts of the asynapsed segments present thickenings.

3) Hybrids showing irregular fertility
These are the hybrids *L.f. fulvus* × *L. macaco* (2n=52).

The karyotypes of these hybrids show that each of the 8 metacentric chromosomes of *L. macaco* corresponds to 2 acrocentric chromosomes from *L. fulvus*. Spermatogenesis is reduced to varying degrees, but in some cases, there are enough spermatozoa to permit reproduction (1 out of 6 animals). These hybrids show 8 trivalents with delayed pairing, as has been seen in the first category of fertile hybrids. In some cells the unpaired arms of adjacent trivalents can come into contact, synapse and give rise to heterosynapsis. In other cases they have end-to-end association, the asynapsed arms being more dense. In a few cells the unpaired arms of trivalents or of the pseudo-chains come into contact with the sex bivalent [Fig. 3a, 3b, 3c]. In one hybrid, in which most germ cells degenerate early, meiosis revealed important disturbances in synapsis in the four cells analysed: segments were unpaired and hypercondensed.

Discussion

The synaptonemal complexes of the trivalents show a delay in pairing of the centromeric region and the short arms of the acrocentric chromosomes involved. The asynapsed heterochromatic arms of the acrocentric chromosomes heterosynapsing later, as has been reported by Moses *et al.* (1979), have the opportunity to come into contact with those of a neighbouring trivalent giving rise to a chain-like configuration.

In all the cells studied that show such a chain-like configuration, study of synaptonemal complexes reveals a total number of bivalents in agreement with the autosomal pairs of the karyotype. These complex patterns thus result from association of 2 or 3 trivalents and no bivalent contributes. Similar findings have been made in hybrid mice heterozygous for 9 trivalents (unpublished personal observations): chains involving 2 or 3 trivalents mediated by the short arms of the acrocentric chromosomes have been found in a certain number of cells, varying from one individual to another. In contrast, the work of Bogdanov *et al.* (1986) on rodents (*Ellobius talpinus*) heterozygous for 10 Robertsonian translocations mentions the involvement of bivalents in the formation of chains. However, in the majority of cells studied the association of 2 or 3 trivalents predominates.

Among the lemur hybrids studied here, the association of several trivalents into a chain-like structure occurred only in the *L. fulvus* × *L. macaco* hybrid category, suggesting a genic origin. This hypothesis also seems probable since there is a large individual variability in the occurrence of those chain-like structures or even in marked delay in synapsis. The latter are regularly seen in the spermatocytes of one hybrid ("Romulus"), the cells resembling the Z cells described by Beaumont & Mandl (1962). However these abnormal patterns are formed, their presence suggests the possibility that they could be able to affect fertility of the individuals.

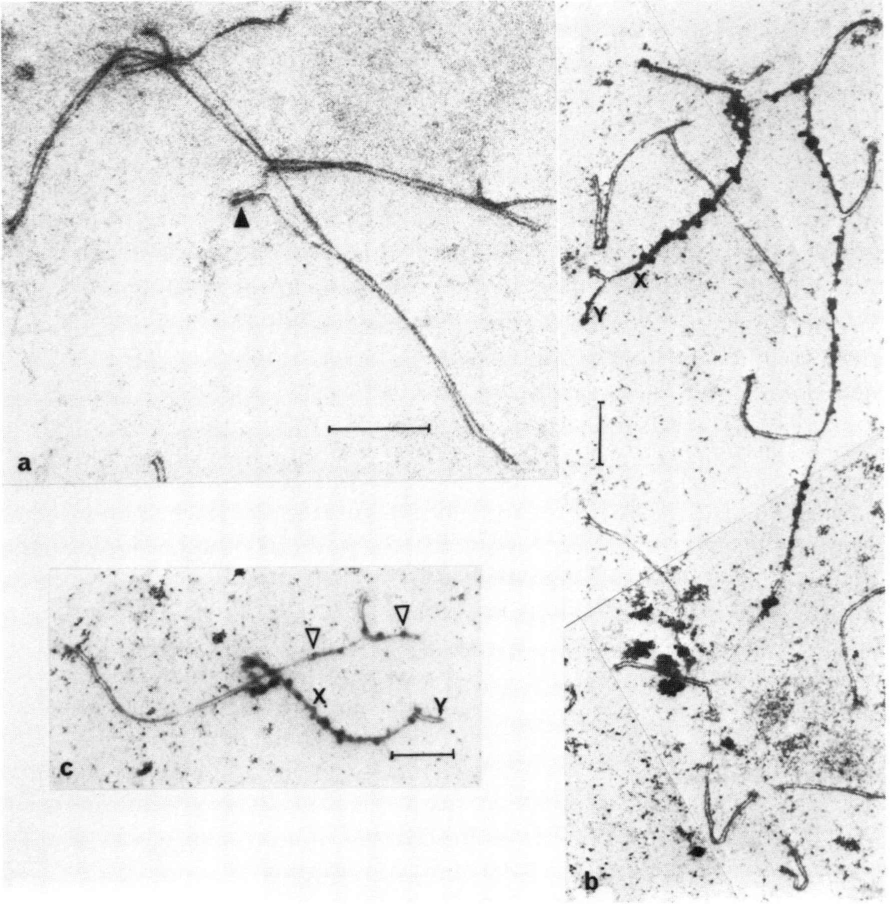

Figure 3a, 3b, 3c. Heterologous associations in hybrids LFF × LMA (bar = 5 μm): **a**: heterosynapsis of the free arms of 2 adjacent trivalents (arrow) forming a pseudochain. **b**: pseudochain involving an end-to-end association of the free arms of trivalents showing thickenings and excrescences. **c**: the free arm of the X chromosome comes into contact with the free arm of a trivalent. Note also excrescences on the unsynapsed arms of the trivalent (arrows).

Several hypotheses have been proposed to explain the relation between abnormal chromosomal configurations and sterility:

1) The hypotheses of Lifschytz & Lindsley (1972) according to which the inactivation of the X chromosome during spermatogenesis is essential and any disturbance to this process hinders the normal development of the spermatocyte, leading to sterility in the male. Forejt (1982) supports this hypothesis in explaining that, in Robertsonian translocations in mice, the unpaired or incompletely paired region of the translocated chromosomes which form non-homologous associations with the sexual bivalent disturb the early inactivation of the X chromosome. The sterilising effect of this rearranged autosome/sexual bivalent association appears to be a general phenomenon not specific to the mouse, since similar cases related to disturbed spermatogenesis have been reported in man (Gabriel-Robez *et al.* 1986). However, the hypothesis of Lifschytz & Lindsley alone does not explain the reduced fertility of the lemur hybrids, in which the sexual bivalent-rearranged chromosome association is rarely observed (Table 2).

2) The hypothesis of Miklos (1974), further developed by Burgoyne and Baker (1984), suggests that there exist chromosomal pairing sites which should normally become saturated during synapsis between homologous elements. A defect in synapsis would keep these sites active and induce a cellular degeneration. In most infertile and hypofertile hybrids (Table 2) many cells showed a marked disorder in synapsis involving the elements entering into chain formation during pachytene. The existence of these large unpaired segments could be taken as support for this latter hypothesis. Also, the unpaired axes of most chains and even of certain isolated trivalents show regions with morphological characteristics such as thickening, hypercondensation and multiple excrescences which are normally observed on the free arms of the sexual bivalent. These abnormalities could reflect metabolic disorders leading to degeneration of the cell. Identical alterations to synaptonemal complexes have been identified in human spermatocytes in cases of inversion (Saadallah & Hultén 1986) and in human and mice XO oocytes (Speed 1986). It should be noted that in the case of chromosomal rearrangements, the germ cells are protected from degeneration if these pairing sites are saturated by homosynapsis (Saadallah & Hultén 1986) and heterosynapsis (Saadallah & Hultén 1986; Gabriel-Robez *et al.* 1987) or even by autosynapsis (Speed 1986).

These results allow us to understand why, if in most cases the existence of multivalents involving more than 5 chromosomes is usually related to male sterility in both mouse (Gropp 1982) and certain lemur hybrids like "Bassam", "Christophe" and "André" (Table 2) or in human pathology (Chandley *et al.* 1986) there are some cases in which multivalents are formed in normal fertile males. Indeed Saadallah & Hultén (1985) reported the case of a human male carrying a complex chromosomal rearrangement leading to a meiotic hexavalent without any defect of spermatogenesis, and Dutrillaux & Rumpler (1977) found pentavalents in a small proportion of spermatocytes in a normal fertile *L. fulvus* × *L.f. albocollaris* hybrid.

On the other hand, the association of these chromosomal multivalents or chain-like structures with a reduced fertility of hybrids supports the role played by Robertsonian translocations in reproductive barriers during speciation of lemurs, as described earlier (Dutrillaux & Rumpler 1977). It seems that these chromosomal rearrangements, as long as they give rise to only small numbers of trivalents with minor delays in synapsis, do not disturb spermatogenesis. Though the increasing of the number of Robertsonian translocations is suggested as having a deleterious effect on gametogenesis, it seems not to be the unique cause of germ cell failure. The genetic background of the hybrids (Ratomponirina *et al.* 1982) is probably involved in the delay in synapsis of the large arms of both trivalents and multivalents during meiosis and could represent another cause of germ cell degeneration.

Conclusion

The study of synaptonemal complexes in hybrids heterozygous for Robertsonian translocations shows a delay in pairing of the trivalents and multivalents. In the trivalents the pairing delay is variable and leads sometimes to the formation of chain-like structures, whereas in multivalents it is always marked and associated with male sterility. Whatever the meiotic configurations, it seems that the extent of the defect in synapsis could be an important cause of reduced fertility in the hybrids. This result favours the hypothesis of Burgoyne & Baker (1984), according to whom long delays in synapsis lead to degeneration of the germ cells. On the other hand, the usual association between hybrid sterility and meiotic multivalents constitutes a good model to explain how speciation could have occurred in some kinds of lemurs.

Acknowledgements

We are indebted to Prof. M. Moses for critically reading the manuscript. We acknowledge the technical assistance of Mr J.L. Maetz, Mr G. Cadiou for his collaboration in photographic work, Mr A. Christmann for grooming the animals well and thank Mrs M. Lavaux for secretarial assistance.

References

Beaumont, H.M. and A.M. Mandl 1962. A quantitative and cytological study of oogonia and oocytes in the foetal and neonatal rat. *Proc. Roy. Soc. Lond.* B155, 557–559.

Bogdanov, V.F., O.L. Kolomiets, E.A. Lyapunova, I.Y. Yanina and T.F. Mazurova 1986. Synaptonemal complexes and chromosome chains in the rodent *Ellobius talpinus* heterozygous for ten Robertsonian translocations. *Chromosoma* 94, 94–102.

Burgoyne, P.S. and T. Baker 1984. Meiotic pairing and gametogenic failure. In *Controlling Events in Meiosis, Symposia of the Society for experimental biology*, XXXVIII, C.W. Evans and H.G. Dickinson, eds, 349–362. Cambridge.

Chandley, A.C., R.M. Speed, S. McBeath and T.B. Hargreave 1986. A human 9;20 reciprocal translocation associated with male infertility analysed at prophase and metaphase I of meiosis. *Cytogenet. Cell Genet.* 41, 145–153.

Dresser, M.E. and M.J. Moses 1980. Synaptonemal complex karyotyping in spermatocytes of the Chinese hamster (*Cricetulus griseus*). *IV*. Light and electron microscopy of synapsis and nucleolar development by silver staining. *Chromosoma* 76, 1–22.

Dutrillaux, B. and Y. Rumpler 1977. Chromosomal evolution in Malagasy lemurs. II. Meiosis in intra-and interspecific hybrids in the genus *Lemur. Cytogenet. Cell Genet. 18*, 197–211.

Evans, E.P., G. Breckon and C.E. Ford 1964. An air-drying method for meiotic preparations from mammalian testes. *Cytogenetics* 3, 289–294.

Forejt, J. 1982. X-Y involvement in male sterility caused by autosome translocations — a hypothesis. In *Genetic Control of Gamete Production and Function*, P.G. Crosignani, B.L. Rubin and M. Fraccaro, eds, 135–151. New-York, London, Academic Press.

Gabriel-Robez, O., C. Ratomponirina, B. Dutrillaux, F. Carré-Pigeon and Y. Rumpler 1986. Meiotic association between the XY chromsomes and the autosomal quadrivalent of a reciprocal translocation in two infertile men, 46,XY,t(19;22) and 46,XY,t(17;21). *Cytogenet. Cell Genet.* 43, 154–160.

Gabriel-Robez, O., C. Ratomponirina, M. Croquette, J.L. Maetz, J. Couturier and Y. Rumpler 1987. Reproductive failure and pericentric inversion in man. *Andrologia* 19, 662–669.

Gropp, A., H. Winking and C. Redi 1982. Consequences of Robertsonian heterozygosity: segregational impairment of fertility versus male-limited sterility. In *Genetic Control of Gamete Production and Function*, P. Crosignani, B. Rubin and M. Fraccaro, eds, 115–134. New-York, London, Academic Press.

Hamilton, A.E. and J. Buettner-Janusch 1977. Chromosomes of Lemuriformes. III. The genus *Lemur*: karyotypes of species; subspecies and hybrids. *Ann. N.Y. Acad. Sci.* 293, 125–159.

Lifschytz, E. and D. Lindsley 1972. The role of X-chromosome inactivation during spermatogenesis. *Proc. Nat. Acad. Sci.* 69, 182–186.

Miklos, G.L.G. 1974. Sex chromosome pairing and male fertility. *Cytogenet. Cell Genet.* 13, 558–577.

Moses, M.J. 1977. The synaptonemal complex and meiosis. In *Molecular human cytogenetics*, vol. II, R.S. Sparkes, D. Commings and C.F. Fox, eds, 101–125. New York: Academic Press.

Moses, M.J., P.A. Karatsis and A.E. Hamilton 1979. Synaptonemal complex analysis of heteromorphic trivalents in *Lemur* hybrids. *Chromosoma* 70, 141–160.

Ratomponirina, C., J. Andrianivo and Y. Rumpler 1982. Spermatogenesis in several intra- and interspecific hybrids of the lemur (*Lemur*). *J. Reprod. Fert.* 66, 717–721.

Rumpler, Y. and B. Dutrillaux 1976. Chromosomal evolution in Malagasy lemurs. I. Chromosome banding studies in the genera *Lemur* and *Microcebus. Cytogenet. Cell Genet.* 17, 268–281.

Saadallah, N. and M. Hultén 1985. A complex three-breakpoint translocation involving chromosomes 2, 4 and 9 identified by meiotic investigations of a human male ascertained for subfertility. *Hum. Genet.* 71, 312–320.

Saadallah, N. and M. Hultén 1986. Electron microscopic investigations of surface spread synaptonemal complexes in a human male carrier of a pericentric inversion inv(13)(p12;q14). The role of heterosynapsis for spermatocyte survival. *Ann. Hum. Genet.* 50, 369–383.

Speed, R.M. 1986. Oocyte development in XO foetuses of man and mouse: the possible role of heterologous X-chromosome pairing in germ cell survival. *Chromosoma* 94, 115–124.

The origin of ring-formation and self-compatibility in *Gibasis pulchella* (*Commelinaceae*)

A.Y. Kenton, S.J. Owens and D. Langton

Jodrell Laboratory, Royal Botanic Gardens, Kew, Richmond, Surrey, TW9 3DS, UK

The genetic system of permanent hybridity maintained by the fixation of heterozygous chromosomal complexes is well known, largely from the extensive studies carried out in *Oenothera* and other species of the *Onagraceae* (reviewed by Holsinger & Ellstrand 1984). Although in *Oenothera*, the presence of true-breeding meiotic rings showed that reciprocal translocations (interchanges) initiated the formation of Renner complexes, much of the early work identified these mainly by the segregation of associated phenotypes. The small size and uniform morphology of the *Oenothera* complement made detailed cytological studies difficult, leaving the nature and size of the interchanges in some doubt. Nevertheless, subsequent studies on permanent hybridity in other genera generally assume the mechanism and significance of complex-heterozygosity to be similar to that proposed for *Oenothera*. (e.g. Lewis & Szweykowski 1964, Tilquin 1981). Recent work has shown that the cytological "preadaptations" thought to predispose certain genotypes to the fixation of multiple heterozygous interchanges (Darlington 1929, Cleland 1972) are not always present in established complex-heterozygotes (e.g. Goldblatt 1980, Kenton & Rudall 1987). It was therefore of interest to make a detailed study of chromosomal rearrangements in a complex-heterozygote favourable to cytological analysis. *Gibasis pulchella*, a perennial species of *Commelinaceae* confined to upland forest habitats in Mexico, admirably suits this purpose because it has very large chromosomes whose well-defined C-band patterns allow them to be identified individually.

We have examined wild-collected material from 12 populations (details in Kenton *et al.* 1987). The basic chromosome number is x=5, with diploids (2n=10) and triploids (2n=15) occurring either separately or in mixed populations. Triploids are all structurally heterozygous and probably arose by non-disjunction in degenerate or unbalanced interchange heterozygotes. Diploids are structural homozygotes, forming bivalents only, or single-interchange heterozygotes, forming three bivalents (II) and one quadrivalent (Θ4), or complex-heterozygotes, forming a complete ring of ten chromosomes (Θ10) at meiosis. No intermediate cytotypes have been found among over 100 plants examined to date.

Structural homozygotes

These genotypes, from 5 populations, have >90% of stainable pollen and exhibit gametophytic incompatibility (Owens 1981). C-band patterns are very similar in each population, and structural homozygotes are characterised by a haploid set (the "A" complex) consisting of five chromosomes with distinct C-banded markers (Figs 1a, 3). Nucleolus organising regions (NOR's) are visible

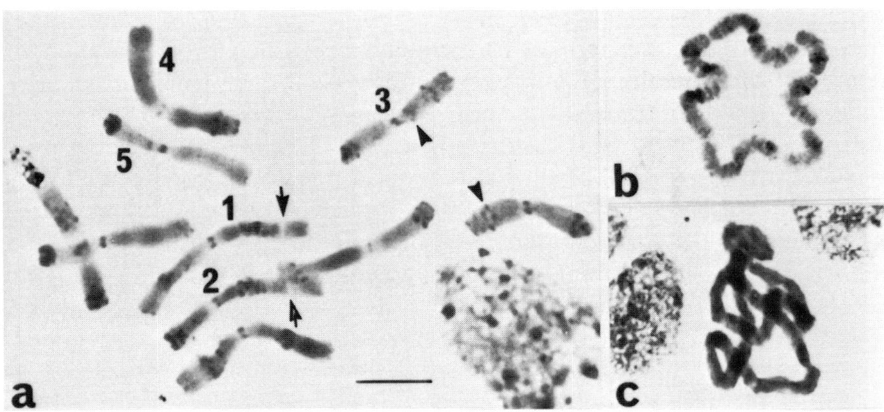

Figure 1. a structural homozygote P25, chromosomes constituting the A complex are numbered from 1-5. Note change in position of bands in chromosome 3 (arrowheads) and constrictions in chromosomes 1 and 2 (arrows). **b** complex-heterozygote 9776, Θ10 at male meiosis (stained with Giemsa). **c** complex-heterozygote P56, Θ10 at female meiosis (stained with orcein).

as heteromorphic secondary constrictions on each of chromosomes 2 and 5 (Fig. 5a). The only visible cytological difference among populations is the position of C-bands in the two arms of chromosome 3. A change of band position usually involves the long arm and may be heterozygous (Fig. 1a) or homozygous. This has not been investigated fully, but is probably due to paracentric inversion. Bivalents contain one or two chiasmata which are usually terminal although unlocalised chiasmata also occur.

Single-interchange heterozygotes

Single-interchange heterozygotes forming a Θ4 in more than 50% of cells are rare. Seven plants showing this behaviour are all self-incompatible like the bivalent-formers, and essentially contain two A complexes. The chromosomes involved in the interchange have been identified in two individuals from one population. In each case, the interchange is different and does not occur in con-populational complex-heterozygotes. In addition to genotypes forming a frequent Ѳ4, a few essentially bivalent-forming plants show a Ѳ4 in only 2 or 3% of cells. Here, the interchanges cannot be identified chromosomally. Pollen stainability in Ѳ4 genotypes ranges from 40 to 75% depending upon the frequency and orientation of the quadrivalent. As yet, the significance, if any, of floating interchanges to complex heterozygosity in *G. pulchella* is uncertain.

Complex-heterozygotes

Self fertile complex-heterozygotes occur in 6 populations. A Ѳ10, with terminal chiasmata only, is formed in all cells at male and female meiosis (Figs 1b, c). The percentage of stainable pollen ranges from 40 to 75%, the remainder being

Table 1. Breeding systems in different cytotypes of G. pulchella

Cytotype	Mean % stainable pollen	Pollen tube growth (self)	Viable embryo sacs	*Seeds/ capsule
II	91.0 ± 1.9	–	3–6	3–5
Ѳ10	57.7 ± 2.6	+/(–)	0–3	0–3

* after random pollination

shrunken and empty. Table 1 compares gametogenesis and breeding systems in structural homozygotes and complex-heterozygotes. Although artificial selfings of ring-formers have always resulted in ring-forming progeny as in *Oenothera*, self pollen shows a mixed response on its own stigma. Of the well-filled, apparently normal grains, 50% or more reach the ovary within 24 hours. Between 10 and 50% do not germinate, and of those which do, up to 25% appear to be inhibited. This suggests that morphologically normal pollen may be of two different genotypes, either differing in S alleles, as in *Oenothera*, where pollen tube competition decides which genotype will accomplish fertilisation, or being selectively inefficient in synthesising proteins for development. The fact that 25–60% of grains are already inviable suggests that more than one factor may be involved in producing defective pollen. One genotype, PO8, seems to be either self-incompatible or sterile. Most of its pollen either does not germinate or does not adhere to the stigma, with the remainder forming only short tubes. Mutations of the gynoecium, and cleistogamous flowers, occur sporadically, and no fruits have so far been set.

In the remaining Ѳ10 genotypes, a maximum of 3 out of a possible 6 viable embryo sacs has been observed. Dissection of immature seed capsules reveals 1 to 3, or very occasionally 4, seeds differing in size, but in about 70% of capsules, only one good seed is set at maturity, compared to 3–5 in bivalent formers after random pollination. The effects of possible unequal segregation of the ring and environmental conditions must be taken into account in considering the very low seed set. Although it seems likely that a gametic balanced lethal system may be involved, artificial hybridisation in either direction between bivalent formers and ring formers has produced ring-forming progeny. Any gametic system must therefore be leaky, at least on the female side.

A few individuals from each population have been C-banded at mitosis and meiosis. Five different C-banded karyotypes have been distinguished (Figs 2, 3),

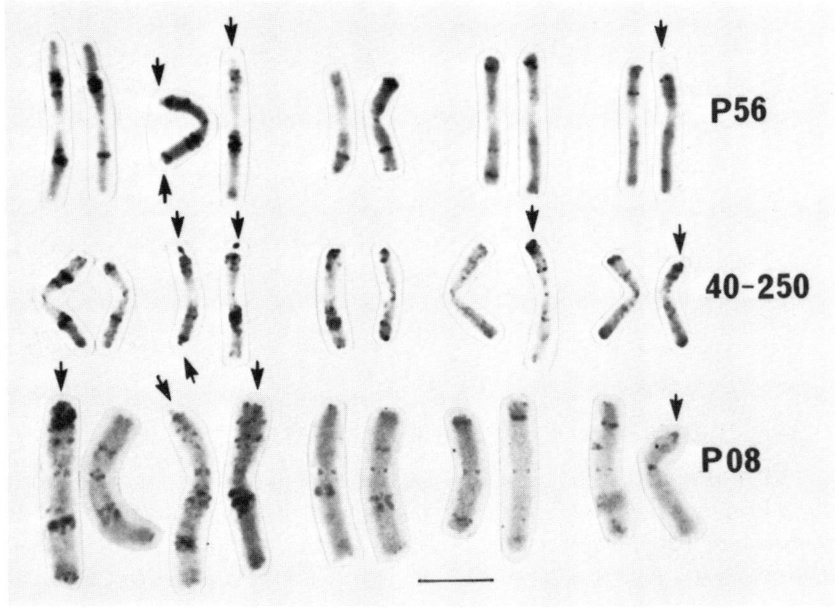

Figure 2. C-banded karyotypes of complex-heterozygotes. Arrows show sites of secondary constrictions.

two of which occur in the same population (Fig. 3). Of these, the karyotype shown by P56 is remarkable in being very similar to that of the bivalent-former (cf. Figs 1a, 2, 3) in that the chromosomes can be arranged in pairs. Characteristic chromosome arms, with strong C-banded markers, remain in their original positions, showing that heterozygous interchanges do not always involve whole chromosome arms, as proposed for *Oenothera* (Cleland 1972), where pericentric heterochromatin was thought to encourage proximal breaks and whole arm exchanges. In *G. pulchella,* strong intercalary and terminal C-bands may influence the position of breaks.

A conspicuous feature of P56 is the presence of a bisatellited chromosome (BS) carrying a NOR at each end (Figs 2, 4, 5). This appears to have its origin in transfer of the terminal heterochromatin and NOR of chromosome 5 to the long arm of chromosome 2, with the break point in chromosome 2 being distal to a major C-block. Chromosome BS appears in five other genotypes, (e.g. 9776 and 40–250), but the satellites distal to the NOR's differ in their staining intensity with Giemsa (Fig. 2). In 8535, BS is smaller, with the break point in the long arm of chromosome 2 probably having occurred near to the centromere, proximally, rather than distally to the major C-block (Fig. 3). The sterile genotype, PO8, is the

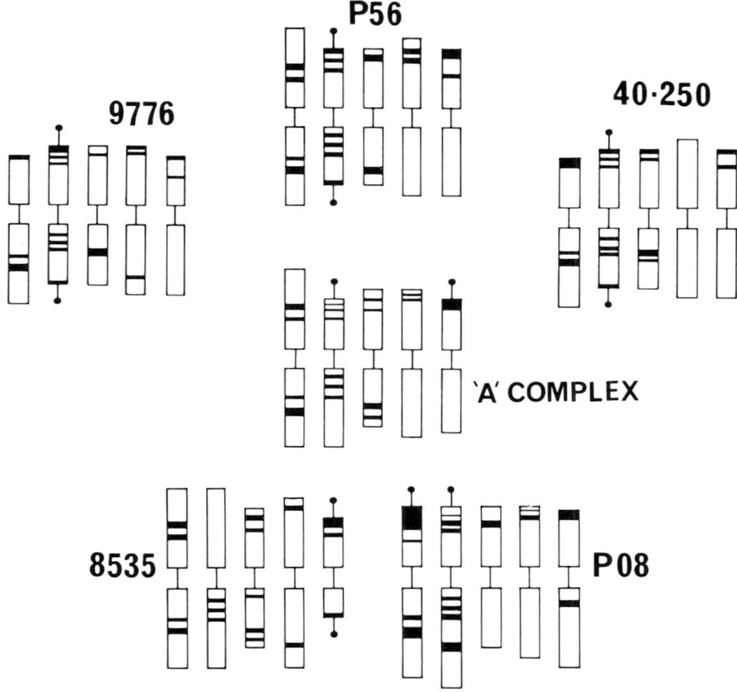

Figure 3. Idiogram summarising banded karyotypes of the standard "A" complex (centre) and "B" complexes from 5 complex heterozygotes.

only one that does not contain BS. Here, the modified chromosome 2 is longer than normal and has probably received material translocated from chromosome 1. Translocations involving a distal segment of chromosome 1 occur in 40–250, 9776 and PO8. In each case, the break point in chromosome 1 is distal to the major intercalary C-block. In 40–250 and 9776, the resulting interchange is markedly unequal, and chromosome material appears to have been lost, while in PO8, the deleted chromosome 1 contains a large C-block, probably from chromosome 5. In PO8, a new chromosome with a highly proximal C-band is probably derived from chromosome 3 (Fig. 2). Non-staining constrictions distal to the major C-blocks in chromosomes 1 and 2 have sometimes been observed in bivalent-formers (Fig. 1a), and may indicate weak points at which breakage is more frequent.

Each of the ring-formers has one standard haploid set in which the marker chromosomes of the A complex can be readily recognised, but the modified chromosomes constituting the B complex can vary among genotypes, even within a population. Fig. 3 summarises the five different B complexes, and shows that rearrangements tend to result in chromosomes having the same relative distributions of heterochromatin. This is also reflected in the presumed paracentric inversions causing a change in position of C-bands in chromosome 3. In PO8, this change makes one chromosome 3 resemble chromosome 1 in its symmetrical distribution of C-bands. The other (rearranged) chromosome 3, with its very proximal C-block, resembles a chromosome 2. It seems likely that the position of breaks and rearrangements may be influenced both by the nature of the DNA adjacent to C-blocks and by the final distribution of C-blocks on the rearranged chromosomes, perhaps related to their original position in the interphase nucleus (cf. Schweizer & Loidl 1987).

Chromosome positions in the Θ10 have been analysed at metaphase I in three genotypes. We have identified an alternate arrangement of chromosomes belonging to the A & B complexes, potentially, at least, resulting in balanced gametes (cf. Kenton *et al.* 1987). Chromosome BS, which appears in four of the five B complexes, is a ready marker, even in unbanded preparations, and its appearance in all pollen mitoses of Θ10 genotypes suggests that the B complex is transferred preferentially through the pollen. Both constrictions are visible at pollen mitosis, but interestingly, the more strongly expressed constriction may be either on the short or on the long arm. (Figs 4c, d).

Chromosome BS

This chromosome appears only in Θ10 plants and some unusual features make it of particular interest. It pairs normally in the Θ10, but in hybrids with bivalent-formers, where the chiasma frequency is reduced, it often forms a ring univalent, which is noticeably smaller in the deleted BS of 8535 (Figs 4a, 5c). Although this association is indistinguishable from a chiasmate one, the standard BS closely resembles a normal chromosome 2 (Figs 2, 3) and certainly does not appear to be an isochromosome. A chiasma would imply the occurrence of complex homologies or a duplication. Each of these is a possibility, particularly as multiple NOR sites may exist in the complement (Kenton *et al.* 1987), but also the possibility that it represents some kind of ectopic pairing in or adjacent to regions of heterochromatin (cf. Kastritsis *et al.* 1986) must at least be considered, particularly since sticky associations have been seen between C-bands at mitosis. An interesting feature of this kind of association is that it could bring together

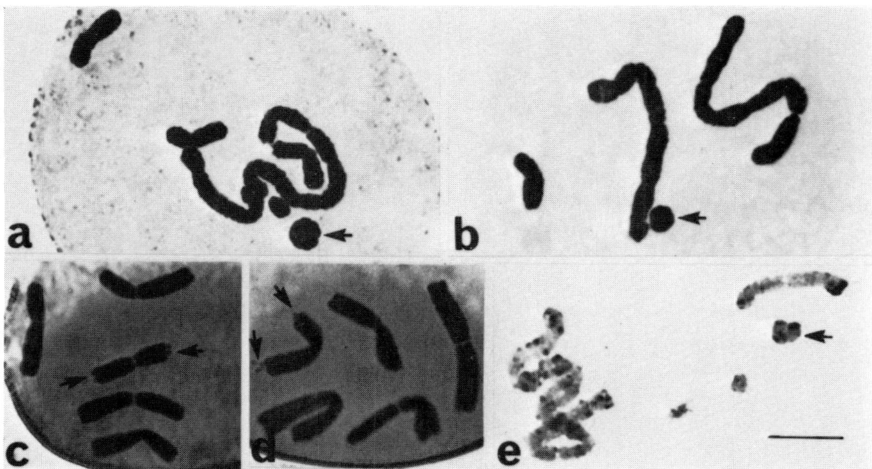

Figure 4. The BS chromosome. **a, e** metaphase I in F1 hybrid (2n=10+B) showing chain of 7 chromosomes and the standard BS as a ring univalent (arrowed). **b** as a, a different hybrid (2n=10) showing 2 chains of 4 chromosomes and the deleted BS as a ring univalent (arrowed). **c** pollen mitosis in 9747, showing BS with larger constriction on the short arm. **d** as c, P56, showing larger constriction in the long arm. All secondary constrictions are arrowed.

chromosomes having only short regions of homology (cf. Chinnappa & Victor 1979; Kenton *et al.* 1987; Sukhadev *et al.* 1985). Silver nitrate staining in Ⓞ10 genotypes shows two active NOR's in BS, (Fig.5c, f), and one on each of chromosomes 2 and 5, corresponding to four prominent nucleoli in interphases of root tip meristems (Fig.5e), while preliminary observations suggest that in the bivalent-formers, two of the NOR's are usually dominant (Fig.5b, d).

The apparently active state of both NOR's in BS is unusual, as in experimental material of barley (Nicoloff *et al.* 1977) and tomato (Quiros 1976), translocated NOR's were always suppressed. Although bisatellited chromosomes do appear in nature (e.g. Strid 1968; Palomino *et al.* 1988), they are rare, and their function and significance has not been fully explored. However, in *Drosophila hydei* with an X–Y translocation, NOR's demonstrated on each end of the same chromosome were associated with sequences exerting position-effect variegation (Van Breugel 1970; Hennig *et al.* 1975), while a "cuckoo" *Aegilops* chromosome causing chromosome breaks when introduced into wheat was also identified by structures resembling subterminal constrictions in each arm (Finch *et al.* 1984). Because of these unusual results and the presence of BS only in ring-forming, self-compatible genotypes, it is tempting to speculate that this chromosome may in some way be implicated in features unique to the complex-heterozygotes, i.e. self-compatibility or lethality, and it seems reasonable to suppose that some interaction between NOR's or adjacent sequences might well result if BS became homozygous.

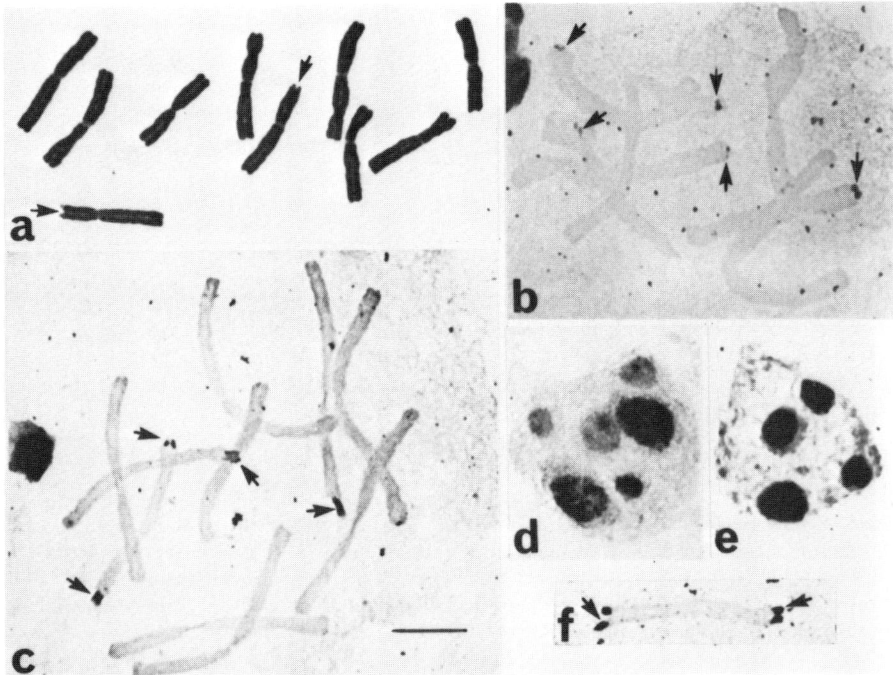

Figure 5. NOR's. **a** Feulgen-stained karyotype of structural homozygote P25, showing prominent secondary constriction in one chromosome 2 and one chromosome 5 (arrowed). **b** Ag-stained structural homozygote P38 showing two prominent and three minor NOR's. **c** complex-heterozygote 40-250, showing 4 NOR's including two on BS. **d, e** nucleoli in root tip interphases of structural homozygote and complex-heterozygote, respectively. **f** BS chromosome from complex-heterozygote P56, showing a large NOR at each end.

The change in breeding system from self-incompatible (II, Θ4) to self-compatible (Θ10) appears to be unaffected by various structural differences among ring-formers, except for P08, which does not contain BS and appears to be sterile. One possible explanation is that the formation of BS involves a structural mutation that results in elimination or suppression of one of the S alleles. This would also explain the mixed response of self pollen on the stigma of the ring-formers. A speculative model (Fig.6) places the S gene near the terminal heterochromatin associated with the NOR on chromosome 5. There, one of the S alleles could be either removed or suppressed by rearrangements that move it to a site on the long arm of chromosome 2 during the formation of BS. In *Nicotiana*, de Nettancourt *et al.* (1975) showed that a duplication of the NOR carried on a centric fragment conferred self-compatibility to self-incompatible genotypes. This was suggested to operate by reinstating nucleolar activity and protein synthesis switched off during the incompatibility reaction.

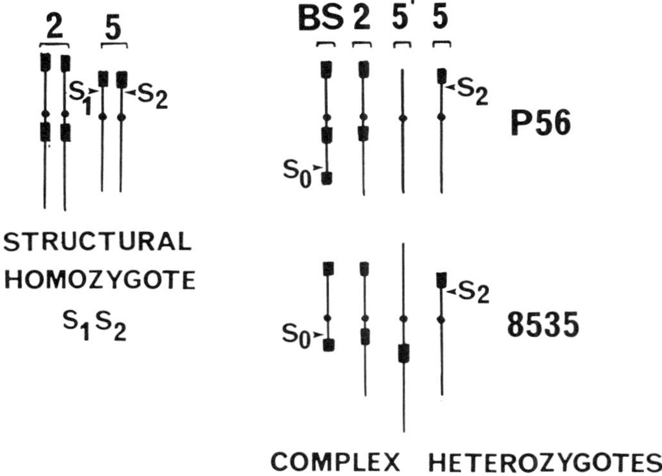

Figure 6. Hypothetical scheme to explain the change in breeding system in complex-heterozygotes in terms of chromosome structural rearrangements (see text for details).

These studies showed that the S gene, or a factor affecting its expression, (a) is located in or near NOR heterochromatin, at least in *Nicotiana*, and (b) might be affected by nucleolar activity.

In summary, *Gibasis pulchella* has proved a useful organism for the study of complex heterozygosity in that the structural rearrangements involved can be visualised. We have shown that the position of breaks is near to intercalary or terminal heterochromatin, and that interchanges may involve relatively small segments. A gametic balanced lethal system probably transfers the complex carrying the interchanges preferentially through the pollen, but is almost certainly leaky on the female side. Structural rearrangement involving NOR's may have some significance in the change in breeding system from self-incompatible to self-compatible. Finally, we suggest that complex-heterozygotes may well be useful as model systems in investigations of gene control, mutation and gene mapping, particularly if molecular biology can be used to complement cytological and genetical approaches.

References

Chinnappa, C.C. and R. Victor 1979. Achiasmatic meiosis and complex heterozygosity in female cyclopoid copepods (Copepoda, Crustacea). *Chromosoma* 71, 227–236.

Cleland, R.E. 1972. *Oenothera: cytogenetics and evolution.* Academic Press, New York.

Darlington, C.D. 1929. Ring-formation in *Oenothera* and other genera. *J. Genet.* 20. 345–363.

Finch, R.A., T.E. Miller and M.D. Bennett 1984. "Cuckoo" *Aegilops* addition chromosome in wheat ensures its transmission by causing chromosome breaks in meiospores lacking it. *Chromosoma* 90, 84–88.

Goldblatt, P. 1980. Uneven diploid chromosome numbers and complex heterozygosity in *Homeria* (Iridaceae). *Syst. Bot.* 5, 337–340.

Hennig, W., B. Link and O. Leoncini 1975. The location of nucleolus organizer regions in *Drosophila hydei*. *Chromosoma* 51, 57–63.

Holsinger, K.E. and N.C. Ellstrand 1984. The evolution and ecology of permanent translocation heterozygotes. *Amer. Naturalist*, 124, 48–71.

Kastritsis, C.D., Z.G. Scouras and M. Ashburner 1986. Duplications in the polytene chromosomes of *Drosophila auraria*. *Chromosoma* 93, 381–385.

Kenton, A., A. Davies and K. Jones 1987. Identification of Renner complexes and duplications in permanent hybrids of *Gibasis pulchella* (Commelinaceae). *Chromosoma* 95, 424–434.

Kenton, A. and P. Rudall 1987. An unusual case of complex-heterozygosity in *Gelasine azurea* (*Iridaceae*), and its implications for reproductive biology. *Evolutionary trends in plants* 1, 95–102.

Lewis, H. and J. Szweykowski 1964. The genus *Gayophytum* (*Onagraceae*). *Brittonia* 16, 343–392.

de Nettancourt, D., M. Devreux, F. Carluccio, U. Laneri, M. Cresti, E. Pacini, G. Sarfatti and A.J.G. van Gastel 1975. Facts and hypotheses on the origin of S mutations and on the function of the S gene in *Nicotiana alata* and *Lycopersicon peruvianum*. *Proc. R. Soc. Lond. B.* 188, 345–360.

Nicoloff, H., M. Anastassova-Kristeva, G. Kunzel and R. Rieger 1977. The behaviour of nucleolus organizers in structurally changed karyotypes of barley. *Chromosoma* 62, 103–109.

Owens, S.J. 1981. Self-incompatibility in the *Commelinaceae*. *Ann. Bot.* 47, 567–581.

Palomino, G., R. Viveros, and R. Bye 1988. Cytology of five Mexican species of *Datura* (*Solanaceae*). *Southwestern Naturalist* 33, 85–90

Quiros, C.F. 1976. Selection for increased numbers of extra heterochromatic chromosomes in the tomato. *Genetics* 84, 43–50.

Strid, A. 1968. Stable telocentric chromosomes formed by spontaneous misdivision in *Nigella doerfleri* (*Ranunculaceae*). *Bot. Notiser* 121, 153–164.

Sukhadev, P., P.S. Janake, J.V.V.S.N. Murthy, M.V.S. Rao and V. Manga 1985. Meiotic behaviour of multiple interchange heterozygotes in pearl millet. *Cytologia* 50, 347–350.

Schweizer, D. and J. Loidl 1987. A model for heterochromatin dispersion and the evolution of C-band patterns. *Chromosomes Today* 9, 61–74.

Tilquin, J.P. 1981. An unusual case of a complex heterozygote presenting no taxonomic problem in *Chelidonium majus* L. (*Papaveraceae*). *Experientia* 37, 341–342.

Van Breugel, F.M.A. 1970. An analysis of white-mottled mutants in *Drosophila hydei*, with observations on X–Y exchanges in the male. *Genetica* 41, 589–625.

Genomic destabilisation and the origin and elaboration of complex hybridity in *Isotoma petraea*

S.H. James

Botany Department, University of Western Australia, Nedlands, W.A. 6009, Australia

Isotoma petraea is a herbaceous perennial member of *Lobeliaceae* which grows on granite outcrops and other rocky habitats throughout the arid interior of Australia. Over most of its distributional range, its populations exhibit seven bivalents (7II) at meiosis, but most populations are polymorphic for small-ring floating interchange heterozygosity. In its most southwestern populations, however, it exhibits complex hybridity with from six to all fourteen chromosomes associated into the two complexes. The complex hybrid populations are essentially monomorphic for ring size, the plants are highly inbreeding, and the complex hybridity is stabilised by balanced lethal systems (James 1965).

The mechanisms involved in the evolutionary origin of complex hybridity remain unclear. Darlington (1931) outlined three possible mechanisms by which complex hybridity could be assembled in natural populations. He favoured a model by which interchange hybridity and lethal systems were sequentially accumulated within a single inbreeding lineage. On the other hand, Cleland (1936) and most other students of *Oenothera* have favoured an alternative model which proposes that the multiple interchange rings in *Oenothera* complex hybrids were assembled at one stroke by the hybridisation of outbreeding strains which were homozygous for appropriately differentiated chromosome rearrangements and subsequently stabilised by the development of inbreeding and balanced lethal systems. The third model envisages the gradual accumulation of interchange hybridity through the hybridisation of a stock with a sequence of populations differentiated by interchange differences. This model was favoured by Catcheside (1932) for *Oenothera* and is similar to one that has been proposed for *Isotoma* (James 1970).

In *Isotoma*, complex hybridity apparently came into being as small ring-of-six (O6) complex heterozygotes on Pigeon Rock, a large granite outcrop 145 km. north of Southern Cross in the goldfields of Western Australia, and has migrated from that site into populations to the south west, territorial gain being associated with an expansion of the ring size and an increase in the system's efficiency.

This paper will describe the breeding behaviour and some aspects of the polymorphisms in evidence within the Pigeon Rock population. The data presented lead to an attractive compromise of all three of the models proposed by Darlington although they strongly support the contention that initiating Θ6 complex hybridity on Pigeon Rock evolved within highly inbreeding lineages. They also encourage the speculation that mobile genetic element activity may be responsible for the diversity which was utilized in the evolutionary assembly of the system.

Interchange polymorphism

Of 331 plants collected from Pigeon Rock, over the years, 223 (67.4%) were Θ6 interchange heterozygotes, four (1.2%) were Θ10 interchange heterozygotes and 104 (31.4%) were structural homozygotes exhibiting 7II at MI in PMC meiosis. The Θ6 interchange heterozygotes may be identified cytologically, by pollen fertility analysis, or by allozyme analysis. The Θ6 plants exhibit around 79% pollen fertility, which is a measure of the frequency of meiotic disjunction, both in the anthers and in the ovules. The 7II's exhibit more than 96% fertile pollen and the Θ10's about 56%. The Θ6's are also invariably heterozygous for the PGM 3 allele and that allele is found only in Θ6 heterozygotes.

Each Θ6 is heterozygous for two complexes, called *N* and *S* respectively. *N* is two interchanges removed from the primitive *ax* chromosome end sequence which characterises the species generally and is found in its eastern Australian relatives, *I. axillaris* and *I. anaethifolia*, as well. *N* also carried the *PGM3* allele and a fully penetrant recessive lethal gene which prevents *N.N* homozygotes existing in the population. *S* has the primitive *ax* chromosome end sequence and carries either *PGM1* or *PGM2*. Its lethal content is discussed below.

Inbreeding

In *Isotoma petraea*, the characteristic Lobeliaceous pollination mechanism is often short-circuited in that the stigma fails to emerge from the syngenesious anther tube, so that self pollination is effected. In a survey of 1955 flowers on 105 Pigeon Rock plants grown in the glasshouse at Perth, the stigma emerged in only 6.2% of flowers on structural homozygotes and in only 3% of flowers on Θ6 interchange heterozygotes. Even when the stigma emerged, it already carried self pollen so that most of the seed produced, even from those flowers with some potential for crossing, must have been the result of self pollination. Comparable but unquantified pollination behaviour has been observed at Pigeon Rock. It is clear that outcrossed seed must constitute something less than 1% in this population.

Seed abortion

Selfed capsules include sterile ovules resulting from non-disjunctional EMC meiosis, unfertilised ovules, normal seeds and seeds exhibiting post-zygotic abortion. Some structural homozygotes and all of the Θ6 interchange heterozygotes exhibit significant proportions of aborted seeds (Fig. 1). The

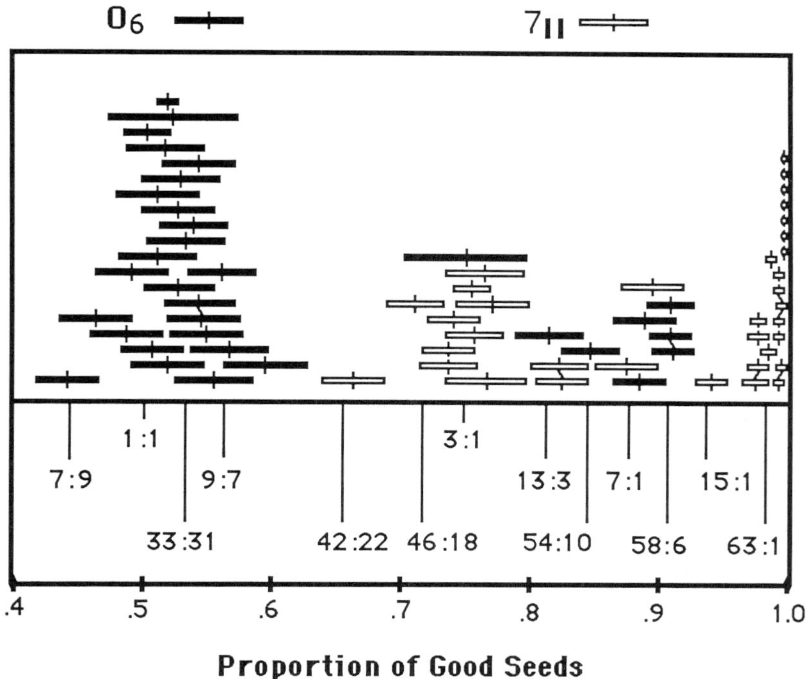

Figure 1. Seed abortion in selfed capsules of a sample of 68 Pigeon Rock plants. Each symbol represents the proportion of good seeds in at least two capsules showing similar behaviour ± 1.96 S.D. of the estimate. A range of mono-, bi- and trifactorial ratios which may be relevant are also indicated.

proportion of aborted seeds in selfed capsules was characteristic for individual plants grown under optimal conditions but varied from a few% through to about 60% between plants. Seed abortion patterns are obviously genetically determined and indicate that most plants in the population are heterozygous for recessive lethal genes which cause homozygotes to abort during seed development. However, the occurrence of seed abortion ratios significantly less than 50% indicates that the *N.N.* and/or *S.S.* homozygotes need not exhibit seed abortion.

Generally, the seed abortion ratio in capsules obtained by articifial cross pollination was either about 50%, in crosses involving allozymically more similar genotypes, or considerably less in crosses between more dissimilar parents. This demonstrates heterosis for seed quality in some intrapopulational crosses and indciates that the allozymically less similar plants may have fewer seed aborting factors in common.

Seed germination

Compared to materials from other populations, Pigeon Rock selfed seeds germinate tardily, following giberellic acid (GA3) breakage of the natural dormancy period. While germination percentages of the good seeds are usually quite high, the seedlings produced are mostly weak so that the yield of adult plants from any bed of selfed seed is low. Seed produced by crossing different Pigeon Rock plants, or by crossing Pigeon Rock plants with material from other populations, however, germinates rapidly and develops into much more robust seedlings. This shows that pronounced heterosis is exhibited by both inter- and intrapopulational cross material at the germination and seedling stages.

An interpretation of the genetic structure of the Pigeon Rock population

The Pigeon Rock *Isotoma* population is not a conventional Mendelian population. It is composed of plants which are highly inbreeding. Each plant represents an individual inbred lineage which has been genetically isolated from other such lineages, on average, for considerable lengths of time. During this isolation, deleterious mutations have accumulated within each lineage, separately, leading to a dramatically reduced performance at germination and in the seedling stages. The independent accumulation of deleterious mutations in separate lineages would provide a basis for the heterosis observed in inter- and intrapopulational crosses. Some of the deleterious mutations would be recessively lethal, with the lethal effect operating at any time from immediately post-zygotic to sub-adulthood. The accumulation of recessive lethals would fix a requirement for parental levels of hybridity in the selfed progeny and result in permanently hybrid inbred lineages. In these permanently hybrid structural homozygotes, which exhibit strongly terminally localised chiasmata in their seven bivalents at meiosis, only one in 128 of their selfed seed might be expected to carry a competent genotype, while all the heterotic products of intrapopulational crosses may well be adequate. Natural selection may be expected to improve the yield of fully fit seeds produced by the surviving lineages. It has not done so by lowering the level of inbreeding in the population.

The efficiency of the inbreeding permanent hybrids, however, has been improved in two ways. Firstly, the N complex was assembled. N has three chromosomes associated together by two overlapping interchanges. This reduces the number of independently segregating units to five and raises the level of selfed seeds having parental levels of hybridity to one in 32. However, the fecundity of these Θ6's is reduced because only about 79% of their meiotic cells are fully disjunctional. The establishment of the interchange rearrangements was also associated with the origin of the *PGM 3* allele and a unique recessive lethal condition which precludes *PGM 3/3* homozygotes from the population.

Iteration of a deterministic model incorporating the theory outlined above indicates that a stable chromosomal polymorphism may be maintained by very high levels of inbreeding. Significant levels of Θ6 interchange heterozygosity will be maintained if the recurrent selfing rate, s, is greater than 90%. Indeed, about one third 7II's and two thirds Θ6's will be maintained if s is about 97%. The Pigeon Rock chromosomal polymorphism is well predicted by this model.

Secondly, the reproductive efficiency of the inbred permanent hybrids has been improved by the accumulation of recessive lethal factors which eliminate the homozygotes at earlier post-zygotic stages in seed development in both the *N* complex and in its structurally primitive counterpart, *S*. These factors are also found, but less abundantly, in the 7II's. In many of the Θ6's, there is a 1:1 seed abortion ratio. These plants breed true for their *PGM (1 or 2)/3* heterozygosity, their interchange hybridity and their seed abortion ratio. Such plants are clearly complex hybrids, but, it is believed, the seed abortion reflects an improved version of the original balanced lethal system, the original being coined with a more primitive requirement for parental levels of hybridity.

Sources of the genetic diversity in the Pigeon Rock population

The genetic diversity in any population ultimately must be derived from mutation, either from within the population, or external to it and introduced into the population by migration. However, the genotypically determined phenotypic diversity generated by the processes of diploid sexuality is usually taken to be the variation exposed to natural selection.

In *Isotoma petraea* at Pigeon Rock, the normal operation of diploid sexuality has been suppressed by chiasma localisation, high levels of autogamy and a strict requirement for parental levels of hybridity in the selfed progeny. The role of migration in contributing to the genetic diversity of the Pigeon Rock population is difficult to assess and at this stage is left as an open question. On the other hand, the divergence of the putatively isolated inbred lineages with respect to factors contributing to seed abortion and to heterotic responses in intrapopulational crosses strongly suggest that the mutational origin of genetic diversity may be of particular importance.

Genetic basis of seed abortion

Analysis of the genetic basis of seed abortion in these materials is rendered somewhat difficult by inconsistency often expressed between capsules on the one plant (Fig. 2). However, this inconsistency can be minimised by care in horticulture, and in these studies, seed abortion ratios were assessed from at least two capsules showing comparable behaviour. Since each capsule contains up to 1000 ovules, the seed abortion estimates should be fairly reliable.

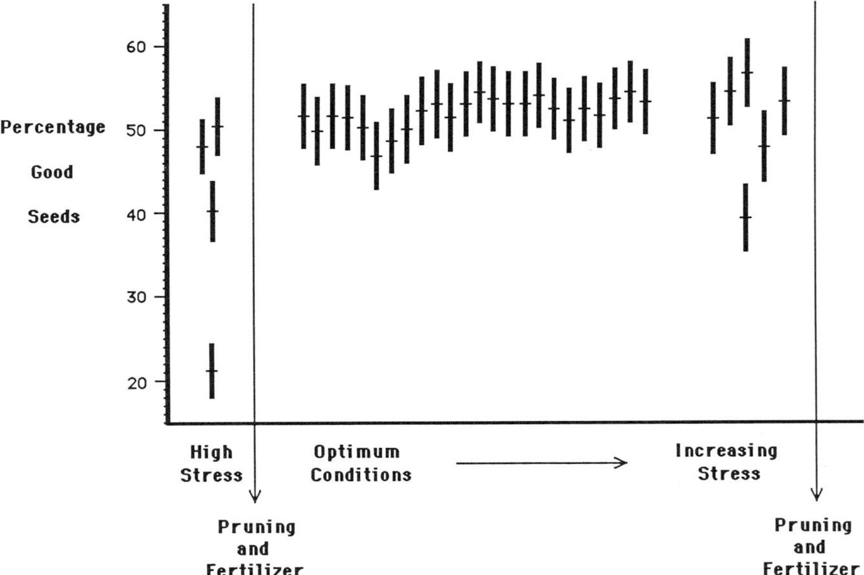

Figure 2.   Seed abortion in individual selfed capsules of a O6 Pigeon Rock plant showing relative homogeneity amongst 24 capsules which formed under optimal growing conditions and heterogeneity amongst capsules which formed under less optimal conditions.

While many of the O6's exhibit about 50% seed abortion and conform with the classical 1:1 ratio characteristic of balanced lethal systems, most do not. They exhibit more than 50% good seeds, suggesting that some of the complex homozygote combinations may survive to seed maturity, at least. This means that one (or both) of the complexes is devoid of recessive seed aborting lethals, which would yield 75% (or 100%) good seed, or that the plants also carry elements outside the ring which interact epistatically to modify the seed abortion ratio. Indeed, all the seed abortion ratios in selfed capsules of Pigeon Rock plants and their derivatives may be modelled in terms of mono-, bi-or trifactorial ratios involving recessive lethal genes and their epistatic modification. Preliminary analyses of seedling progenies from plants exhibiting departures from the conventional 1:1 seed abortion ratio have yielded *PGM* homozygote frequencies appropriate for the models proposed.

When Pigeon Rock Θ6's are crossed to 7II's from other populations, Θ6 (*N.ax*) and 7II (*S.ax*) twin hybrids are produced. 7II twin hybrids derived from parental Θ6's exhibiting a 1:1 and 9:7 seed abortion ratio respectively exhibited a seed abortion ratio generally conforming with a 9:7 ratio (Fig. 3). This indicates that

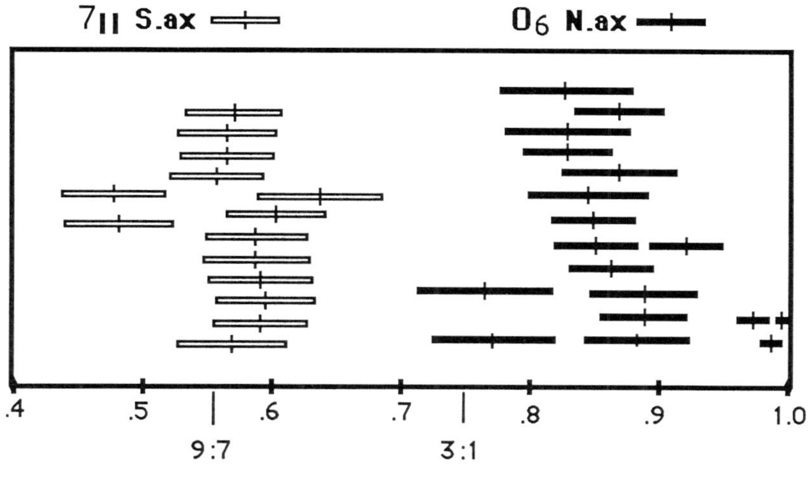

Proportion of Good Seeds

Figure 3. Seed abortion in individual selfed capsules of several twin hybrid plants derived from crosses between Pigeon Rock Θ6 and "495" (Yackeyackine Soak) 7II plants.

the chromosomes of the *S* complex carried two separate recessive lethals, now independently assorting in the 7II twin hybrid. The relatively high level of good seeds produced by the Θ6 twin hybrid, which might have been expected to exhibit a 3:1 ratio, was not so easily explained.

The inference that the *S* complex carries two recessive lethal seed aborting elements is supported by other observations. Thus, in a sample of some 15000 seeds from 24 capsules harvested under optimal growing conditions, one Θ6 plant was shown to exhibit seed abortion which agreed only with a 33:31 trifactorial ratio (Fig. 2). This ratio suggests that one complex, *N*, carries one recessive lethal while the other complex, *S*, carries two, each of which is modified by independently assorting elements outside the ring (Fig. 4). With this model, it would be expected that one in 33 of the good seeds would be an *S.S* homozygote, the remainder *N.S* heterozygotes.

		N				S			
		AB	Ab	aB	ab	AB	Ab	aB	ab
N	AB	×	×	×	×	✓	✓	✓	✓
	Ab	×	×	×	×	✓	✓	✓	✓
	aB	×	×	×	×	✓	✓	✓	✓
	ab	×	×	×	×	✓	✓	✓	✓
S	AB	✓	✓	✓	✓	×	×	×	×
	Ab	✓	✓	✓	✓	×	×	×	×
	aB	✓	✓	✓	✓	×	×	×	×
	ab	✓	✓	✓	✓	×	×	×	✓

Ring of 6

$$N : \quad -x \quad +y \quad +z$$
$$S : \quad +x \quad -y \quad -z$$

First Bivalent

$$A : \quad -x \quad -y \quad -z \quad (normal)$$
$$a : \quad -x \quad +y \quad -z$$

Second Bivalent

$$B : \quad -x \quad -y \quad -z \quad (normal)$$
$$b : \quad -x \quad -y \quad +z$$

Figure 4.   A trifactorial model which may explain the 33:31 seed abortion phenotype of the plant illustrated in Fig. 2.

Transpositions?

Recessive lethals and their modifying elements may be simply modelled as deficiencies and duplications of chromosome segments respectively, i.e., as the consequences of transposition. This approach has been adopted previously, in an analysis of the balanced lethal system in a Θ12 complex heterozygote from another population (Beltran & James 1970). It is reasonable to expect, also, that a duplication would be able to exhibit a dominant or recessive ability to modify its appropriate deficiency according to the site into which it is inserted and the accuracy with which it was inserted. This additional freedom provides all the flexibility necessary to model any trifactorial ratio in terms of duplications and deficiencies.

In addition, duplications inserted into non-homologous chromosomes may provide a mechanism for interchange "mutation" to occur through homologous recombination. Thus, the duplication of segments from two chromosomes into a third would provide the possibility of generating the rearrangement necessary to construct the N complex. Similarly, the occurrence of occasional $\Theta10$'s in the population may be accounted for by this mechanism.

Mobile genetic elements?

The occurrence of complementary deficiencies and duplications would suggest that a mechanism promoting the transposition of chromosome segments within the genome is of particular importance in the Pigeon Rock population. Such a mechanism can be modelled in terms of homologous recombinational events between segments in otherwise non-homologous chromosomes, in terms of mobile genetic element activity, or as a combination of both mechanisms. Indeed, something akin to the second mechanism would be required before the first could operate.

Mobile genetic element activity may be inferred from mutational instability. This instability may be induced by physiological stress, genomic shock or by genetic complementation. The seed abortion ratio in at least some of the Pigeon Rock plants or their derivatives is an aspect or the phenotype which exhibits considerable instability. Thus, the seed abortion ratios in 24 capsules collected from one plant growing under optimal conditions were homogenous ($Chi^2_{(23)}$ = 30.3) while significant heterogeneity was exhibited between capsules collected in periods when the plant was under increased stress (Fig. 2). Secondly, seed abortion in the 06*N.ax* (06*PR* × 7II*495*) twin hybrid was particularly variable (Fig. 3), ranging from about 25% (the simplest expectation) down to less than 1%. It is not easy to invent a simple Mendelian model based on duplications and deficiencies which would account for less than 25% seed abortion in an interpopulational hybrid, let alone the range of values exhibited. Alternatively, it may be suggested that in these twin hybrids, an element which induces the production of modifiers of the lethal component of N is activated so that many of the good seeds carry germinal *de novo* modifying mutations.

It should be noted that in this population, the normal systems for the generation of diversity, i.e, those associated with normal diploid sexuality, are strongly suppressed by autogamy, terminal chiasma localisation, a low chromosome number and fixed hybridity. Under such conditions, an alternative mechanism for the generation of diversity would be adaptive. Mobile genetic elements are the obvious candidates to provide this alternative. Continuing research on this material will seek evidence of mobile genetic elements and their activity using molecular techniques in association with conventional genetic analysis.

The elaboration of complex hybridity in *Isotoma*

Analysis of the transmission of allozymically marked complexes in crosses between different populations of large ringed complex heterozygotes in *Isotoma petraea* demonstrate substantial identity among components of the lethal systems characterising different populations (Lavery & James 1987). This makes it unlikely that complex hybridity has evolved separately in all the various populations. It is, perhaps, more probable that this unlikely genetic system has spread from a single source. The geographical distribution of complex hybrids in terms of their ring sizes, the increasingly earlier times at which their seed lethals operate, and the progressive reduction in flower size, all suggest that Pigeon Rock was the source of the genetic system. Migration of the genetic system from Pigeon Rock towards the south west also conforms with the direction of the prevailing summer wind which may be expected to be a major determinant of migrational pathways in this small-seeded species.

The selfed progeny of Pigeon Rock complex hybrids exhibit pronounced inbreeding depression with respect to seed germination capabilities, seedling vigour and establishment capabilities. At Pigeon Rock, the Θ6 complex heterozygotes coexist with structural homozygotes (and rare Θ10's) in about a 2:1 ratio; they have not been able to displace their competitiors. However, intra-and, more importantly, interpopulational hybrids involving Pigeon Rock Θ6's are relatively more vigorous. Such hybrids are likely to be heterozygous for a number of transpositions which may involve the chromosomes both outside and inside the ring. These hybrids may also be destabilised, like the *N*-containing twin hybrid described above, generating duplications throughout its genome. Homologous recombination processes acting upon duplicated segments in otherwise non-homologous chromosomes could then generate the interchanges necessary to assemble a larger ring. It may well be that this sequence of events was involved in the establishment of complex hybridity in the 3-mile Rock population, some 100 km SSW of Pigeon Rock. In this population, complex hybrid Θ10's, and alethal 7II's or Θ4's occur in a ratio of about 9:1 (James 1970). The selfed progeny of the 3-mile Rock Θ10 complex hybrids are clearly more vigorous than those of the Pigeon Rock Θ6, and they have reduced, but not excluded, their non-complex hybrid competitors from the population. All other complex hybrid populations examined to date are composed entirely of complex hybrids. These observations are in agreement with the hypothesis that complex hybridity originated on Pigeon Rock and has spread from that site, sequentially, to other populations. Interpopulational hybridizations at these new sites created forms able to conserve increasing levels of hybrid vigour in sequentially enlarged rings. This model provides an attractive synthesis of all the three models proposed by Darlington in 1931.

Acknowledgements

This paper incorporates a quantity of preliminary information provided by my students, Reno Beltran, Jane Sampson and Julia Playford, which is yet to be consolidated and formally published; I am very appreciative of their contribution. Support by the Australian Research Grants Committee for much of the work is gratefully acknowledged.

References

Beltran, I.C. and S.H. James 1970. Complex hybridity in *Isotoma petraea*. III. Lethal system in Θ12 Bencubbin. *Aust. J Bot.* 18, 223–232.

Catcheside, D.G. 1932. The chromosomes of a new haploid *Oenothera*. *Cytologia* 4, 68–113.

Cleland, R.E. 1936. Some aspects of the cytogenetics of *Oenothera*. *Jour. Genet.* 48, 99–110.

Darlington, C.D. 1931. The cytological theory of inheritance in *Oenothera*. *Jour. Genet.* 24, 405–474.

James, S.H. 1965. Complex hybridity in *Isotoma petraea*. I. The occurrence of interchange heterozygosity, autogamy and a balanced lethal system. *Heredity* 20, 341–353.

James, S.h. 1970. Complex hybridity in *Isotma petraea*. II. Components and operation of a possible evolutionary mechanism. *Heredity* 25, 53–77.

Lavery, P.A. and S.H. James 1987. Complex hybridity in *Isotoma petraea*. VI. Distorted segregation, gametic lethal systems and population divergence. *Heredity* 58, 401–408.

Karyotypic variation and evolution in the common shrew, *Sorex araneus*

J.B. Searle*

School of Biological Science, University of East Anglia, Norwich NR4 7TJ, UK

** Present address: Department of Zoology, University of Oxford, South Parks Road, Oxford OX1 3PS, UK*

The common shrew (*Sorex araneus*) has one of the most variable karyotypes of any species of mammal. The variation occurs both within and between populations (polymorphism and polytypy) and it appears that the predominant source of such variation is the Robertsonian fusion mutation (centric fusion, Robertsonian translocation; Searle 1984, 1986).

Of the chromosomes within the common shrew karyotype, the sex chromosomes (X, Y1, Y2) and autosomes *af, bc* and *tu* are invariant. However, the autosome arms *g–r* may occur as acrocentrics (which appears to be the ancestral state) or fused together in various combinations as metacentrics. Over the northern Palaearctic range of the common shrew many hundreds of animals have now been examined using banding methods. From these studies, which have been particularly detailed in western Siberia, Poland, Czechoslovakia, Finland, Sweden, Britain and Switzerland, it is apparent that the range of the species is subdivided into a patchwork of karyotypic races, each distinguished from neighbouring races by the combination of metacentric chromosomes it possesses (for recent reviews see: Zima & Král 1985, Hausser *et al.* 1986, Halkka *et al.* 1987, Fedyk & Leniec 1987). Workers in the field disagree as to what criteria should be used to define a karyotypic race, but it emerges from our recent meeting ('The Population and Evolutionary Cytogenetics of *Sorex araneus*', Oxford, 30–31 August 1987) that somewhere between 20 and 40 karyotypic races have been adequately described (abstracts to meeting available from me). Using all but the most conservative criteria, there are likely to be at least 100 karyotypic races of common shrew over the entire range of the species.

So far, considering all races, 36 different metacentrics (out of the 66 possible) have been described. Also, all the variable chromosome arms (*g–r*) may occur in the acrocentric condition and indeed in some karyotypic races certain chromosome arms are always found in the acrocentric state. In parts of the range of a race both particular metacentrics and the homologous acrocentrics are found in the same populations, giving rise to Robertsonian polymorphism.

In 1984 I presented a phylogeny based on a cladistic analysis of the chromosomal characteristics of the 12 karyotypic races of common shrew that had been described at that time. I identified a number of phylogenetic groups

which shared certain combinations of metacentrics: in particular, the West
European, the East European and the Siberian phylogenetic groups. The
constituent races of these groups occupied discrete geographic regions, lending
support to the cladistic scheme. However, since publication of the phylogeny,
the situation in eastern Europe has become considerably more complex (Wójcik
1986, Fedyk & Leniec 1987, Halkka *et al.* 1987) and a reappraisal of the
relationships of the races from that region is now necessary. However, the
concept of the West European phylogenetic group has been strengthened by
new data (Zima & Král 1985, Hausser *et al.* 1986, Wójcik 1986, Fedyk & Leniec
1987, Searle & Wilkinson 1987, Fredga 1987, Wójcik 1987).

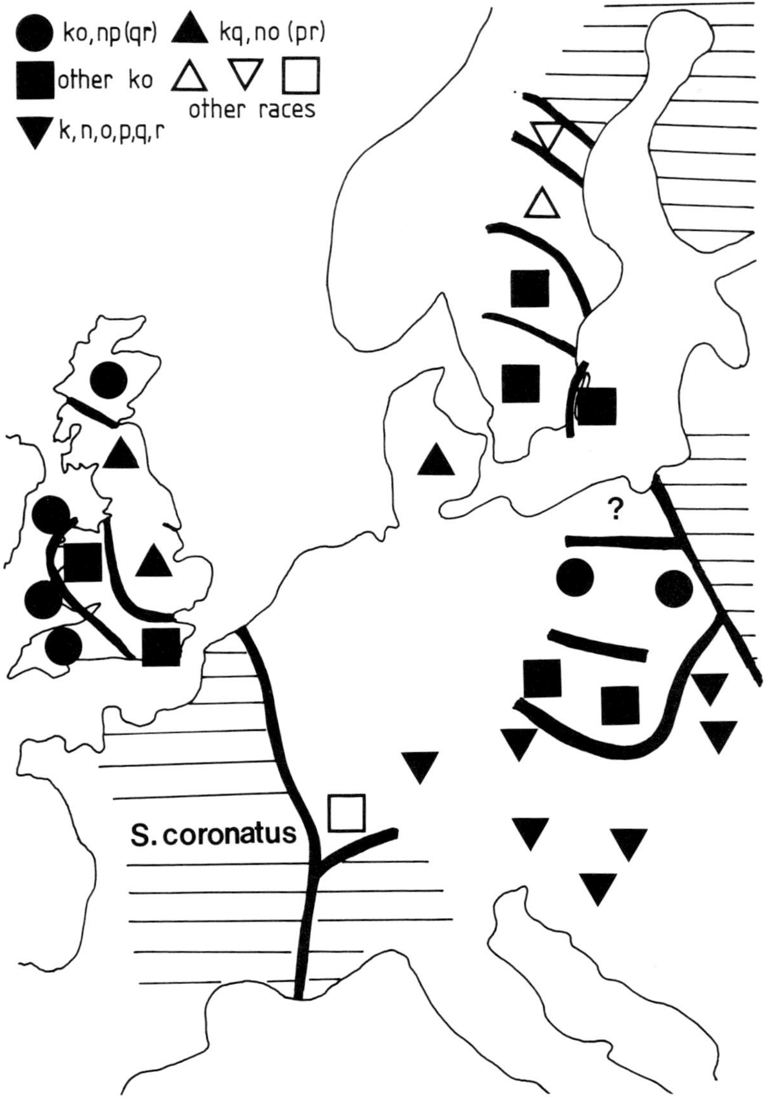

Figure 1. The distribution of the West European phylogenetic group of common shrews and
constituent races (see text). The approximate boundaries between these races are shown. Shading
indicates the distribution of different phylogenetic groups of common shrew and, also, the sibling
species *Sorex coronatus*.

Three metacentrics characterise the West European phylogenetic group: *jl, hi, gm*. In addition nine other metacentrics occur within the karyotypes of particular races within the group: *ko, kp, kq, no, np, nq, pr, pq, qr*; *k, n, o, p, q, r* may also occur as acrocentrics. (Data used in the description of the East European phylogenetic group comes from Wójcik 1987, Fredga 1987 and reviews previously cited.)

The distribution of the West European phylogenetic group is shown in Fig. 1. Races belonging to this phylogenetic group have been found in Sweden, Denmark, Britain, West Germany, Switzerland, Poland, Czechoslovakia, Hungary, Austria and Yugoslavia (see Table 1). However, in eastern Poland,

Table 1. The basic karyotypes within the West European phylogenetic group of the common shrew.

Karyotype	'Type' site	Distribution
k, n, o, p, q, r	Ulm	C & SE Europe
kr, no, p, q	Vaud	N Switzerland
kp, nr, o, q	Stugun	C Sweden
kq, n, o, p, r	Sidensjö	C Sweden
kq, no, p, r	Wrentham	E England (Britain)
		Jylland (Denmark)
kq, no, pr	Cothill	C Britain
		Sjaelland (Denmark)
ko, n, p, q, r	Öland	Öland (Sweden)
ko, nr, p, q	Drnholec	N & C Czechoslovakia
ko, nr, pq	Hallefors	S Sweden
ko, pr, n, q	Hermitage	S Britain
ko, pr, nq	Åkarp	S Sweden
ko, np, q, r	Arreton	Isle of Wight (Britain)
		Anglesey (Britain)
		SW Wales (Britain)
		SW England (Britain)
		C Poland
ko, np, qr	Tarty	NE Scotland (Britain)
		SW Wales (Britain)
		SW England (Britain)

'Oxford race': kq, no (pr); 'Aberdeen race': ko, np (qr)

northern Sweden and Finland and further east other forms are found. Similarly, the Valais race of southern Switzerland (and presumably northern Italy) does not belong to the West European phylogenetic group. *Sorex coronatus* replaces *S. araneus* over most of France. From this distribution I suggested (Searle 1984) that the West European phylogenetic group occupied a refugium in south-eastern

Europe at the last glacial maximum (20–15,000 b.p.), when northern Europe would have been climatically inhospitable to common shrews. It is suggested that the West European phylogenetic group spread north-westwards on amelioration of the climate. Apparently the phylogenetic group could not expand or persist further to the east or west because of alternative phylogenetic groups which occupied refugia in Italy (Valais race), Spain (*S. coronatus*) and eastern Europe and Asia (eastern European forms).

Table 1 shows the 13 basic karyotypes so far identified in the West European phylogenetic group. Note that some of these karyotypes are found in several regions that are isolated from each other (in the sense that other races and/or geographical barriers occupy intermediate territory). This table illustrates the problem of how to define a karyotypic race. For the ultimate 'splitter' each isolate should be described as a separate karyotypic race (i.e. 21 races). In contrast, Searle & Wilkinson (1987) combine populations with a basic karyotype of *ko, np* and those with a basic karyotype of *ko, np, qr* as Aberdeen race; both populations where metacentrics *ko, pr* are fixed and those where *pr* occurs at a very low frequency are considered to belong to the Hermitage race; and both populations characterised by metacentrics *kq, no, pr* and those with *kq, no* are classified as Oxford race. This, as we will see, makes sense in a British context. Taking this approach of 'lumping' of populations where chromosome arms are not combined into different metacentrics to its extreme, the West European phylogenetic group could be deemed to be represented by as few as 6 races.

What can we say about the constituent races of the West European phylogenetic group? In particular, what can we say about their origin and how they came to have their present distribution? To consider these questions I would like to present my own set of data on karyotypic variation of common shrews in Britain. This represents a geographic survey involving 1412 animals from 151 sites from 41 (out of 54) counties in England and Wales and 6 (out of 9) regions in Scotland. This project has run for the last 9 years, and I am extremely grateful for the assistance and advice of many people. In particular, I would like to single out the invaluable technical help of Ms Ann Reilly over the past 3 years.

Included within this data set are 170 individuals from 22 sites collected by C.E. Ford and co-workers in the 1950's and 1960's (reviewed by Ford & Hamerton 1970), wherever reinterpretation has been possible (Searle 1984, Searle & Wilkinson 1987, Searle & Reilly unpublished results).

Results

As already described, within Britain there are three basic karyotypic races: the Oxford, Hermitage and Aberdeen races. The distributions of these races are indicated in Fig. 2. Most widespread of these is the Oxford race which occupies central and northern England and southern Scotland. The Aberdeen race occurs in northern Scotland, south-western England and south-western Wales and on the Isle of Wight and Anglesey. The Hermitage race occupies northern and central Wales and southern England and a narrow connecting region. There is also some evidence that it occurs in the extreme south-west of Scotland (see below).

The approximate areas of contact between these races are shown in Fig. 2.

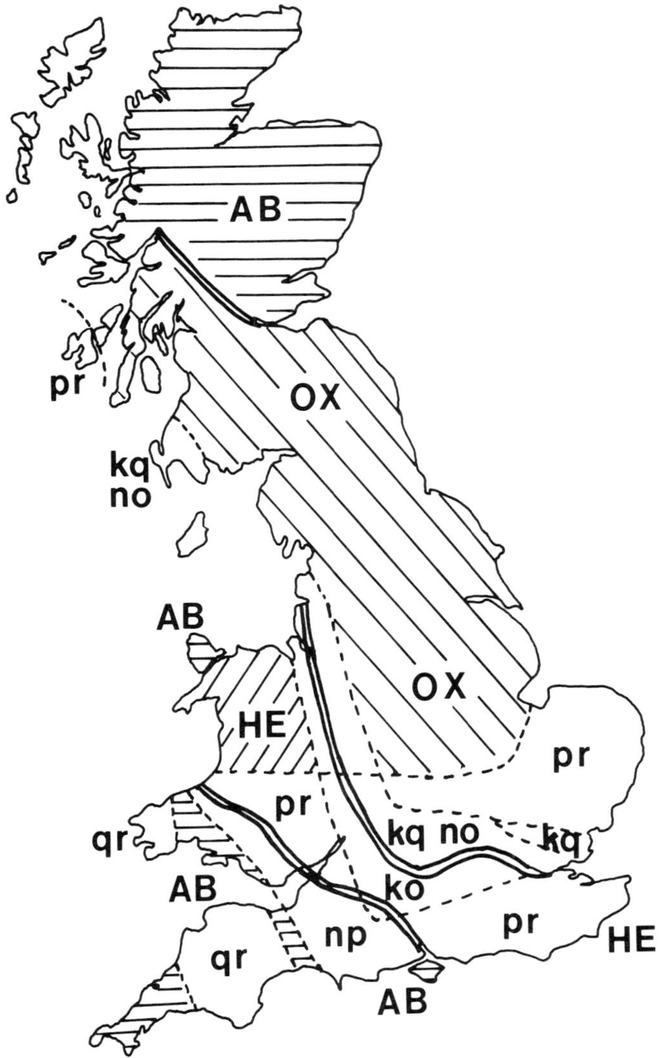

Figure 2. The distribution of the Oxford (OX), Aberdeen (AB) and Hermitage (HE) races of common shrew in Britain and areas of monomorphism (shaded) and polymorphism. The boundaries to the area of polymorphism of a particular arm combination are indicated by double lines (inter-racial hybrid zones) and dashed lines.

Inter-racial hybridisation has been demonstrated at several locations between the Oxford and Hermitage races and the Hermitage and Aberdeen races in southern Britain. Due to poorer sampling much less is known about the inter-racial contact areas in Scotland; no hybrids or racially-mixed populations have been located yet.

The contact area between the Oxford and Hermitage races in southern Britain is the best-studied to date. To the extreme east the Thames estuary separates the two races. Nothing is known about the interactions within London as this area is not very amenable to study. However, west of London, we know that the Oxford-Hermitage contact area is a hybrid zone. The course of this zone relates to various geographical features. Along its westward course it runs along the chalk ridge of the Chilterns and Berkshire Downs. The zone then veers northwards following the Severn valley for part of its course. The course of the Hermitage-Aberdeen hybrid zone may relate to ranges of hills and mountains: in Wales it appears to run over the Brecon Beacons and it runs along or close to the Mendips in south-western England. One may expect hybrid zones and other clines of heterozygote disadvanges to be trapped by geographical barriers and areas where the species occurs at low density (Barton & Hewitt 1985), as may be the case in upland areas, which would have less lush vegetation and hence a smaller invertebrate food supply than that of lowland alluvial areas.

The Oxford-Hermitage hybrid zone is characterised by a narrow area (approximately 15 km wide) where both Oxford and Hermitage race-specific metacentrics (*kq* and *no* and *ko*) occur, sometimes in the same individual (Searle 1986, Searle & Reilly unpublished results). In this area there is a high frequency of individuals that are homozygous for acrocentric chromosomes k, n, o, q. This is presumed to be a result of selection against individuals with a hybrid karyotype, which would have long chain configurations and suffer high frequencies of non-disjunction at meiosis (see Searle 1986). Clearly, in the region between the pure race distribution (either Oxford or Hermitage) which is metacentric-dominated and the hybrid zone which is acrocentric-dominated, one would expect an area of Robertsonian polymorphism, where individuals are homozygous metacentric, heterozygous or homozygous acrocentric for the race-specific arm combinations. As indicated in Fig. 2 there is indeed Robertsonian polymorphism for *kq* and *no* on the Oxford side of the hybrid zone (extending some 40 km) and for *ko* on the Hermitage side (extending some 20 km: see Searle 1986). This area of polymorphism is likely to be limited because the Robertsonian heterozygotes are presumably at a selective disadvantage relative to homozygotes (as a result of an increased frequency of non-disjunction associated with the presence of a trivalent configuration at meiosis). The width of the area of polymorphism will be determined by the degree of selection against these heterozygotes and the vagility of common shrews.

Polymorphism for *pr* and *np* is likewise associated with the Hermitage-Aberdeen hybrid zone which is dominated by a high frequency of acrocentrics *n, p, r*. Additionally to these hybrid-zone associated polymorphisms,

polymorphisms for *qr* (in Devon and in Pembrokeshire) and *pr* (in south-eastern England and on Islay) have been demonstrated. The polymorphism of *kq* and *no* in south-western Scotland may be associated with the occurrence of the Hermitage race in that region; more sampling is needed.

Interpretation

The colonisation of Britain

The Aberdeen race clearly forms a 'Celtic fringe' to the west and north of Britain (Fig. 2) and like the Celtic peoples this peripheral distribution is attributed to subsequent partial displacement by the other karyotypic races (Searle & Wilkinson 1987).

Thus, during the late glacial and early post-glacial period, when there was a land bridge between Britain and Europe and the climate was suitable for common shrews (sometime during the period 15,000–9,000 b.p.), Britain became occupied by successive waves of invasion of different karyotypic races. Of the extant races, the first to colonise was the Aberdeen race. This was partially displaced by the Hermitage race which was in turn partially displaced by the Oxford race. On the basis of present-day sea-level data the land bridge between Britain and the continent would have been over land at present covered by the North Sea. The relative distribution of the races is compatible with this place of invasion (see Fig. 2).

Whether the 3 races colonised Britain as closely successive waves of invasion or as distinct temporally-separated pulses (analogous to the human colonisation of Britain by first the Celtic peoples and later the Anglo-Saxons) cannot be said with any certainty. However, it is thought that the Menai Straits between Anglesey and mainland Britain were formed at about the same time that the land bridge between Britain and the continent was flooded. Therefore, if the Menai Straits were the barrier which stopped the Hermitage race displacing the Aberdeen race from Anglesey (the Aberdeen race is fully displaced from the mainland north Wales), then either: (a) the Hermitage race (closely followed by the Oxford race) entered Britain shortly before its separation from Europe and completed displacement shortly afterwards (the 'closely successive waves model') or (b) the races entered Britain independently but there were further range changes subsequent to the separation of Britain (the 'pulse model'). The closely successive waves model would presumably mean the colonisation of Britain between 10,000 and 9,000 b.p., that is after the Younger Dryas (the cold period that marked the final stage of the last ice age) and before Britain and Anglesey became islands, with racial displacement completed shortly afterwards.

Although it is suggested that the positions of inter-racial hybrid zones of common shrew in Britain are related to minor geographical barriers and local areas of low density where the zones are now trapped, presumably when these zones travelled through other parts of Britain and continental Europe they crossed other such barriers, sometimes more substantial. It is therefore a major challenge to determine why the zones are trapped at these particular barriers.

The most likely reason for hybrid zones to move is in response to large changes in population structure (Barton & Hewitt 1985), as would have occurred at the end of the last glaciation. If a gradient in population density is created, hybrid zones will tend to move down it. All other things being equal a race moving into a region previously unoccupied by the species will have a lower density than one following it because individuals will be 'lost' from the main body of the race into the virgin territory. However, once the first colonising race of a species reaches a coastline its density will increase. Individuals of this race can disperse only into territory that is already occupied rather than into new territory. This will increase the density of this race and slow down and maybe stop the hybrid zone that follows it. In this way, hybrid zones between karyotypic races of the common shrew may have moved through Europe (and ultimately Britain) when northern Europe became available for colonisation at the end of the ice age, and have halted when there was no new territory for the species to expand into. Thus, the positions of the hybrid zones of the common shrew in Britain may reflect the numbers of individuals constituting each colonising race and the influences of geographical features in slowing down and stopping hybrid zones.

Metacentrics pr *and* qr

Metacentric *qr* occurs in the Aberdeen race karyotype from NE Scotland, S Wales and SW England. This metacentric may have arisen by independent Robertsonian fusion of acrocentrics *q* and *r* in each of these localities (with subsequent spread) or within the Aberdeen race wave of invasion there was a sub-wave consisting of individuals with *qr* in their karyotypes. The polymorphisms involving *qr* in south-west Britain reflect the contact of areas dominated by metacentric *qr* and areas dominated by acrocentrics *q* and *r*. As with polymorphisms connected with hybrid zones these polymorphisms vary clinally, the width of the clines presumably being a reflection of the degree of heterozygote disadvantage and vagility of common shrews.

In the case of metacentric *pr* it would appear more likely that it arose and spread within Britain, although other explanations are possible. Fig. 3 shows the approximate 50% frequency isocline for this metacentric. From the position and shape of this isocline, I suggest that this metacentric arose within central or northern Britain and spread south. To the west the moving cline would have been halted by the Aberdeen-Hermitage hybrid zone (to which metacentric *pr* is now an important component, see Results section). As clines of heterozygote disadvantage tend to accumulate (Barton & Hewitt 1985), in south-eastern England the moving cline of *pr* may have been trapped by the clines of polymorphism for *kq* and *no* to which the *pr* cline is now nearly coincident. To the east the 50% isocline of metacentric *pr* follows the chalk ridge that runs into

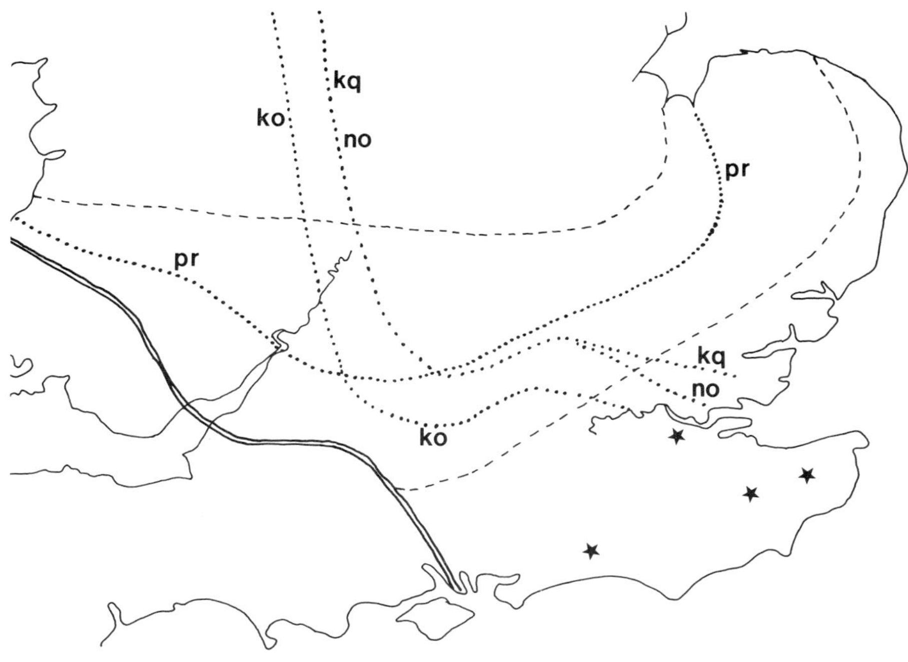

Figure 3. The 50% frequency isocline for metacentric *pr* (dotted line) in comparison with similar isoclines for other metacentrics. The main distribution for polymorphism of arm combination *pr* is indicated by dashed lines (the distribution to the south-west is bounded by the Hermitage-Aberdeen hybrid zone, as indicated by double lines). However, metacentric *pr* has also been found at low frequency in sites (stars) in an area otherwise dominated by acrocentrics *p* and *r*.

East Anglia, which may act as a local density trough.

It is of interest that there is polymorphism for arm combination *pr* on Islay with the acrocentric condition predominating. Thus, in the extreme north-west of the distribution of the Oxford race, as in the extreme south-east, the metacentric *pr* condition has failed to become fixed.

A European context; relationships between British and Continental races
If the Aberdeen, Hermitage and Oxford races spread from refugia in southern Europe as waves of invasion one may expect to see them elsewhere in Europe. Most significantly, the Aberdeen race (Table 1) is found in Poland (Fig. 1) where the West European phylogenetic group makes contact with other (eastern) phylogenetic groups (i.e. as may be expected if it was one of the earliest waves of invasion to spread north and west from a refugium in SE Europe) (Wójcik 1986, 1987, Fedyk & Leniec 1987).

Furthermore, Fredga (1987) has found the Oxford race in Denmark (Table 1, Fig. 1). This location is compatible with a fairly late invasion north. There are five other races in southern Sweden which presumably represent earlier invasions through Denmark. Their phylogenetic position (i.e. within the West European phylogenetic group) and the orientation of the 3 known hybrid zones in Sweden (i.e. they run approximately east-west) suggests an invasion from the south. Furthermore the 3 most southerly races in Sweden carry metacentric *ko* and therefore may represent part of the same wave of invasion as the Hermitage race in Britain.

In general terms the occurrence of the Hermitage race in mainland Europe may be more problematical than that of the Oxford and Aberdeen races. If the scenario described for Britain is correct, this race on invasion was characterised by metacentric *ko* with *n, p, q, r* acrocentrics. In Britain, metacentric *pr* could spread into this background but in other parts of Europe other metacentrics (composed of arms *n, p, q, r*) may have become incorporated into the racial karyotype. Thus I have lumped together all races with *ko* in their karyotype in Fig. 1, except the Aberdeen race which appears to be a distinct entity.

It may be significant that not only does a *ko* race occur adjacent to and northerly of an Oxford race in Scandinavia, but also two such races are apparently contiguous with and to the south of the Aberdeen race in Poland.

The other group of races, indicated in Fig. 1, which apparently occur extensively in south-eastern Europe are those with *k, n, o, p, q, r* in the acrocentric state (see Table 1). These presumably occupied a refugium to the south of the *ko*, Aberdeen and Oxford races during the last ice age. Races in the northern and south-westerly extremes do not fit into the four categories so far described. The situation in north-western Poland needs clarification.

Clearly, more sampling is needed within the West European phylogenetic group to clarify the relationships and tease out cases of parallel evolution. The recent discovery of shrews belonging to the West European phylogenetic group near Moscow (Ivanitskaya 1985), well to the east of other such races, opens up a completely new insight into this phylogenetic group.

Acknowledgements

In recent times, this work has received support from N.E.R.C. and the Royal Society. I thank Dr. G.M. Hewitt for laboratory facilities. I am grateful to Prof. K. Fredga and Dr. J.M. Wójcik for unpublished information.

References

Barton, N.H. and G.M. Hewitt 1985. Analysis of hybrid zones. *Ann. Rev. Ecol. Syst.* 16, 113–148.

Fedyk, S. and H. Leniec 1987. Genetic variability of Polish populations of *Sorex araneus* L. I. Variability of autosome arm combinations. *Folia Biol., Kraków.* 35, 57–68.

Ford, C.E. and J.L. Hamerton 1970. Chromosome polymorphism in the common shrew, *Sorex araneus. Symp. Zool. Soc. Lond.* 26, 223–236.

Fredga, K. 1987. Distribution of the chromosome races in Sweden and Denmark. *The Population and Evolutionary cytogenetics of Sorex araneus. An International Meeting, Oxford.* (Abstracts of meeting.)

Halkka, L., V. Söderlund, U. Skarén and J. Heikkilä 1987. Chromosomal polymorphism and racial evolution of *Sorex araneus* L. in Finland. *Hereditas* 106, 257–275.

Hausser, J., E. Dannelid and F. Catzeflis 1986. Distribution of two karyotypic races of *Sorex araneus* (Insectivora, Soricidae) in Switzerland and the post-glacial recolonization of the Valais: first results. *Z. f. zool. Systematik u. Evolutionsforschung* 24, 307–314.

Ivanitskaya, E.J. 1985. Taksonomiczeskaja i citogeneticzeskaja analiz transberingijskich swjazej zemlerojek-burozubok (*Sorex*: Insectivora) i piszczuch (*Ochotona*: Lagomorpha). Ph.D. thesis, Univ. Moscow. (In Russian.)

Searle, J.B. 1984. Three new karyotypic races of the common shrew *Sorex araneus* (Mammalia: Insectivora) and a phylogeny. *Syst. Zool.* 33, 184–194.

Searle, J.B. 1986. Factors responsible for a karyotypic polymorphism in the common shrew, *Sorex araneus. Proc. R. Soc. Lond.* B229, 277–298.

Searle, J.B. and P.J. Wilkinson 1987. Karyotypic variation in the common shrew (*Sorex araneus*) in Britain — a "Celtic Fringe". *Heredity*, in press.

Wójcik, J.M. 1986. Karyotypic races of the common shrew (*Sorex araneus* L.) from northern Poland. *Experientia* 42, 690–962.

Wójcik, J.M. 1987. Karyotypic races of common shrew from Poland. *The Population and Evolutionary Cytogenetics of Sorex araneus, An International Meeting, Oxford.* (Abstracts of meeting.)

Zima, J. and B. Král 1985. Karyotype variability in *Sorex araneus* in central Europe (Soricidae, Insectivora). *Folia Zool., Brno* 34, 235–243.

Differences in the nucleolar organisers on sex chromosomes and Haldane's Rule in a hybrid zone

*G.M. Hewitt, †J. Gosalvez, †C. Lopez-Fernandez, *M.G. Ritchie, *W. Nichols and *R.K. Butlin

*School of Biological Sciences, University of East Anglia, Norwich, and †Genetics Department, Universidad Autonoma, Madrid, Spain

Hybrid zones are relatively narrow regions where two different parapatric genomes meet, mate and hybridise. These genomic differences may be recognised phenotypically in a variety of ways, from obvious characters such as bird plumage or plant morphology through less accessible features such as chromosome structure and allozyme variation to the recent field of variation in the DNA itself. They are of interest for two primary reasons, firstly, because they are common [a recent survey (Hewitt 1986) listed over 150 well established cases and this is now in excess of 170] with many more possible examples; secondly, because they are natural laboratories in which we can study the interaction of two partly differentiated taxa and thereby gain some insight into the process of speciation and the genes involved in divergence.

Types of zone

We may, in principle, distinguish two modes of origin for hybrid zones. Following Mayr's (1970) terminology, a primary zone of intergradation is where the differences evolved in a continuous distribution, and a secondary zone of intergradation is where the differences evolved while the two populations were geographically isolated and then formed a steep cline when their ranges expanded and met. Both are possible. Thus an environmental gradient can impose selection for modifier genes which will progressively steepen the gene clines between two internally coadapted genotypes (Clarke 1966, Endler 1977); while it is also clear that hybrid zones occur in regions not inhabited by the taxa during the ice age and these have made secondary parapatric contact after being allopatric (Hewitt 1986). This secondary contact model has been more favoured (White 1968, Carson 1975, Hewitt 1975) but this does not mean that the differences arose because of or during allopatry. This could have happened by selection, drift or their interaction over many range fluctuations and a long period of time. For example, biochemical differences betweeen the toads *Bombina bombina* and *B. variegata* that currently form a long hybrid zone in central Europe provide evidence for a divergence time of several million years (Szymura *et al* 1985), long enough for several ice ages. It is difficult, if not impossible, to deduce the evolutionary history of two taxa now meeting in a hybrid zone and to assign differences to primary or secondary models (Barton & Hewitt 1985). This is particularly true of the tropics (Endler 1982), but in the more temperate regions understanding the recent history of a zone may be somewhat simpler because of ice age clearances and a growing knowledge of Quaternary biology.

Current hybrid zones may have a number of proximate explanations. Following secondary contact, if the allelic differences are selectively equal there will be a gradual diffusion to produce a progressively shallower cline. If one form is advantageous then the alleles for this will advance as a wave replacing the other alleles through the species range. If there is selection against the heterozygotes or recombinants produced by hybridisation then a tension zone (Key 1968) will result, the width of which will be a product of the dispersal of the organism and the lower fitness of hybrids. If the two genotypes are adapted to different environments then a hybrid zone will be maintained in the ecotonal transition. It is also theoretically possible that a hybrid zone could be maintained if selection favoured hybrids in a narrow ecotone (Moore 1977). Evidence has been produced in support of all these possibilities; however, from a review of the literature, it appears that hybrid unfitness is most often involved, with ecological differentiation possibly also a major factor (Hewitt 1986). The category of near neutral mixing after secondary contact is also not to be overlooked, but there is little convincing evidence yet for the hybrid superiority model except in local isolates rather than hybrid zones (Barton & Hewitt 1985).

Distinguishing causes

How may we distinguish these various possibilities for any particular zone, what data do we need and what experiments should we do? Fortunately each model has certain properties which can be measured and tested against each other and in conjunction. A tension zone will have hybrids that are less fit than the two pure forms. This may be demonstrated with some endeavour in both laboratory and field, as has been done for the sex chromosome zone in *Podisma pedestris* (Barton & Hewitt 1981a, Nichols & Hewitt 1987). The width of a tension zone will be determined by the dispersal of the organism and the selection against the heterozygote for the locus in question, so both width and dispersal should also be measured to see if the three parameters agree (e.g. Hewitt & Barton 1980). Where there are coincident clines for a number of loci, and this is the case in most zones (Barton & Hewitt 1985, Hewitt 1986), then measuring linkage disequilibrium and clinal shapes allows one to produce estimates of dispersal, selection and the number of genes selected, which may then be compared with actual measures. Recent practical and theoretical work by Szymure & Barton (1987) in *Bombina* reveals the power of this approach.

The position of a tension zone will not necessarily be determined by a change from one ecological community to another, but it will tend to lie in regions of consistent low density (Hewitt 1975, Barton 1979, Barton & Hewitt 1981b, Nichols & Hewitt 1986). If the hybrid zone is, however, produced by adaptation of the two parapatric genotypes to different ecologies then it should be possible to demonstrate this. One would expect the width of the zone to vary with the steepness and patchiness of the ecological transition, and also that clines for different genes would not all change at the same place — the null point of selection for the adaptation of a particular allele need not be the same location in an ecotone as for alleles of another gene. Consequently, detailed investigations of population density and ecology in relation to the zone are required.

Historical evidence is required to decide if a zone is moving or there is neutral mixing. Clines several orders of magnitude greater than the dispersal distance may well be examples of the latter.

A range of questions should be asked of any hybrid zone in order to understand its status and thereby proceed to the more general questions of genome differentiation, racial distribution and speciation. To this end we have been studying a zone in the grasshopper *Chorthippus parallelus* in the Pyrenees.

Zone in *Chorthippus parallelus*

This flightless meadow grasshopper is widely distributed in Western Europe. The French subspecies *C.p. parallelus* differs from the Spanish subspecies *C.p. erythropus* in lacking the reddish tinge of the hind tibiae of *erythropus*, in lacking its distinctive courtship song and in having fewer stridulatory pegs on average (Reynolds 1980). A few intermediate individuals had been found in the Pyrenees (Kruseman 1982) and further investigation has revealed a complex hybrid zone between these two taxa (Butlin & Hewitt 1985 a, b). Whilst this zone was discovered on the basis of morphological and behavioural characters, it is now clear that it involves chromosomal, allozymic and DNA differences as well. By contrast, many of the best studied zones were first discovered from their chromosomes and other characters followed [e.g. *Podisma pedestris* (Hewitt 1975), *Caledia captiva* (Shaw 1981), *Warramaba* (White 1978), *Sorex araneus* (Searle 1984), *Spalax ehrenbergii* (Nevo 1982)]. The two subspecies meet along the Pyrenees (Fig. 1) but since their upper altitudinal limit is around 1900 m they can do this only in the low cols in the central part of the range and around the western and eastern ends. During the ice ages the Pyrenees were covered by an ice sheet, and *C. parallelus* would have retreated south into Spain and probably east through lowland Provence and south into Italy. As the climate ameliorated some 10000 BP the two races would have expanded from their refuges along with many other organisms and met each other first in the lowlands at either end of the Pyrenees and later in the higher cols.

Our studies have concentrated on two regions of secondary contact, one near Mont Louis in the Eastern Pyrenees at Col de la Quillane and the other near Gabas in the Western Pyrenees at Col du Pourtalet.

Hybrid testis dysfunction

Laboratory crosses between pure populations of French *C.p. parallelus* and Spanish *C.p. erythropus* from either side of the zone reveal that male F1 hybrids have atrophied testes and are virtually sterile, while female F1 hybrids mated and layed eggs which hatched apparently normally (Hewitt *et al.* 1987). F1 females backcrossed to pure *C.p. parallelus* or *C.p. erythropus* males produced viable BX progeny, but the BX males showed variable intermediate levels of testis dysfunction. These data provide yet another clear example of Haldane's Rule that "when in the F1 offspring of a cross between two animal species or races one sex is absent, rare or sterile, that sex is always the heterozygous sex" (Haldane 1922). Furthermore, they clearly demonstrate that the genomes of the two races have diverged so that hybrid males are completely unfit and the F1 generation as a whole is probably only half as fit as the parentals at the most. This is clear evidence of a tension zone.

Accurate location of the zone

In order to locate the zone accurately and thereby facilitate the study of its genetics, ecology and dynamics, transects were made through Col de la Quillane and Col du Pourtalet (Fig. 1). These began with samples every few kilometres over some 60 km, and, in later years as the position of the zone became clearer, samples were taken every few hundred metres in critical regions. We measured a whole suite of morphological characters which showed variable amounts of change from pure *parallelus* to *erythropus*. When all are combined in a discriminant function (first canonical variant) it provides a clear step cline through the zone and separates the two subspecies (Butlin & Hewitt 1985a). The single best discriminator is the number of pegs on the stridulatory file of the male. The width of the cline for this character at Col de la Quillane is about 5 km, where width is defined as 1/maximum slope (Bazykin 1969). At Col du Pourtalet it appears to be wider at some 12 km (Fig. 2). Both appear to be fairly smooth sigmoidal clines although we require more data at the edges of the Col du Pourtalet zone.

The calling and courtship songs of the two subspecies differ in both structure and usage (Butlin & Hewitt 1985b). Various measures of these characters through Col de la Quillane show that they change in the same place as the morphological characters but the clines appear to be broader. The second principal component of the Fourier transform of courtship song changes over the zone with a cline width of about 10 km.

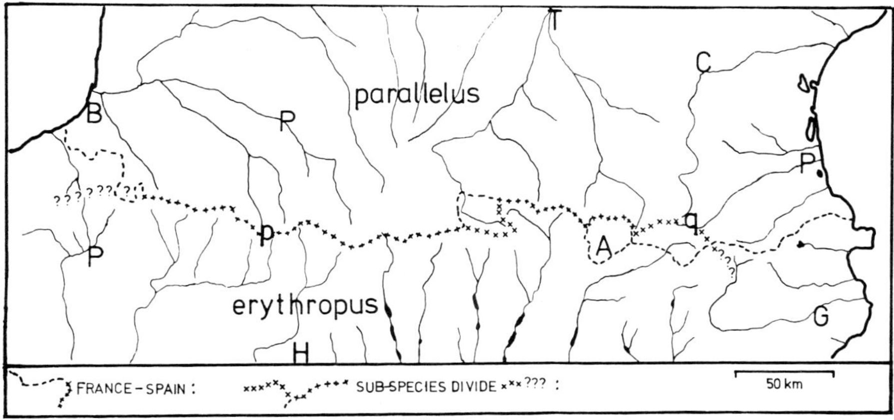

Figure 1. The subspecies *C.p. parallelus* and *C.p. erythropus* are separated by the high mountains of the Pyrenees. Below 1900 m they meet and hybridise. Hybrid zones studied in detail are marked at Col du Pourtalet (p) and Col de la Quillane (q). This division follows the higher natural divide rather than the political boundary, but the course of hybridisation is not yet accurately determined at the ends of the mountain range. Major cities are marked by capital letters.

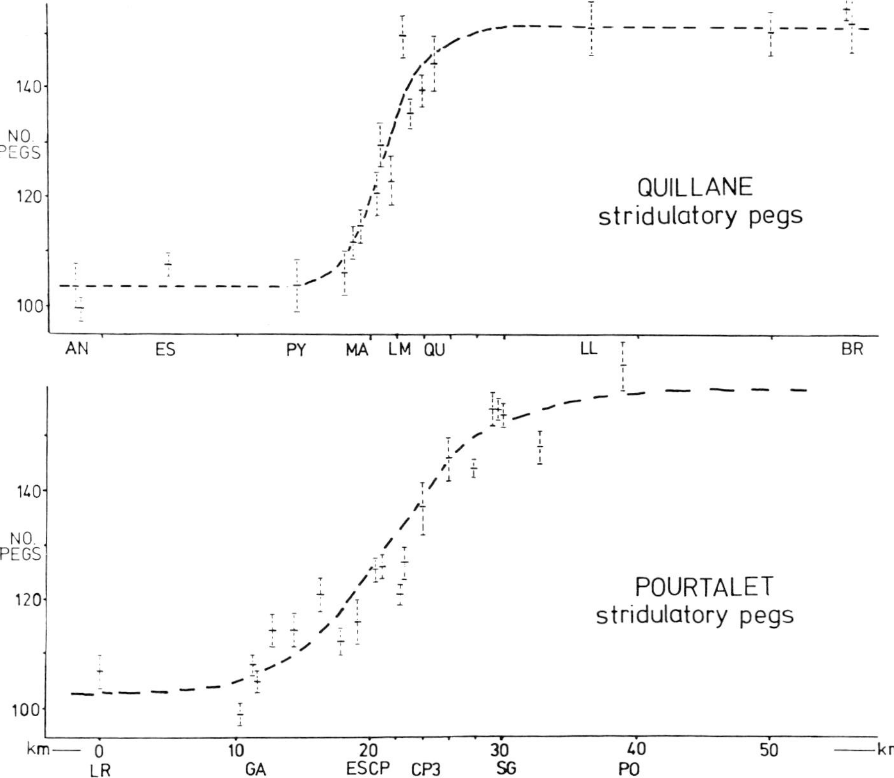

Figure 2. Clines for stridulatory pegs on male hind femurs across Col de la Quillane and Col du Pourtalet. Means and standard deviations are marked. Some key populations are located by letters, e.g. AN = Aunat, LL = Llo, LR = Laruns, PO = Polituara.

The two subspecies also differ for a number of enzyme systems (Butlin & Hewitt 1985a, Duijm personal communication); at Col de la Quillane the cline for Esterase-2 is about 15 km wide and occurs in the same place as those for morphological and song characters.

Thus the clines for the various behavioural, morphological and allozyme differences between *C.p. parallelus* and *C.p. erythropus* are coincident in this zone which is located near the high point of two disjunct cols on the Pyrenean ridge. This suggests that the hybrid zones for genes controlling these characters are not primarily ecologically determined, otherwise they might change in different places. Clines may be held together by hybrid unfitness caused by heterozygote disadvantage or recombination between co-adapted alleles of the two pure races. They may also be in density troughs. Genes for characters of ecological significance may be linked into this system. Detailed studies of the habitat and ecology of these zones similar to those in *Podisma pedestris* (Nichols & Hewitt 1988) are needed to detect any ecological correlation and these are in hand.

Nucleolar organisers in *Chorthippus parallelus*

Using a silver staining technique (Rufas *et al* 1983) to locate the nucleolar organising region in testes samples, we found that *C.p. parallelus* has three primary NORs, interstitially on chromosome L2, terminally on L3 and terminally on the X. In contrast, *C.p. erythropus* does not have the NOR on the X (Fig.3). This provides a very clear chromosomal marker and we have scored its frequency through the hybrid zone (Fig. 4).

Col de la Quillane
A transect sampled every 300 metres at Col de la Quillane shows the zone for the X-NOR to be only some 600 m wide, centred near the middle of the eastern side of Lac de Matemale and coincident with the cline for peg number. The narrowness of this zone argues strongly that it is maintained by selection against hybrids balanced by the insects' dispersal each generation (Barton & Hewitt 1985). The effective selection on the X-NOR cline is given by the relationship $w = \sqrt{8\sigma^2/s}$, where w = width, σ^2 is dispersal expressed as variance of parent offspring distance over one generation, s is selection. If we take $\sigma = 30$ m from a simple mark-recapture experiment, then selection is about 2%. We clearly need good estimates of dispersal in the zone and these are in hand; it is quite likely that dispersal varies between different places in the zone but is unlikely to be an order of magnitude different.

Since this NOR is on the X chromosome, only females (XX) may be hybrid for its presence or absence; males are XO and either have one or do not. Heterozygote disadvantage for the X-NOR may operate only in the female. Selection may also act on this region in the males since in the hybrid zone they will have various mixtures and recombinations of *parallelus* and *erythropus* alleles at loci throughout the complement. The X-NOR may not function properly in the presence of a hybrid background genome. We have some preliminary evidence that suggests this may be the case.

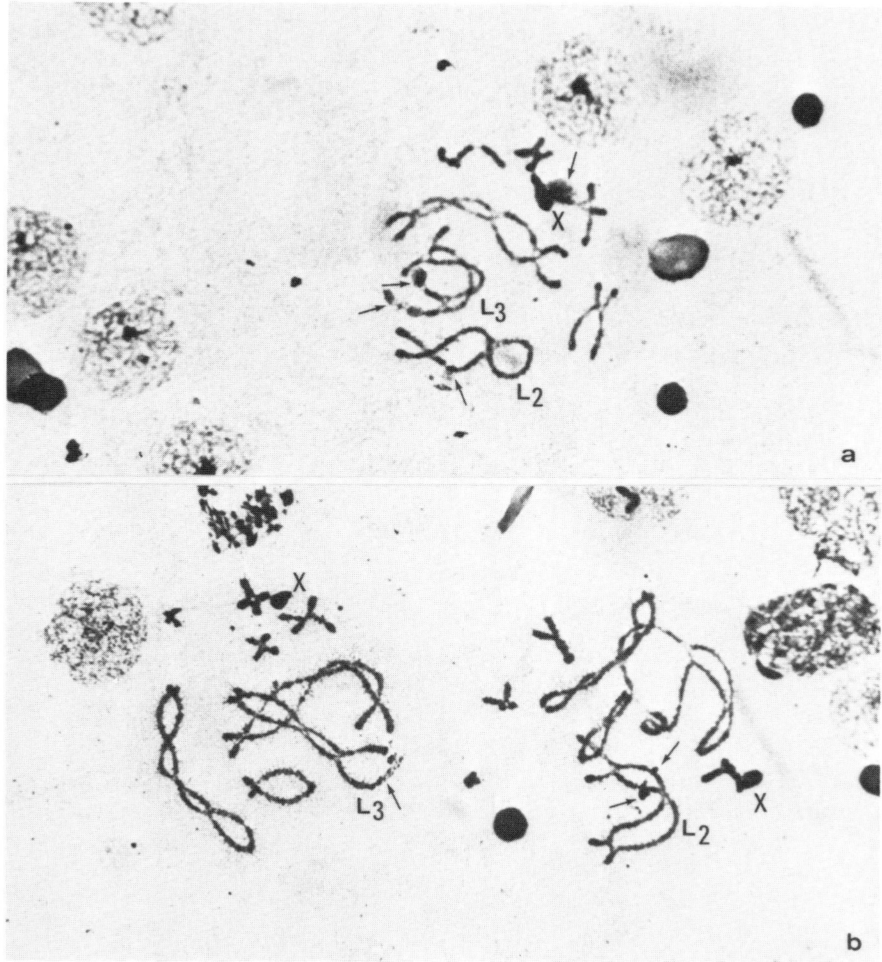

Figure 3. Nucleolar organising regions in *C.p. parallelus* (a) and *C.p. erythropus* (b). *C.p.e.* lacks a NOR on the X.

It was noticed when scoring the X-NORs in the transect that some males had nucleoli which were smaller than normal and this seemed consistent in all cells examined in each individual. Such individuals were found only in the samples between Matemale and Quillane where there is the cline in X-NOR frequency. Pure samples of *C.p. parallelus* from Escaloubre had only normal-sized X nucleoli while pure samples of *C.p. erythropus* from Eyne had nothing on the X. Furthermore, individuals with normal-sized X nucleoli were found only in the samples on the *parallelus* side of cline. Those on the *erythropus* side were all small or showed none at all.

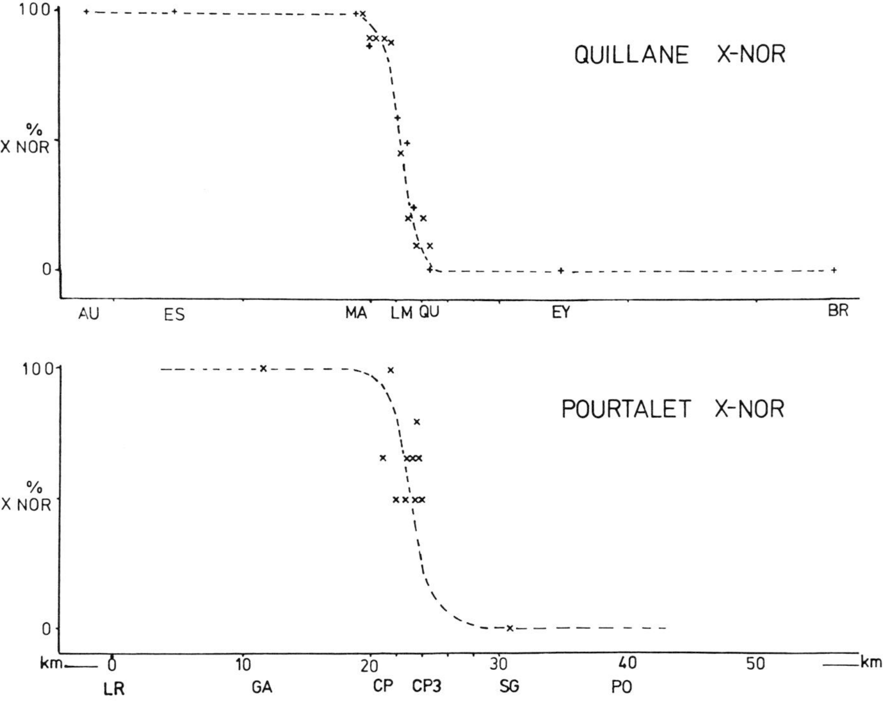

Figure 4. Clines for the X chromosome nucleolar organising region (X-NOR) across Col de la Quillane and Col du Pourtalet. Most points comprise 8–10 individuals but a few are smaller samples. The cline marked at Pourtalet is thus for the present an indication only.

Col du Pourtalet

Our samples from Gabas (*C.p.p.*) to Sallent de Gallego (*C.p.e.*) also show that the X-NOR changes across the Col du Pourtalet (Fig. 4), but the picture is at present less clear than at Col de la Quillane. This is partly because our intensive sampling covered only some 4 km at the presumed centre of Pourtalet compared with 6 km at Quillane, but it is also possibly an indication of a wider zone for the X-NOR over Quillane. There is also variation between adjacent samples, which may in some cases be due to sampling variance; but it may also reflect a more patchy distribution, which itself could lead to a wider zone. Clearly this requires further research.

The cline for peg number is also wider at Pourtalet than Quillane, as would be expected since those factors that are affecting the insects' dispersal, be they patchy distributions, density troughs, more viscous habitats or genes for jumping, will diffuse all the genes, from nucleolar DNA sequences to polygenes for peg number or song. As at Quillane, the clines for X-NOR and peg number are coincident, being centred on the Col du Pourtalet. This argues for the hybrid zone's being maintained by selection against heterozygotes or recombinants between the coadapted genomes of *C.p. parallelus* and *C.p.erythropus*. It does not rule out the possibility that the two subspecies may have different ecological adaptations, but environmental differences are not likely to be the only or the main factors producing the present zone.

In two individuals from the hybrid zone at Col du Pourtalet the L3 chromosome showed NOR silver staining near the centromere in addition to the standard nucleolus at the distal end of its short arm. This indicates the existence of cryptic rRNA which may possibly be expressed in hybrid genomes that have unbalanced control. Also some individuals from this zone which had active X-NORs did not have attached nucleoli at diplotene.

Discussion

It seems the clines for peg number and X-NOR at Col du Pourtalet are different in width from those at Col de la Quillane and the most likely explanation is a difference in the structure of the habitat that produces a more effective dispersal and diffusion over the former col. However, there are other possible reasons. For example, it could be that the genomes in the Eastern and Western Pyrenees are different in one or both subspecies, so that selection on the hybrids for the characters in question is different. It is also possible that one or both zones is not yet at equilibrium, but this seems a less likely explanation since Col du Pourtalet is both higher and narrower and yet the zone appears wider. The fact that song and enzyme clines are wider than peg and X-NOR clines indicates that selection on them is less intense and may be very small at each locus.

There is evidence both from crosses between the two subspecies and from the zone between them that hybrids are probably less fit. Obviously this must be investigated in the field. Section against hybrids has recently been shown in the zone in *Podisma pedestris* (Nichols & Hewitt 1987). The X-NOR would appear to contribute to this hybrid unfitness, it forms a narrow cline and there are indications of variable nucleolar expression within the zone which suggest that the control of rDNA is imperfect in hybrid genomes. A number of factors should be considered here. Firstly, the number of active NORs is different between *C.p. parallelus* (3) and *C.p. erythropus* (2) and since one would expect a correlation of overall NOR size and the number of rDNA copies (Flavell & Martin 1982) the NORs in *erythropus* may be bigger than in *parallelus*. Secondly, because of this there may be over- or under-production of rDNA in the zone which would affect fitness. Thirdly, however, compensation in the production of NOR is documented in *Drosophila* (Cullis 1982); this may ameliorate the obvious affects of hybridisation on nucleolar size but may still affect the organism's fitness.

We noted earlier that crosses between the two subspecies were an example of Haldane's Rule with only the males' fertility severely reduced. The difference in X-NOR may be a component of this, since the most tenable fundamental explanation of this rule is that the sex chromosomes evolve difference more rapidly than autosomes because of genetic drift and hemizygosity (Charlesworth *et al.* 1987). The more proximate cause may be that the XO male has more problems in controlling rRNA production than the XX female. Clearly this is an avenue worth pursuing.

Acknowledgement

This work was supported by grants from the SERC.

References

Barton, N.H. 1979. Gene flow past a cline. *Heredity* 43, 333–339.

Barton, N.H. and G.M. Hewitt 1981a. Hybrid zones and speciation. *In Evolution and Speciation: Essays in Honour of M.J.D. White.* W.R. Atchley and D.S. Woodruff, eds, 109–145, Cambridge. Cambridge Univ. Press.

Barton, N.H. and G.M. Hewitt 1981b. The genetic basis of hybrid inviability in the grasshopper *Podisma pedestris. Heredity* 47, 367–383.

Barton, N.H. and G.M. Hewitt 1985. Analysis of hybrid zones. *Ann. Rev. Ecol. Syst.* 16, 113–148.

Bazykin, A.D. 1969. Hypothetical mechanisms of speciation. *Evolution* 23, 685–687.

Butlin, R.K. and G.M. Hewitt 1985a. A hybrid zone between *Chorthippus parallelus parallelus* and *Chorthippus parallelus erythropus* (Orthoptera: Acrididae): morphological and electrophoretic characters. *Biol. J. Linn. Soc.* 26, 269–285.

Butlin, R.K. and G.M. Hewitt 1985b. A hybrid zone between *Chorthippus parallelus parallelus* and *Chorthippus parallelus erythropus* (Orthoptera: Acrididae): behavioural characters. *Biol. J. Linn.* 26, 287–299.

Carson, H.L. 1975. The genetics of speciation at the diploid level. *Amer. Nat.* 109, 83–92.

Charlesworth, B., J.A. Coyne and N.H. Barton 1987. The relative rates of evolution of sex chromosomes and autosomes. *Amer. Nat.* (in press).

Clarke, B.C., 1966. The evolution of morph ratio clines. *Amer. Nat.* 100, 389–400.

Cullis, C.A. 1982. Quantitative variation of the ribosomal RNA genes. In *The Nucleolus*, W.G. Jordan and C.A. Cullis, eds, 103-112. Cambridge. Cambridge Univ. Press.

Endler, J.A., 1977. *Geographic variation, speciation and clines.* Princeton. Princeton Univ. Press.

Endler, J.A., 1982. Problems in distinguishing historical from ecological factors in biogeography. *Am. Zool.* 22, 441, 452.

Flavell, R.B. and G. Martini 1982. The genetic control of nucleolus formation with special reference to common bread wheat. In *The Nucleolus.* W.G. Jordan and C.A. Cullis, eds, 113-128. Cambridge. Cambridge Univ. Press.

Haldane, J.B.S., 1922. Sex ratio and unisexual sterility in hybrid animals. *Journal of Genetics*, 12, 101–109.

Hewitt, G.M. 1975. A sex chromosome hybrid zone in the grasshopper *Podisma pedestris* (Orthoptera: Acrididae). *Heredity* 35, 375–387.

Hewitt, G.M. 1986. The structure and maintenance of hybrid zones. In *Orthoptera* 1, J. Gosalvez, C. Lopez-Fernandez and C. Garcia de la Vega, eds 15–54, Madrid, Fundacion Ramon Areces.

Hewitt, G.M. and N.H. Barton 1980. The structure and maintenance of hybrid zones as exemplified by *Podisma pedestris.* In *Insect Cytogenetics.* R.C. Blackman, G.M. Hewitt and M. Ashburner, eds, 149–170. Oxford, Blackwell.

Hewitt, G.M., R.K. Butlin and T.M. East 1987. Testicular dysfunction in hybrids between parapatric subspecies of the grasshopper *Chorthippus parallelus. Biol. J. Linn. Soc.* 31, 25–34.

Key, K.H.L. 1968. The concept of stasipatric speciation. *Syst. Zool.* 17, 14–22.

Kruseman, G. 1982. Materiaux pour le faunistique des Orthopteres de France. Fascicule II. Les acridiens des Musees de Paris et d'Amsterdam. *Verslagen en Technische Gegerens* 36. Institute voor Taxonomische Zoologie (Zoologisch Museum) Universiteit van Amsterdam.

Mayr, E. 1970. *Populations, species and evolution.* Cambridge, Mass. Harvard Univ. Press.

Moore, W.S. 1977. An evaluation of narrow hybrid zones in vertebrates. *Q. Rev. Biol.* 52, 263–278.

Nevo, E. 1982. Speciation in subterranean mammals. In *Mechanisms of Speciation*, 191–218. Academia Nazionale Dei Lincei Rome 1981. Allan Liss Inc., New York.

Nichols, R.A. and G.M. Hewitt 1986. Population structure and the shape of a chromosomal cline between two races of *Podisma pedestris* (Orthoptera: Acrididae). *Biol. J. Linn. Soc.* 29, 301–316.

Nichols, R.A. and G.M. Hewitt 1988. Genetical and ecological differentiation across a hybrid zone. *Ecological Entomology* 13, (in press).

Rufus, J.S., J. Gosalvez, C. Lopez-Fernandez and H. Cardoso 1983. Complete dependence between Ag-NORs and C-positive heterochromatin revealed by simultaneous Ag-NOR C-banding method. *Cell. Biol. Int. Rep.* 7, 275–281.

Searle, J.B. 1984. Three new karyotypic races of the common shrew *Sorex araneus* (Mammalia: Insectivora) and a phylogeny. *Syst. Zool.* 33, 184–194.

Shaw, D.D. 1981. Chromosomal hybrid zones in Orthopteroid insects. In *Evolution and Speciation. Essays in Honour of M.J.D. White*, W.R. Atchley and D.S. Woodruff, eds, 146–170. Cambridge. Cambridge Univ. Press.

Szymura, J.M. and N.H. Barton 1987. Genetic analysis of a hybrid zone between the fire-bellied toads *Bombina bombina* and *B. variegata*, near Cracow in Southern Poland. *Evolution* 40, 1141–1159.

Szymura, J.M., C. Spolsky and T. Uzzell 1985. Concordant change in mitochondrial and nuclear genes in a hybrid zone between two frog species (genus *Bombina*). *Experientia* 41, 1469–1470.

White, M.J.D. 1968. Models of speciation. *Science* 158, 1065–1070.

White, M.J.D. 1978. Chain processes in chromosomal speciation. *Syst. Zool.* 27, 285–298.

Chromosomal rearrangements, ribosomal genes and mitochondrial DNA: contrasting patterns of introgression across a narrow hybrid zone

D.D. Shaw, A.D. Marchant, M.L. Arnold and N. Contreras

Population Genetics Group, Research School of Biological Sciences, Australian National University, Canberra, Australia

We dedicate this paper to Bernard John upon his retirement as an acknowledgement of his exemplary contribution to the study of chromosomes

Homozygous chromosomal differences between closely related and parapatrically distributed taxa have now been documented in a wide variety of organisms. It has been proposed that they can act as major barriers to gene flow between the taxa and hence, initiate speciation (White 1978). However, definitive empirical evidence to support this proposal is lacking. Recent developments in molecular biology have provided an array of independent genetic markers, both nuclear and cytoplasmic, which now permits detailed comparative assessments of the significance of chromosomal rearrangements as barriers to gene flow at the interface between chromosomally differentiated and hybridising taxa.

We have used several of these molecular markers to assess the evolutionary status of a hybrid zone between two genomically distinct taxa within the grasshopper *Caledia captiva*. The markers used include restriction fragment length variation in the mitochondrial genome (mtDNA) and in the ribosomal DNA (rDNA), a highly repeated 168 bp satellite sequence, allelic variation at four enzyme-coding loci and a series of pericentric rearrangements involving most members of the genome.

These analyses reveal that, although the hybrid zone is extremely narrow, with concordant chromosomal and highly repeated DNA clines spanning only 1 km, the other diagnostic nuclear and cytoplasmic markers are found at high frequencies up to 400 kilometres beyond the chromosomal limits of the hybrid zone.

These data suggest that the hybrid zone is very old and capable of moving while still retaining its very narrow chromosome structure. They also suggest that the chromosomal structural differences are the principal genetic components in maintaining the narrowness of the hybrid zone and that strong selective processes are involved, acting at the genomic level.

Material and methods

Three transects along coastal Queensland were analysed: (a) a northern transect from Peregian Beach (Moreton) to Insulator Creek (Torresian), a distance of 1500 kilometres, (b) a southern transect between Kilcoy (Moreton) and Gregors Creek (Torresian) in south-east Queensland — a distance of 35 km which includes (c) a detailed transect across the hybrid zone where samples were taken at 200 metre intervals from six localities (TA, TA1–TA4, TB-transect 1, Fig. 1 and Fig. 3). The individuals analysed for both mitochondrial and ribosomal

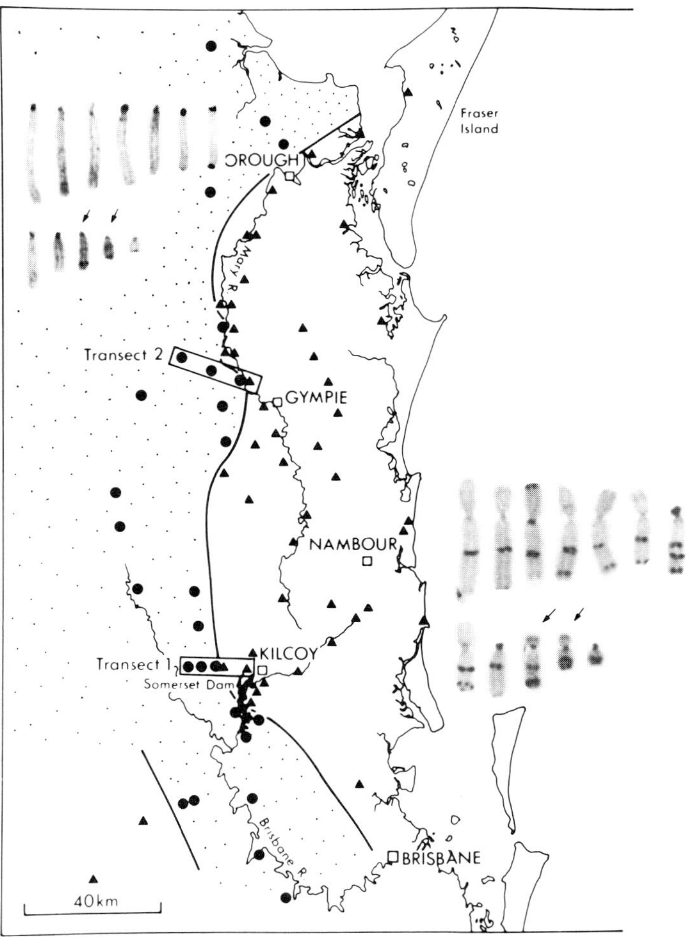

Figure 1. The distribution of the Moreton and Torresian taxa in S.E. Queensland where they are parapatrically associated over a distance of 250 km and form a very narrow chromosomal hybrid zone. The C-banded haploid karyotypes of the Torresian & Moreton taxa are shown. They differ by nine pericentric rearrangements and a series of interstitial and terminal C-bands. The arrows depict the locations of the ribosomal genes on chromosomes 10 and 11 of both the Torresian and Moreton.

DNA were the same as those previously analysed for chromosomal, C-band and allozymic variation by Shaw *et al.* (1985).

Total DNA was isolated from individual grasshoppers using the isolation procedure described by Arnold *et al.* (1987a). Total nuclear DNA was digested with the restriction enzyme ClaI, gel-electrophoresed, the restricted DNA transferred to Gene Screen (New England Nuclear) and hybridised to a ^{32}p nick-translated 0.8 kb rDNA probe from the *C. captiva* rDNA cistron. (Arnold *et al.* 1987b).

mtDNA, digested with four restriction enzymes (Hae III, Hind III, MspI and Xba I), was probed with three pUC-18 derived plasmids containing SacI fragments of *C. captiva* mtDNA, comprising the whole molecule.

Cloning of the highly repeated DNA sequence from the Moreton taxon was performed using DNA renaturing at a Cot <0.02 (mols × secs)/litre for the synthesis of radioactive probes. A Taq I digestion of total DNA was probed with the ^{32}p labelled Cot fraction and a 168 bp fragment was shown to be a subset of this fraction. The Taq I fragments were inserted into the Cla I site of pBR322 and then used to transform *E. coli* ECR 291.

To determine the chromosomal locations of the 168 bp repeat family and the rDNA clones, *in situ* hybridisation was performed on mitotic cells prepared as air-dried slides from 8-day-old embryos. The chromosomal locations of the 168 bp sequences and of the 18S + 28S ribosomal genes were determined by hybridisation of ^{3}H-cRNA synthesised using cloned sequences from these repeated families (Appels *et al.* 1978).

Electrophoretic variation was analysed at four loci (*Got-2, Idh-1, Mpi* and *Pgi*) using cellulose acetate gels and standard staining procedures (Daly *et al.* 1981).

Chromosomal preparations were obtained from mid-gut caecal cells of field-collected adults or from 8-day-old embryonic tissue. Cells were C-banded according to standard procedures and stained in giemsa.

Results

(a) mtDNA restriction fragment length polymorphisms.
Using four restriction enzymes, 10 mtDNA clones were identified in the Moreton taxon in populations sampled from across its entire distribution, from Peregian to Lakes Entrance, a distance of 1500 km (Fig. 2). 13 mtDNA clones were found to characterise those sampling locations known to be chromosomally and allozymically Torresian (S.W. Papua, Northern Territory and North Queensland).

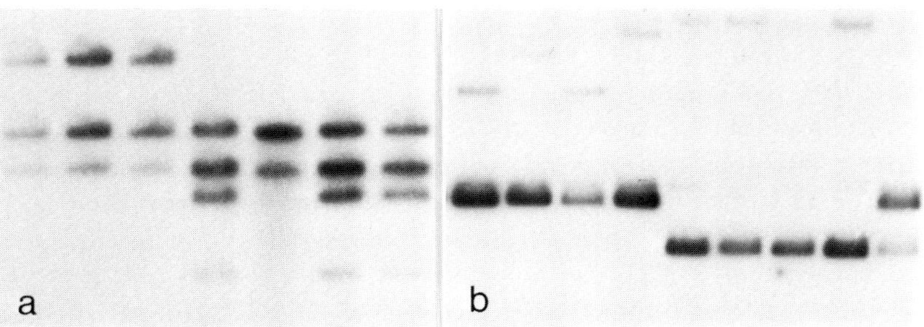

Figure 2. Autoradiographs of Southern hybridisations after **a** digestion of mtDNA with *XbaI* and probed with 3 pUC-18 derived plasmids containing *Sac I* fragments of the total *C. captiva* mtDNA genome. Left to right: 3 Torresian and 4 Moreton individuals. Using *Xba I, Hae III, Msp I* and *Hind III*, 10 Moreton and 13 Torresian clones have been identified. **b** digestion of total nuclear DNA with *Cla I* and probed with a 0.8 kb ^{32}P nick translated rDNA fragment. Left to right: 4 Moreton 2.9 kb restriction variants, 4 Torresian 2.1 kb fragment variants and a Torresian individual contains both the Moreton 2.9 kb and the Torresian 2.1 kb fragments from a population located 3 km to the west of the hybrid zone.

However, those samples taken close to the hybrid zone at its southern end (Fig. 3) and from the coastal region up to 200 km north-east of the hybrid zone

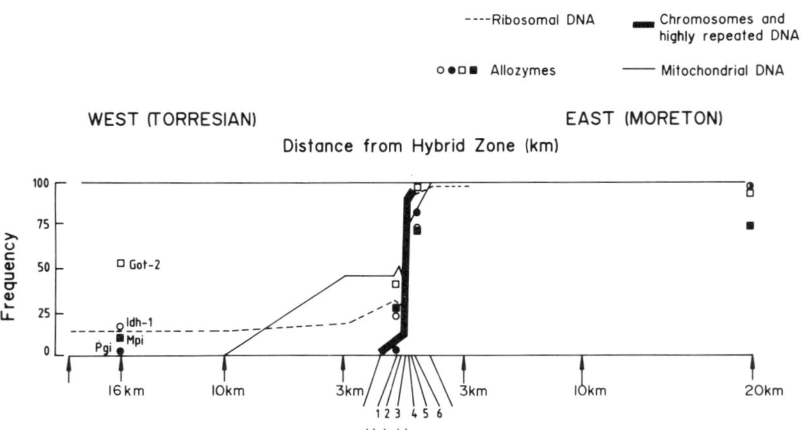

Figure 3. The frequencies of the diagnostic markers along a 40 km transect which includes Transect 1 (Fig. 1). Note the very abrupt change in all chromosomal rearrangements and C-bands over a distance of only 1 km. This contrasts markedly with the introgression of Moreton mtDNA, rDNA and allozymes up to distances of 20 km into Torresian populations.

(Fig. 4) were found to contain high frequencies of 2 Moreton mtDNA clones. In the coastal populations, these two Moreton clones characterise every individual which has been examined from this region (Fig. 4). None of the 13 Torresian clones has been found beyond 1 km to the east of the hybrid zone (Fig. 3).

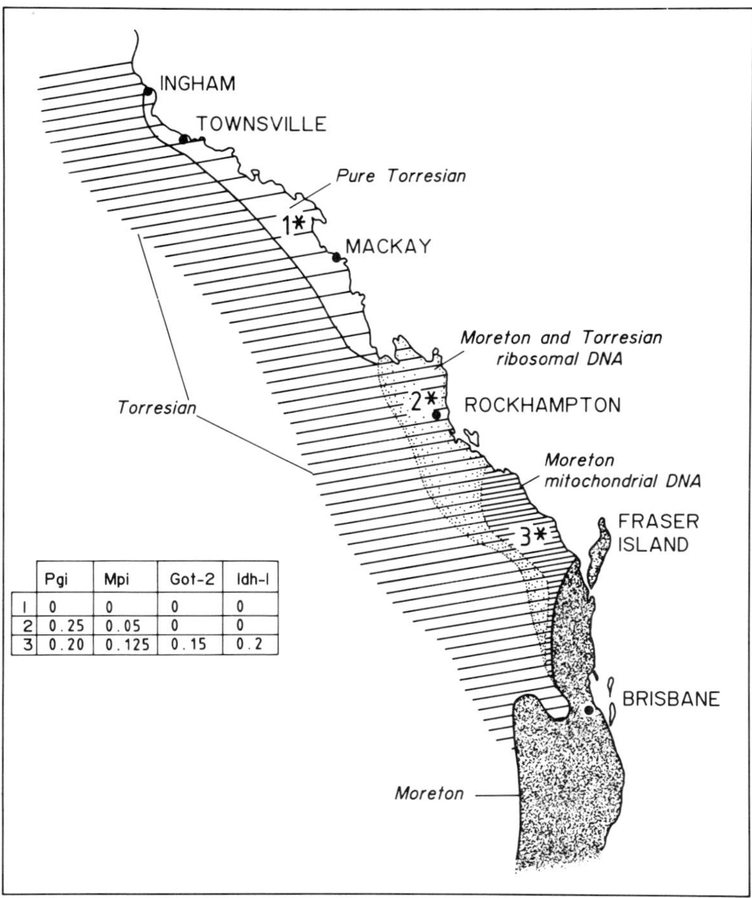

	Pgi	Mpi	Got-2	Idh-I
1	0	0	0	0
2	0.25	0.05	0	0
3	0.20	0.125	0.15	0.2

Figure 4. The distribution of Moreton mtDNA, rDNA and allozymes along the northern, coastal transect of Queensland. Two Moreton mitochondrial DNA clones characterise all individuals which are chromosomally Torresian up to 200 km north of the hybrid zone. Similarly, the 2.9 kb Moreton rDNA fragment pattern is polymorphic in all Torresian populations up to 400 km to the north of the zone. Moreton allozymes have been found in Torresian individuals 300 km to the north (samples 2 & 3). None of these Moreton markers has been detected in Torresian populations from north Queensland, the Northern Territory or Papua New Guinea.

(b) rDNA restriction fragment length variation
The ribosomal genes are located on autosomes 10 and 11 of both the Torresian and Moreton genomes (Fig. 1).

Cla I digestion of individual DNA samples, followed by hybridisation to the 28S/non-transcribed spacer probe produces a number of fragments ranging from 2.1 kb to 11.5 kb in size. Of those fragments which show homology to the probe, the 2.1 kb and 2.9 kb fragments are diagnostic for the Torresian and Moreton taxa, respectively. However, in the southern and central coastal region of Queensland between 120 km and 400 km to the north of the hybrid zone, the Moreton 2.9 kb fragment is present at frequencies ranging from 0.11-0.87 (Fig. 4). Across the southern transect, the Moreton 2.9 kb fragment is found in Torresian populations up to 30 km to the west of the hybrid zone. The Torresian 2.1 kb fragment is found at extremely low frequencies in samples up to 6 km to the east of the hybrid zone (Fig. 3).

(c) Patterns of allozyme variation
Northern Torresian populations are diagnostically distinct from Moreton populations for alleles at the *Got-2, Idh-1* and *Pgi* loci. The *Mpi* locus is polymorphic in both Torresian and Moreton populations.

Across the southern transect, the *Got-2* Moreton allele, which is almost fixed 20 km east of the zone (as it is in all other Moreton populations), was found at a frequency of 56% 13 km west of the zone, even though it is almost or completely absent in other Torresian populations (Daly *et al.* 1981). The *Idh-1* Moreton allele is similarly almost fixed 20 km east of the zone, but is present at 12% frequency to the west. Variation at the *Mpi* locus shows a pattern different from these two. The Torresian allele, which is fixed or almost so in all populations, exists at polymorphic frequencies in all Moreton populations, including Fraser Island, and may represent a primary polymorphism. The *Pgi* locus differs from the other three by showing a complete replacement of Torresian and Moreton alleles over distances of 15 km and 20 km respectively on either side of the hybrid zone. This locus also shows the greatest frequency change (77%) of all the allozymes across the centre of the zone.

In the northern transect, Moreton alleles have been found in Torresian populations up to 300 km north of the hybrid zone, at frequencies up to 25%.

Pericentric rearrangements and C-band variation

The Torresian genome is made up exclusively of acro- and telocentric chromosomes. All populations from Papua, the Northern Territory and Queensland possess this pattern of karyotypic organisation (Fig. 1).

The Moreton genome is fixed for pericentric rearrangements involving chromosomes 1–6, 10 and 11. Autosomes 7, 8, 9 and 12 are polymorphic in S.E. Queensland; the X chromosome is metacentric in inland populations and acrocentric in the northern and coastal populations. The Moreton chromosomes also carry one or several interstitial and terminal C-bands, each of which occupies a characteristic position on the chromosome that permits every chromosome within the genome to be identified unambiguously (Fig. 1). The Moreton genome also carries 1.5×10^5 copies of the 168 bp repeat sequence and *in situ* hybridisation has revealed that these sequences are located within the C-banded regions of every Moreton chromosome (Arnold *et al.* 1985). The Torresian genome contains only 3.5×10^3 copies of this sequence which are restricted to the telomeric regions of autosomes 10, 11 and 12.

The chromosomal structure of the hybrid zone

An analysis of samples taken at 200 metre intervals across the detailed southern transect (transect 1, Fig. 1) has shown that there is a change in frequency of at least 50% for each diagnostic chromosome (1–10) over only 200 metres. The changes in frequency are concordant for all chromosomes and there is total genomic replacement over a distance of 1 km (Fig. 3). This suggests that either the hybrid zone is very young so that non-homologous chromosomes have not had sufficient time to equilibrate independently or there is strong selection acting in concert on all chromosomes (i.e. selection is favouring a particular genomic pattern). Previous analyses have shown that the F_1 hybrid between the Moreton and Torresian taxa is both fully viable and fertile (Shaw & Wilkinson 1980). However, the F_2 generation is totally inviable and backcrosses suffer approximately 50 inviability, both due to embryonic breakdown during early development. The analysis of chromosomal segregation and recombination in the backcrosses reveals that the majority of the embryonic mortality is due to intrachromosomal effects with no evidence of interactions between chromosomes. In complete contrast, there is evidence of strong linkage disequilibrium between chromosomes within the hybrid zone itself and that the direction of disequilibrium can change over short time periods, favouring an acro- or metacentric genome according to prevailing environmental conditions (Shaw *et al.* 1985).

The C-band differences between the Torresian and Moreton taxa show patterns of change very similar to those observed for the pericentric rearrangements. There is no evidence of penetration of either Moreton pericentric rearrangements or C-bands in Torresian populations beyond 500 metres to the west of the null point.

Conclusions

There are 3 possible explanations for the presence of these Moreton markers at high frequencies in Torresian populations well beyond the limits of the chromosomally defined hybrid zone.

1. They represent shared, ancestral polymorphisms which were present prior to the chromosomal separation of the Moreton and Torresian taxa. This possibility is highly unlikely because the Torresian populations to the north and west of the introgressed area, and in Papua New Guinea and the Northern Territory do not carry any of these characters. Yet all of these Torresian populations have the same karyotype and are more closely related to each other than to any other taxon of *Caledia*, on the basis of allozyme patterns.

2. The mtDNA, rDNA and allozymes have introgressed across the hybrid zone into the Torresian populations up to 400 kms. Such high levels of introgression over large distances would require the involvement of extremely strong selective processes favouring all three marker systems to explain their deep penetration of the zone. Again this seems highly improbable.

3. The presence of these Moreton components represents a relict of a past occupation of the Torresian territory by the Moreton taxon. The Moreton type has subsequently retreated east and south to its present location, leaving behind a series of essentially neutral markers in its wake. This is the most parsimonious explanation and one which readily accounts for the observed patterns. It is envisaged that the hybrid zone, as defined by its chromosomal characteristics, was located close to the present day limit of the Moreton rDNA marker, about 400 km to the north. Climatic and vegetational analyses of this region suggest that the initiation of zone movement would have occurred around 8000 yrs bp at the onset of the most recent arid phase (Nix & Kalma 1972).

During this climatic change the Moreton taxon would have retreated southward with its immediate replacement by the Torresian taxon as a moving front with persistently high levels of hybridisation. The present location of the zone would then represent a finely tuned equilibrium state maintained by environmental factors. More importantly, however, this scenario exposes the fundamental role played by chromosomal rearrangements in the stability of the hybrid zone despite its movement. The karyotypic autonomy of both taxa is also clearly seen by the total exclusion of either Moreton or Torresian chromosomes in populations located only 500 metres on either side of the hybrid zone, despite the high frequencies of other nuclear and cytoplasmic components.

We conclude that chromosomal rearrangement of the kind that we have described in *Caledia* is capable of generating an extremely effective isolating mechanism between karyotypically differentiated and hybridising taxa. We propose that these two contrasting patterns of genomic organisation represent adaptations which are maintained by strong selective processes acting directly upon chromosome structure.

References

Appels, R., C. Driscoll and W.J. Peacock 1978. Heterochromatin and highly repeated DNA sequences in rye (*Secale cereale*). *Chromosoma* 70, 67–89.

Arnold, M.L., R. Appels and D.D. Shaw 1985. Evolution and conservation in a highly repeated DNA family from the grasshopper *Caledia captiva. Cytobios* 43, 149–157.

Arnold, M.L., P. Wilkinson, D.D. Shaw, A.D. Marchant and N. Contreras 1987a. Highly repeated DNA and allozyme variation between sibling species: evidence for introgression. *Genome* 29, 272–279.

Arnold, M.L., D.D. Shaw and N. Contreras 1987b. Ribosomal RNA-encoding DNA introgression across a narrow hybrid zone between two subspecies of grasshopper. *Proc. Natl. Acad. Sci. USA* 84, 3446–3450.

Daly, J.C., P. Wilkinson and D.D. Shaw 1981. Reproductive isolation in relation to allozymic and chromosomal differentiation in the grasshopper *Caledia captiva. Evolution* 35, 1164–1179.

Nix, H.A. and J.D. Kalma 1972. Climate as a dominant control in the biogeography of northern Australia and New Guinea. In *Bridge and Barrier: The Natural and Cultural History of Torres Strait*, D. Walker, ed., 61–91. Australian National University.

Shaw, D.D. and P. Wilkinson 1980. Chromosomal differentiation, hybrid breakdown and the maintenance of a narrow hybrid zone in *Caledia. Chromosoma* 80, 1–31.

Shaw, D.D., D.J. Coates, M.L. Arnold and P. Wilkinson 1985. Temporal variation in the chromosomal structure of a hybrid zone and its relationship to karyotypic repatterning. *Heredity* 55, 293–306.

White, M.J.D. 1978. *Modes of Speciation*. W.H. Freeman, San Francisco.

Chromosome stability and instability in plants

J.S. Parker, A.S. Wilby and S. Taylor

School of Biological Sciences, Queen Mary College (University of London), Mile End Road, London E1 4NS

The observation that the majority of closely related species are chromosomally differentiated has led to attempts to assess the role of chromosome changes in the speciation process. Consideration has been given to the processes by which chromosome variants showing negative heterosis become fixed during speciation. There are problems with the approach employed. Firstly, King (1987) points out that estimates of rates of chromosome change in lineages indiscriminately mix those which lead to negatively heterotic effects with those which do not and indeed are polymorphic in the species considered. Secondly, although the factors which influence fixation have been enumerated (rates of chromosome mutation, selection pressures on hetero- and homokaryotypes, inbreeding, genetic drift, and meiotic drive) the values of these parameters in natural populations are ill-understood.

Study of the cytological structure of populations has revealed variation which represents unique mutation events and polymorphisms, both transient and balanced. Data from a number of plant species are given here concerning (i) rates of chromosome variation and (ii) meiotic drive, which are pertinent both to population cytology and to chromosome fixation during speciation. These data reinforce the idea that selection pressures maintain karyotype stability despite processes tending to rapid change.

Rates of chromosome change

White (1978) states that newly-arisen chromosome rearrangements are found in about 1 in 500 individuals, estimates derived from organisms such as humans and grasshoppers. Specific rearrangements occur at much lower frequencies, between 10^{-3} and 10^{-4} being suggested for reciprocal interchanges (Lande, 1979). How representative are these figures for plants? Studies of *Scilla autumnalis* and *Rumex acetosa* suggest that rates are an order of magnitude higher than this, perhaps two orders of magnitude for specific chromosomes or chromosome segments.

The bulbous species *Scilla autumnalis* (*Liliaceae*) is chromosomally diverse, with 11 cytological races (Taylor, unpublished; Ainsworth *et al.*, 1983). All populations have chromosome polymorphisms, some extremely widespread but the majority population-limited. Remarkably, between 5 and 10% of mature population plants have unique rearrangements (Fig. 1), most frequently centric shifts, although this bias may simply reflect the lack of discrimination of standard feulgen staining of mitotic chromosomes. Thus this is a minimum estimate of frequency.

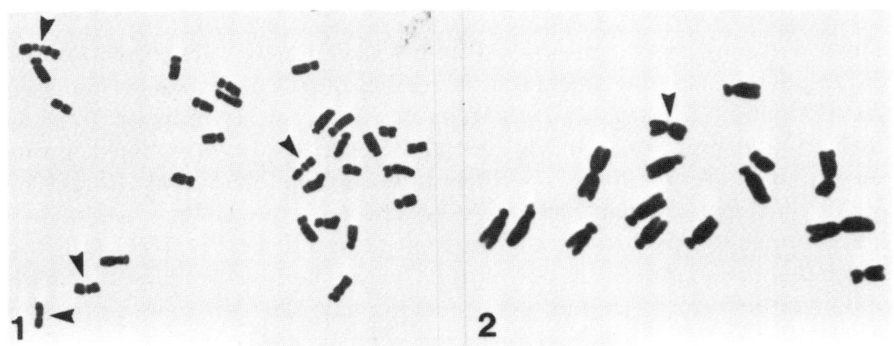

Figure 1. Unique duplication of chromosome 3 in autotetraploid *Scilla autumnalis*. All chromosomes 3 arrowed. **Figure 2.** Unique centric shift of chromosome 2 in male *Rumex acetosa* (arrowed).

Rumex acetosa has a well-differentiated sex-chromosome system with $2n = 12 + XX$ in pistillate (female) plants and $2n = 12 + XY1Y2$ in staminate (male) plants. The Y-chromosomes represent 20% of the diploid complement and are heterochromatic at all stages of the mitotic cycle (Wilby and Parker, 1986). Chromosome variation can be considered in the euchromatic and heterochromatic components of the genome.

The frequency of novel rearrangements in the euchromatin — the autosomes and the X — is 1 in 50, mainly centric shifts (Fig. 2). With numerical changes this rises to 1 in 33. These data are for plants grown from wild-collected seed and from controlled crosses (Table 1). This level of variation, although high, is moderate when compared with that of the Y-chromosomes.

Table 1. The nature and frequency of chromosome variants in plants of <u>Rumex</u> <u>acetosa</u> grown from seed samples collected in the wild and from controlled crosses.

Natural Seed Sample

Numerical Variant	No.	Structural Variant	No.	Total plants
Sex chromosome	4	Interchange	6	
Trisomy	5	Centric shift	21	1254
Triploidy	1	Deletion	1	
Totals	10(0.8%)		28(2.2%)	1 in 33

Controlled Crosses

Numerical Variant	No.	Structural Variant	No.	Total plants
Sex chromosomes	1	Interchange	1	
Trisomy	5	Centric shift	1	931
Triploidy	10	Fission	1	
Totals	16(1.7%)		3(0.3%)	1 in 49

The Ys are hypervariable in centromere position. The centromeres occur anywhere within the central 40% of the chromosome with no preferred locations, but are strictly excluded from the two distal 30% regions. The Ys are independent in behaviour so an enormous number of Y1Y2 variants can be distinguished (Fig. 3). In 10-male samples from natural populations between 4 and 8 variants are found, and on average every second male is identifiable (Wilby and Parker, 1987). Y-variation appears selectively neutral and must be maintained by a high mutation rate. In controlled crosses, centromere relocation occurs with a rate of 1 in 80. In addition, Y–Y interchange occurs at a very high rate with about 1 in 70 males affected (Fig. 4).

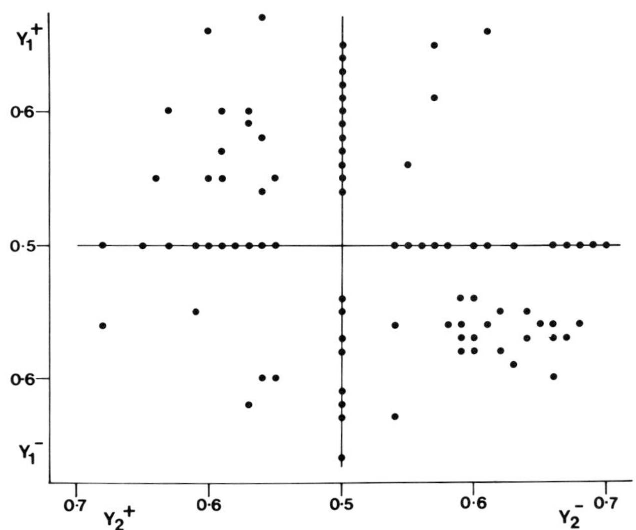

Figure 3. Arm ratios of 87 morphologically distinguishable Y1Y2 variants found in 280 males of *Rumex acetosa* in which at least one Y was acrocentric. The variant in which both Ys are metacentric is omitted (see Wilby and Parker 1986).

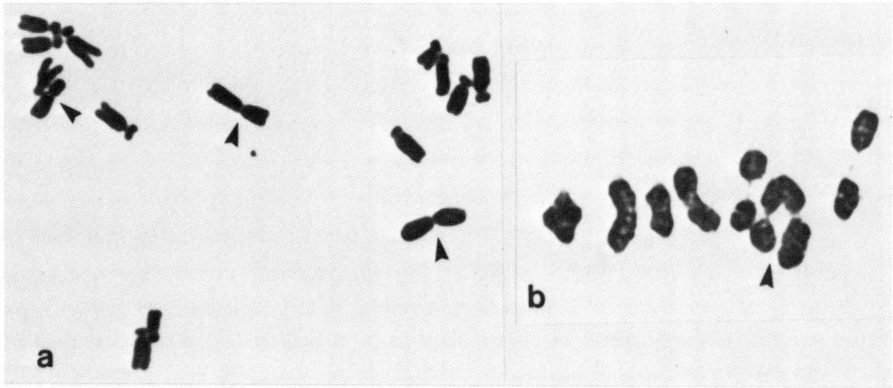

Figure 4. a. Male *R. acetosa* with a Y-Y interchange. Sex-chromosomes arrowed. **b.** Metaphase-1 in a male with a Y-Y interchange; sex-trivalent arrowed.

Non-Mendelian inheritance

Many examples of non-Mendelian recovery of gametes and zygotes in crosses have been recorded involving alleles of single loci. Although meiotic drive has been postulated by White (1978) as of potential importance in karyotype evolution no evidence in support of this has been presented. In addition, the role of non-Mendelian inheritance in the maintenance of chromosomal polymorphisms other than B chromosomes has received little attention. Evidence for both is presented.

a. Polymorphisms

Supernumerary segments are found in both *S. autumnalis* and *R. acetosa*, the former euchromatic and the latter heterochromatic (Figs. 5 and 6). In the

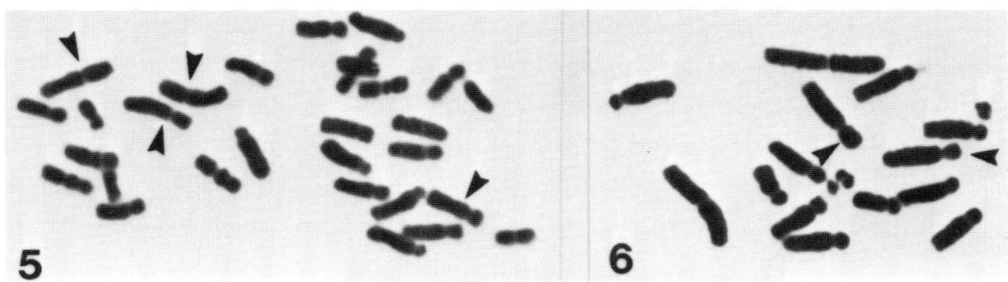

Figure 5. Mitotic chromosomes of plant of *Scilla autumnalis* duplex for supernumerary segment on chromosome 1. Chromosomes 1 arrowed. **Figure 6.** Female *Rumex acetosa* heterozygous for a supernumerary segment on chromosome 1 (arrowed) and two B-chromosomes.

autotetraploid race of *S. autumnalis* a segment on the short arm of chromosome 1 reaches polymorphic proportions (Ainsworth *et al.*, 1983). In nulliplex × simplex crosses the segment is transmitted through both the pollen and eggs in excess (0.65 and 0.66 per gamete; Table 2). In *R. acetosa* a segment on the short arm of chromosome 1 forms a low frequency polymorphism in British

Table 2. The inheritance of supernumerary segment 1 in the autotetraploid race of Scilla autumnalis in simplex × nulliplex crosses.

Female		Male	Offspring			
			++++	+++1	++11	$\bar{k}$
+++1	×	++++	22	30	3	0.66
++++	×	+++1	15	28	–	0.65

populations. The segment-bearing chromosome is preferentially transmitted through the egg ($\bar{k} = 0.63$) but not the pollen (Table 3), indicative of true meiotic drive. Interestingly, drive is not characteristic of all segments in *Rumex* since that on chromosome 6 shows standard inheritance.

Table 3. The inheritance of supernumerary segments on chromosomes 1 and 6 in *Rumex acetosa*.

Cross			Offspring		Total	$\bar{k}$	x^2
Female		Male	++	+S	plants		
+1	×	++	52	88	140	0.63	9.25 **
++	×	+1	103	79	182	0.43	3.16 ns
+6	×	++	56	48	104	0.46	0.61 ns
++	×	+6	34	40	74	0.54	0.48 ns

Non-Mendelian inheritance, then, may be a factor in the maintenance of supernumerary segment polymorphisms and this has recently been proposed by Cabrero and Camacho (1987) for a *Chorthippus vagans* segment. No crossing data are yet available.

b. Unique variants

If meiotic drive is to play a role in the fixation of chromosome rearrangements then at least some newly-arisen variants should display non-Mendelian inheritance. In *R. acetosa* inheritance of seven centric shifts and two reciprocal interchanges has been examined in the plant in which they arose (Fig. 7). Additionally, two reciprocals were examined using the offspring generation.

Mean inheritance ranged from 0.31 to 0.85 per gamete in backcrosses, with six of the eleven crosses giving significant deviations from expectation (Table 4).

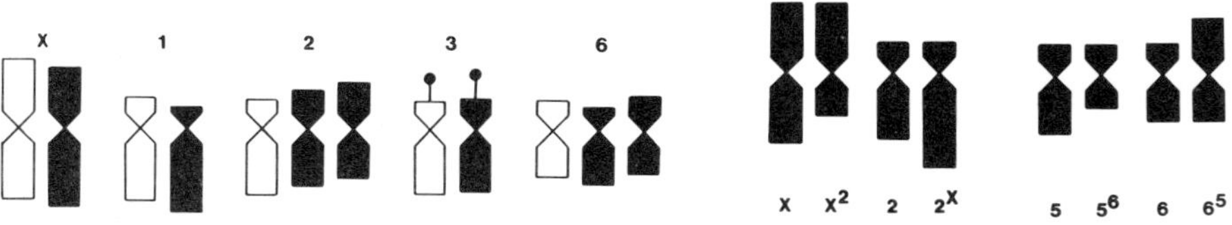

Figure 7. Karyotypes of seven centric shifts and two reciprocal interchanges whose patterns of inheritance have been studied (see Table 4).

Table 4. Transmission rates established from backcrosses for seven centric shifts and two reciprocal interchanges in <u>Rumex acetosa</u>. All rearrangements were novel and were tested in the individual in which they were found. Two were also tested in plants of the offspring generation.

| | Centric Shift | | | | | Interchange | | | |
| | Female | | Male | | | Female | | Male | |
	$\bar{k}$	P	$\bar{k}$	P		$\bar{k}$	P	$\bar{k}$	P
X	0.78	**	–	–	X/2	0.64	***	0.63	ns
1	0.66	**	0.31	*					
2	–	–	0.69	*					
2	–	–	0.85	**					
3	0.40	ns	–	–					
6	0.37	ns	–	–	5/6	–	–	0.36	ns
6	0.55	ns	–	–					

The novel chromosome was inherited in excess in five crosses and loss was observed only once. Non-Mendelian inheritance was seen in both pollen and egg transmission, indicating that a variety of mechanisms is operating; meiotic drive, pollen competition and possibly zygotic death may be implicated, but all can contribute to fixation of novel variants.

Spontaneous variants in other species may also show non-Mendelian inheritance. For example, centric fission products of chromosome 1 in *Hypochoeris radicata* are preferentially inherited through the pollen ($\bar{k}$ = 0.74; Table 5) (C.G. See, pers. com.).

c. Non-Mendelian inheritance in other systems
Non-Mendelian inheritance contributes to maintenance of many B-polymorphisms, depending on directed segregation during meiosis, somatic mitosis or pollen grain mitosis (Jones & Rees, 1982). In maize, abnormal segregation is associated with the presence of K10 which shows neocentric activity and induces it in other knobbed chromosomes in the complement (Rhoades & Dempsey, 1966). Again in maize, directed segregation has been implicated in selective recovery of interchange products from T6-9b (Dempsey, 1961). These examples suggest intimate involvement of centromeric activity in the generation of non-Mendelian inheritance. Genic evidence from maize points in the same direction. Wendel *et al.* (1987) report an F2 involving eighteen

Table 5. The inheritance of a spontaneous centric fission of chromosome 1 in Hypochoeris radicata.

Cross		Offspring		
++ female × telo het male		++	telo het	$\bar{k}$
1		9	26	0.74
2		6	17	0.74
Totals		15	43	0.74***

enzyme loci distributed across seven chromosomes which showed significant deviations from expectation at nine loci. Centromeric linkage of the deviating loci is particularly clear on chromosome 3 which was marked by five loci (Fig. 8).

It is possible that deviations from expectation are the result of differential behaviour of homologous centromeres during segregation, and loci simply respond because of centromeric linkage. Interestingly in *R. acetosa* the pattern of

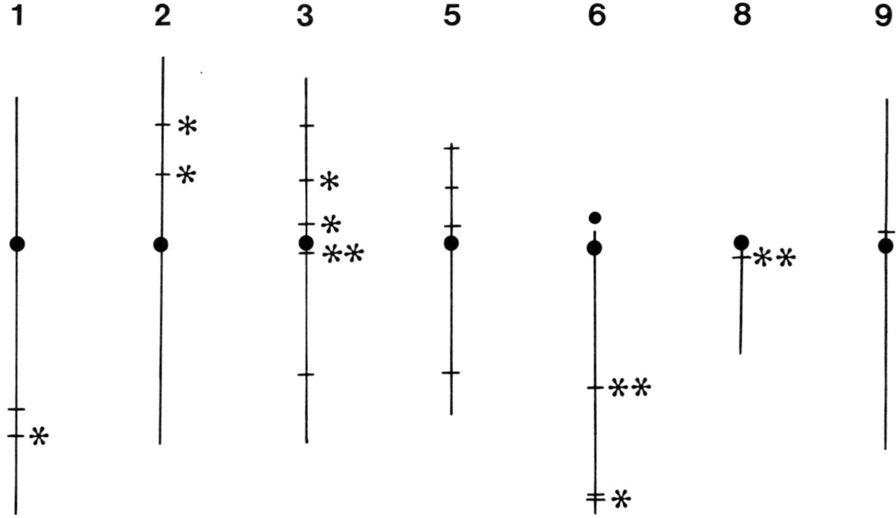

Figure 8. The inheritance of 18 loci on seven chromosomes segregating in a single F2 of maize. Deviations from expectation significant at the 5% (*) and 1% (**) level are marked. Centromeres are marked by circles. Data from Wendel *et al.* 1987.

inheritance does not reflect the nature of the structural change itself. The rearrangements may simply allow identification of centromeres with particular segregation characteristics.

Conservation of the karyotype

The genomes of these two quite distinct flowering plants are highly labile, giving rise in each generation to a wide spectrum of variants at rates at least an order of magnitude greater than that generally accepted. In addition some novel variants carry inherent drive systems which should aid the spread of variants through the population even when exhibiting negative heterosis. Drive should especially enhance changes such as centric shifts which often show no fertility decline in heterozygotes.

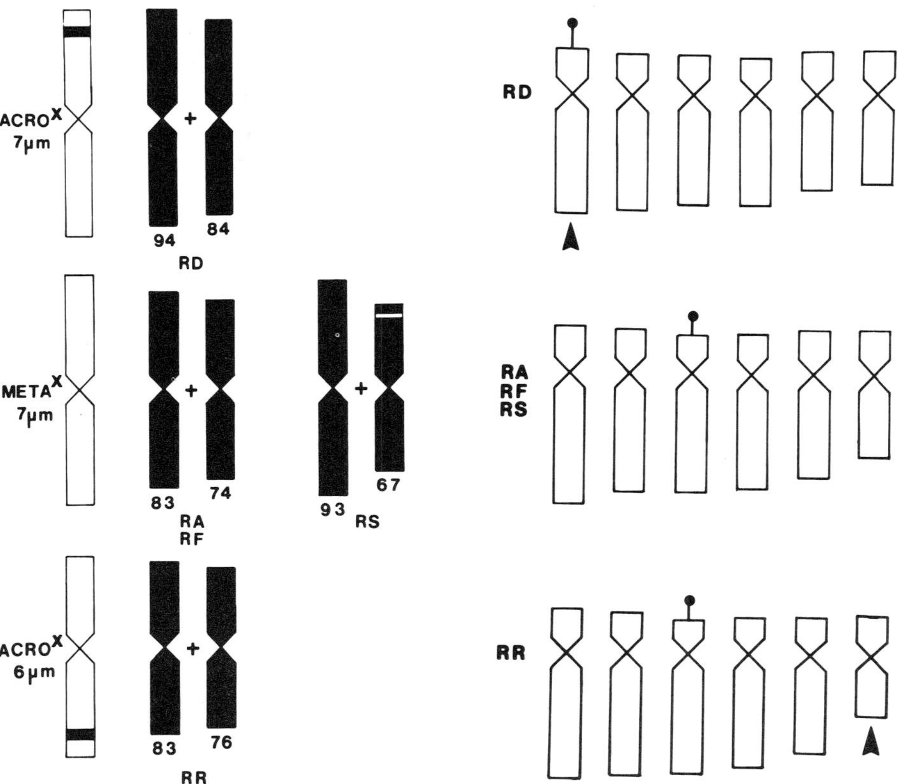

Figure 9. Sex-chromosome and autosome complements of five species in the *R. acetosa* group. Autosome changes arrowed. The sizes of the Y-chromosomes (solid) are expressed as percentages of the X-length. (RA-*acetosa*, RF-*arifolius*, RS-*thyrsiflorus*, RD-*thyrsoides*, RR-*rothschildianus*).

The karyotypes of five species in the *R. acetosa* group have been examined and among the autosomes only two rearrangements have been found: transference of the nucleolar organiser region from chromosome 3 to 1 in a single species and a presumptive deletion of part of the long arm of chromosome 6 in another species. In the sex chromosomes, three types of X and four Y1Y2 morphs characterise these five species. Thus there is underlying stability of the karyotype in this species group, particularly in the autosomes (Fig. 9).

Similar stability is inherent to the cytological races of *Scilla autumnalis*. At the diploid level, three fixed differences are found in the five races while the six polyploids are summations of the diploids with no further rearrangements. Despite the massive variability of this species, karyotypic constancy underlies the species across the whole geographical range from Britain to Israel.

Within the diploids of *S. autumnalis* two quantum changes in total DNA amounts have occurred while maintaining the karyotype morphology. The same phenomenon has been observed in the genus *Hypochoeris* (C.G. See, pers. com.). Karyotype morphology itself then is a character which is subject to stabilising selection in the face of forces such as mutation and meiotic drive potentially disrupting the pattern.

References

Ainsworth, C.C., J.S. Parker and D.M. Horton 1983. Chromosome variation and evolution in *Scilla autumnalis*. *Kew Chromosome Conference II*, P.E. Brandham & M.D. Bennett, eds, Allen & Unwin, London.

Cabrero, J. and J.P.M. Camacho 1987. Population cytogenetics of *Chorthippus vagans*. *Genome* 29, 280–284.

Dempsey, E. 1961. Further studies of T6-9b. *Maize Genet. Coop. Newsletter*.

Jones, R.N. and H. Rees 1982. *B-chromosomes*. Academic Press, London.

King, M. 1987. Chromosomal rearrangements, speciation and the theoretical approach. *Heredity* 59, 1–6.

Lande, R. 1979. Effective deme sizes during long-term evolution estimated from rates of chromosomal rearrangement. *Evolution* 33, 234–251.

Rhoades, M.M. and E. Dempsey 1966. The effect of abnormal chromosome 10 on preferential segregation and crossing-over in maize. *Genetics* 53, 989–1020.

Wendel, J.F., M.D. Edwards, and C.W. Stuber 1987. Evidence for multilocus genetic control of preferential fertilisation in maize. *Heredity* 58, 297–301.

White, M.J.D. 1978. *Modes of Speciation*. W.J. Freeman, San Francisco.

Wilby, A.S. and J.S. Parker 1986. Continuous variation in Y-chromosome structure of *Rumex acetosa*. *Heredity* 57, 247–254.

Wilby, A.S. and J.S. Parker 1987. Population structure of hypervariable Y-chromosomes in *Rumex acetosa*. *Heredity*, 59, 135–143.

B chromosomes and supernumerary chromosome segments in *Liliaceae*: selfish or heterotic DNA?

C. Ruiz Rejón, R. Lozano, F.J. Ortega Nieto and M. Ruiz Rejón

Dpto. de Biologia Animal, Ecologia y Genética, Facultad de Ciencias, Universidad de Granada, 18071 Granada, Spain

Extra chromatin is normally present in eukaryotic organisms in the form of B chromosomes or supernumerary segments (s.s.). B-chromosomes have now been recorded in at least a thousand flowering plants and in more than 260 animals (Jones & Rees 1982). Supernumerary segments, on the other hand, have been recorded only in animal species, notably in the Orthoptera (John 1973; Hewitt 1979). In plants there are intraspecific structural polymorphisms for certain classes of heterochromatin, such as C-bands (e.g. in *Scilla sibirica*, Vosa 1973), knobs (e.g. in maize, Longley 1938) and H-segments (e.g. in *Trillium*, Darlington & La Cour 1940), but s.s. as such are not cited.

In recent years our research has been directed towards detecting and analysing both types of extra chromatin in plants and to do this we we have worked with certain *Liliaceae* species, which are suitable for studies at a cytogenetic level. B-chromosomes have already been found in some species, such as *Allium sphaerocephalon* (Guillén & Ruiz Rejón, 1984) and *Scilla autumnalis* (Ruiz Rejón *et al.* 1980; Oliver *et al.* 1982). We have also cytogenetically analysed various species of *Liliaceae* in order to detect the possible existence of s.s., and have identified this type of extra chromatin in *Tulipa australis* (Ruiz Rejón & Ruiz Rejón 1985), *Allium subvillosum* and *Scilla autumnalis* (unpublished data and C. Ruiz Rejón *et al.* 1987).

Although many uncertainties remain concerning this extra genetic material, the question of the maintenance mechanisms of the Bs and s.s. is of current interest because we are now aware of the widespread distribution of such redundant genetic material and of the apparently redundant nature of a large proportion of the nuclear DNA in eukaryotes. Thus, in this paper we will comment on the parasitic or selfish nature of the Bs of *Allium sphaerocephalon* and on the heterotic nature of a supernumerary segment in *Tulipa australis*.

The B chromosome system of *Allium sphaerocephalon* L.

There are two types of Bs in *A. sphaerocephalon*: the first is submetacentric and the second is a smaller acrocentric (see Fig. 1 and Guillén & Ruiz Rejón 1984). Both types are widely distributed in natural populations; but in this case we have

"

analysed one population from Padul (Granada, Spain) in which the submetacentric Bs are absent. The acrocentric B in *A. sphaerocephalon* is euchromatic and stable during mitosis.

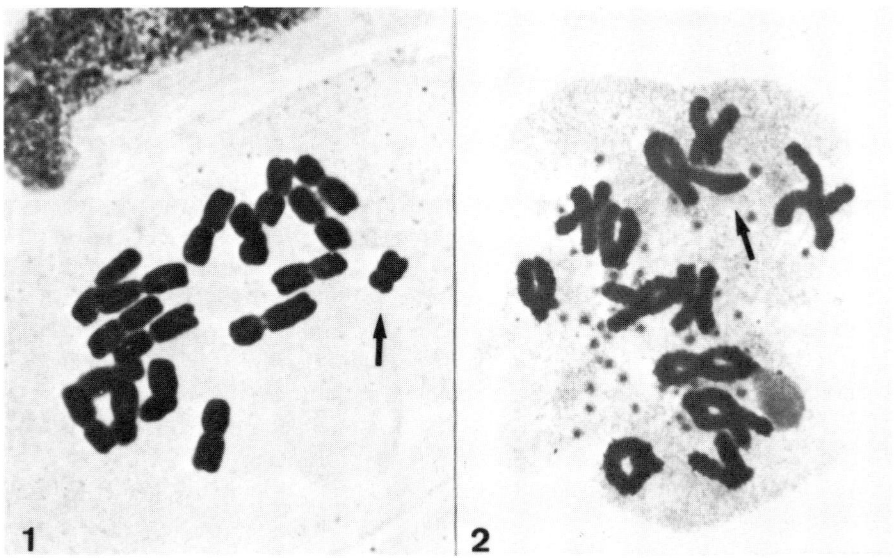

Figure 1. Mitotic metaphase of *Allium sphaerocephalon* with 1B (acrocentric B arrowed). **Figure 2.** Diplotene cell of *Tulipa australis* heterozygous for a supernumerary segment (arrow).

To discover the maintenance mechanism of the acrocentric B we have first of all analysed the frequency of the Bs in bulbs during several years of sampling this population (see Table 1).

Table 1.   Frequencies of bulbs with and without B chromosomes.

Year	N	0B	1B	2B	Mean B
1984	230	0.617	0.300	0.083	0.466
1985	180	0.667	0.233	0.100	0.433
1986	110	0.600	0.300	0.100	0.500
Totals	520	0.631	0.277	0.092	0.467
Means	–	0.628	0.278	0.094	0.433

As can be seen, the frequencies have remained stable over the three years of sampling.

We then analysed the population dynamics by comparing the frequencies and the distribution of B classes in some bulbs and in their progenies: in the year 1986 a sample of 40 bulbs was taken from the wild with their seeds and the B frequencies of both generations were counted cytologically. The results are shown in Table 2.

Table 2.   Summary of the analysis of 40 bulbs and 20 seeds from each.

	No. bulbs	Frequency of karyotypes					No. seeds	Mean B
		0B	1B	2B	3B	other		
Seeds of 0B bulbs	26	0.763	0.138	0.099	0.0	0.0	520	0.336
Seeds of 1B bulbs	10	0.374	0.437	0.184	0.005	0.0	200	0.810
Seeds of 2B bulbs	4	0.0	0.726	0.212	0.050	0.012	80	1.327
Seeds of all bulbs	40	0.589	0.277	0.127	0.006	0.001	800	0.545
Adult bulbs	40	0.650	0.250	0.100	–	–	–	0.450
Survival zygotes to adult bulbs		1	0.818	0.713	–	–	–	–

Bearing in mind that *A. spaerocephalon* is a strictly allogamous plant (Pastor & Valdés, 1983), the frequency of the Bs in the pollen pool of the population can be estimated from the frequency of seeds with Bs among the offspring of the 0B bulbs. On the basis of the data from 520 analysed seeds from each of 26 0B bulbs (20 seeds to each bulb), 23.7% (sum of seeds with 1B and 2Bs) of the pollen carried Bs. This frequency was similar to that expected from the observed frequency of 1B bulbs (in this case 50% of the pollen will have B) and 2B bulbs (all the pollen with B) in the year 1985: $(0.233 \times 0.5) + (0.1 \times 1) = 0.217$. Apparently the acrocentric B of *A. sphaerocephalon* has little or no effect on male fertility and fecundity. However, the frequency of the Bs among the offspring of the 0B bulbs (0.336), and thus the frequency of Bs in the pool, was higher than would be expected if the Bs were sexually transmitted in a Mendelian fashion on the pollen side (0.217). This may be due to the presence of unexpected 2B seeds resulting from a non-disjunction of Bs at the first pollen grain mitosis (see Ruiz Rejón, M. *et al.* 1987).

The average rate at which the Bs were transmitted by the egg cells can be estimated by subtracting the frequency of the Bs in the pollen pool (0.336) from the frequency of the Bs among the seeds of the 1B bulbs (0.810) and 2B bulbs (1.372). These calculations give a rate of transmission of 0.474 and 1.036 respectively, rates which differ insignificantly from the 0.5 and 1 to be expected if the Bs were sexually transmitted in a Mendelian fashion by the female side.

As can be seen in Table 2, the frequency of the Bs was about 0.45 among bulbs and about 0.545 among zygotes. Thus, the B increased in frequency from the adults to the zygotes of the next generation by about 0.095. As we have seen in Table 1, however, the frequency of Bs is stable. This might be due to the low survival of 3B individuals and to the reduction in the survival of 1B and 2B individuals. The relative survival of 3B individuals can be estimated in the following manner: if the expected frequency of 3B individuals is 0.006 the probability of not finding 3B bulbs among 520 analysed bulbs is $0.994^{520} = 0.043$. When the probability of survival of 3B individuals is 0.5 (relative to that of the other individuals) the probability of not finding any 3B individuals is $0.997^{520} = 0.209$. Thus the absence of 3B bulbs suggests that the viability of 3B bulbs is less than 0.5. Nevertheless, the selection against the 3B individuals could not have removed more than about $0.006 \times 2 = 0.012$, or only $0.012/0.095 = 12.5\%$ of the increase in the frequency of the B due to the high transmission by the pollen. Thus, the rest of the elimination must have been through the reduction in the survival of at least some of the 1B and 2B individuals. In fact, on the basis of the frequency of 1B and 2B individuals among the adults (0.25 and 0.1) and among seeds (0.277 and 0.127) it was calculated that the survival of 1B individuals was only 0.818 compared to that of 0B individuals and that the survival of 2B individuals in relation to 0B individuals was 0.713.

The three parameters that we have investigated are the effect of Bs on pollen and seed fertility and vegetative reproduction. As can be seen in Table 3 the individuals without Bs have values similar to those of the 1B and 2B individuals.

Table 3. Pollen and seed fertility and vegetative reproduction of individuals with various numbers of Bs.

No. of Bs	No. of ind.	Pollen Fert. %	Seed Fert. %	Total No. of seeds	Mean No. of bulbils after 5 months of cult.
0	26	96.32	70.54	2006	3.52
1	10	95.45	69.18	758	3.40
2	4	95.23	67.14	275	3.41

Since there was no evidence that the B chromosomes of *A. sphaerocephalon* increased the survival, fertility or fecundity of either 1B or 2B individuals relative to 0B individuals it points to the conclusion that this B-chromosome may be of the "parasitic" type, being maintained in nature by its efficient accumulation mechanism in the pollen grains.

The supernumerary segment of *Tulipa australis* Link.

The supernumerary segment of *T. australis* is located terminally on the short arm of a submetacentric chromosome; it shows dark C-banding and is heteropycnotic at meiotic prophase (see Fig. 2 and Ruiz Rejón & Ruiz Rejón 1985). The segment is widespread in all the analysed Andalousian populations of *T. australis*, but we have centred our analyses on the population in which the segment was first detected (Cruz Atalaya-Padul, Granada, Spain).

The frequency of the segment over 3 years of sampling is shown in Table 4.

Table 4. Frequency of bulbs of <u>Tulipa australis</u> with and without supernumerary segment.

Year	N	St/St	St/s.s.	s.s./s.s.	Mean s. s.
1. 1985	47	0.894	0.106	0.0	0.106
2. 1986	83	0.890	0.110	0.0	0.110
3. 1987	240	0.796	0.204	0.0	0.204
Totals	370	0.830	0.170	0.0	0.170
Means	–	0.860	0.140	0.0	0.140

In 1986 a sample of 83 bulbs was taken from the wild with all their seeds and the supernumerary segment frequency in both generations was counted cytologically (Table 5).

Table 5. Summary of the analysis of 83 bulbs and their seeds.

	No. bulbs	St/St	St/s.s.	s.s./s.s.	No. seeds	Mean s.s.
Seeds of St/St	74	0.948	0.052	0.0	4469	0.052
Seeds of St/s.s.	9	0.357	0.638	0.005	635	0.643
Seeds of all bulbs	83	0.883	0.115	0.002	5104	0.119
Adult bulbs	83	0.890	0.110	0.0	–	0.110
Survival zygotes to adult bulbs	–	1	0.930	–	–	–

The bulbs without the segment had two types of seeds: without-segment seeds (0.948) and heterozygotes for the segment (0.052). Bearing in mind that *T. australis* cross-fertilises, the frequency of seeds with the segment among the offspring of the homozygous standard bulbs (St/St) indicates the frequency of s.s. in the pollen pool of the population (= 0.052). This frequency was similar to that expected from the observed frequency of heterozygous bulbs in 1985 (0.106), with a Mendelian transmission of the segment by the pollen (0.5) = 0.106 × 0.5 = 0.053. That is to say, it seems that the s.s. of *T. australis* was sexually transmitted in a Mendelian fashion by the pollen grains, having no effect on male fertility or fecundity.

The bulbs heterozygous for the segment have three types of seed: standard seeds (0.357), heterozygotes (0.638) and homozygotes for the segment (0.005). The average rate at which the segment was transmitted by the egg cells can be estimated by subtracting the frequency of the s.s. in the pollen pool (0.052) from the frequency of s.s. among the seeds of the heterozygous bulbs (0.643). This calculation gives a rate of transmission of 0.591, which is significantly higher than 0.5, to be expected if the s.s. were sexually transmitted in a Mendelian fashion by the female side. That is to say, the s.s. of *T. australis* exhibits an accumulation mechanism in the egg-cell side.

We have calculated the relative survival of homozygous standard individuals and heterozygous ones for the segment in a way similar to that which we did for *Allium sphaerocephalon*: by comparing the frequencies of these two cytotypes among adults and seeds we have determined that the relative survival of heterozygous individuals compared to the homozygous standard individuals is 0.93. The relative survival of bulbs homozygous for the segment was also estimated in the following way: when the expected frequency of individuals homozygous for the segment is 0.002 the probability of not finding homozygous individuals among 376 bulbs is $0.998^{376} = 0.471$; when the probability of survival of the homozygous individuals for the segment is 0.5 (relative to that of the other individuals) the probability of not finding any homozygous individuals is $0.999^{376} = 0.6861$. Thus, the absence of bulbs homozygous for the segment suggests that the viability of this cytotype is probably less than 0.5.

Finally, we have determined the effect of the segment on pollen and seed fertility. As can be seen in Table 6, the heterozygous bulbs have a pollen fertility

Table 6. Pollen and seeds fertility of bulbs with and without segment.

Karyotype	No. of indiv.	Pollen Fert. %	No. ind.	Seeds Fert. %	Seeds
St/St	10	66.18	45	73.93	4357
St/s.s.	10	65.83	4	89.17	478

similar to that of the homozygous standard individuals. It is important to mention, however, that the heterozygous bulbs have a seed fertility significantly higher than that of the homozygous individuals.

Bearing in mind that the individuals heterozygous for the segment have a lower survival rate but have similar pollen fertility and higher seed fertility than the homozygous standard individuals we have estimated the total fitness of homozygous bulbs to be 0.89 ($1 \times 1 \times 0.82$) and 1 ($0.93 \times 0.99 \times 1$) for heterozygous bulbs.

It is possible then that the supernumerary segment of *T. australis* might be maintained in natural populations by its heterotic nature, that is, overdominance, plus possibly its accumulation mechanism of transmission in the egg cells. These two factors would lead to the belief that the frequency of the segment might increase over generations, and, in fact, the frequency of the segment is not stable in the Cruz Atalaya population, it is increasing (Table 4). We have also found some populations (e.g. Cumbres de Beas, Jaén, Spain) in which the frequency of the segment is very high indeed (0.48).

Discussion

In the light of our results, we believe that the acrocentric B of *Allium sphaerocephalon* is "parasitic" or "selfish" DNA, while the extra segment of *Tulipa australis* is "heterotic" DNA.

The parasitic B of *A. sphaerocephalon* should be added to other cases of parasitic Bs, such as those of rye, *Lilium callosum*, *Melanoplus femur-rubrum*, mealy bug, mottled grasshopper, etc. (for an overall discussion of these cases see Jones 1985). Other Bs, on the other hand, such as those of *Centaurea scabiosa*, *Allium cernuum* and *Allium schoenoprasum*, have no accumulation mechanism, and a third class of Bs, such as those from *Lolium perenne*, enhance the survival chances of individuals. These latter may well be considered as being heterotic (cf. Jones 1985).

The extra segment of *T. australis* is the first to our knowledge that has been analysed with regard to its maintenance mechanism. Our results cannot therefore be compared with previous studies but it is worth noting that the s.s. of *T. australis* shows an accumulation mechanism in the egg cells similar to, for example, maize knobs (Rhoades 1942).

Why should some chromosomes and supernumerary segments be parasitic and others heterotic?

We believe that both the supernumerary chromosomes and segments may have originated as selfish genetic material through non-phenotypical selection (cf. M. Ruiz Rejón *et al.* 1987), but that it may well be that a later evolution of these extra chromatin types leads, in some cases, to their attaining a heterotic or functional rôle. That is to say that once a B or s.s. system has established itself, just as the A genome might develop anti-B (Nur & Brett 1985, 1987) or anti-s.s.

genes, the extra chromatin systems might equally well evolve and diverge in such a way as to take on their own particular rôle and thus become evolutionarily favoured by the phenotype selection. As far as the B chromosomes are concerned, from all the effects which have been observed, it would seem that the existence of some favourable effects of the Bs on the rate of germination, growth, vigour and on survival would have to be classed as heterotics. With regard to the s.s., from the effects that we have been able to see, such as the influence on pairing, recombination and the NORs, and, in *T. australis*, on seed fertility, we think that the s.s. can as yet be regarded only as heterotic in seed fertility.

One final question which remains to be answered is how this evolutionary leap from parasitic to heterotic chromatin might occur. We suggest that, as far as the s.s. are concerned, they may suppress the formation of chiasmata adjacent to their sites and this effect of the elimination of chiasmata at a specific chromosome site leads us to believe that, as in the case of inversions, the s.s. may also maintain favourable allelic combinations in linkage disequilibrium. Furthermore, if the heterozygotes are fitter than the two homozygotes the presence of an s.s. would constitute a system that would prevent the break-up of the linkage disequilibrium and would support itself by heterosis.

In this way, the differences in the parasitic (selfish) or heterotic nature of the B chromosome of *A. sphaerocephalon* and the s.s. of *T. australis* would be due to their evolutionary stage. Nevertheless, it will be essential in the future to find out whether "selfishness" is commoner in B chromosomes than in the s.s. and if so, why? We think, for example, that it is essential to carry out many more experiments, such as comparing the Bs and s.s. from the same species, and also to investigate further to see if the chromosomes with extra segments differ in their allelic combination from those without them.

This study was supported by the Comisión Asesora de Investigación Cientifica y Técnica of Spain, grant 1211–84 C02–01.

References

Darlington, C.D. and L.F. La Cour 1940. Nucleic acid starvation of chromosomes in *Trillium. J. Genet.* 40, 183–213.

Guillen, A. and M. Ruiz Rejón 1984. The B-chromosome system of *Allium sphaerocephalon* L (Liliaceae): Types, effects and origin. *Caryologia* 37, 259–267.

Hewitt, G.M. 1979. *Grasshoppers and crickets. Animal Cytogenet. vol 3: Insecta, Orthoptera.* Berlin-Stuttgart: Gebrüder Borntraeger.

John, B. 1973. The cytogenetic systems of grasshoppers and locusts. II The origin and evolution of supernumerary segments. *Chromosoma* 44, 123–146.

Jones, R.N. 1985. Are B-chromosomes "selfish". In *The Evolution of Genome Size*. T. Cavalier-Smith, ed. 397–425, John Wiley & Sons.

Jones, R.N. & H. Rees 1982. *B chromosomes*. Academic Press, New York.

Longley, A.E. 1938. Chromosomes of maize from North American Indians. *J. Agric. Res.*, 56, 835–862.

Nur, U. and B.L.H. Brett 1985. Genotypes suppressing meiotic drive of a B chromosome in the mealybug, *Pseudococcus obscurus*. *Genetics* 110, 73–92.

Nur, U. and B.L.H. Brett 1987. Control of meiotic drive of B chromosomes in the mealybug, *Pseudococcus affinis* (*obscurus*). *Genetics* 115, 499–510.

Oliver, J.L., F. Posse, J.M. Martinez-Zapater, A.M. Enriquez and M. Ruiz Rejón 1982. B-chromosomes and E-1 Isozyme activity in mosaic bulbs of *Scilla autumnalis* (Liliaceae). *Chromosoma* 85, 399–403.

Pastor, J. and B. Valdés 1983. Revisión del Género *Allium* (Liliaceae) en la Peninsula Ibérica e Islas Baleares. *Anales de la Universidad Hispalense.*

Rhoades, M.M. 1942. Preferential segregation in maize. *Genetics* 27, 395–407.

Ruiz Rejón C., R. Lozano and M. Ruiz Rejón 1987. *Segmentos cromosómicos supernumerarios en vegetales* (*en homenaje al Prof. A. Sañudo*). Fundación R. Areces. Madrid.

Ruiz Rejón C. and M. Ruiz Rejón 1985. Chromosomal polymorphism for a heterochromatic supernumerary segment in a natural population of *Tulipa australis* Link (Liliaceae). *Can. J. Genet. Cytol.* 27, 633–638.

Ruiz Rejón, M., F. Posse and J.L. Oliver 1980. The B chromosomes system of *Scilla autumnalis* (Liliaceae): Effects at the isozyme level. *Chromosoma* 79, 341–348.

Ruiz Rejón, M., C. Ruiz Rejón and J.L. Oliver 1987. La evolución de los cromosomas B: ? existen cromosomas egoistas?. *Investigación y Ciencia* (*Spanish version of Scientific American*) Barcelona.

Vosa, C.G. 1973. Heterochromatin recognition and analysis of chromosome variation in *Scilla sibirica*. *Chromosoma* 43: 269–278.

Nucleolar competition in *Triticeae*

J.R. Lacadena[1], M.C. Cermeño[2], J. Orellana[3] and J.L. Santos[1]

[1]*Departamento de Genética, Facultad de Ciencias Biológicas, Universidad Complutense, 28040 Madrid, Spain*
[2]*Lehrstuhl für Pflanzenbau und Pflanzenzüchtung, Technische Universität München, Freising-Weihenstephan, Federal Republic of Germany*
[3]*Departamento de Genética Agraria, Escuela Técnica Superior de Ingenieros Agrónomos, Universidad Politécnica, 28040 Madrid, Spain*

Ag-NORs

Nucleolus-organiser regions (NORs, the term introduced by McClintock) which are associated with secondary constrictions of satellited chromosomes (SAT-chromosomes), have been shown to be the sites of ribosomal RNA genes (rDNA clusters) in many animal and plant species (*Drosophila*, Ritossa & Spiegelman 1965; Hennig *et al.* 1982; *Xenopus*, Wallace & Birnstiel 1966; *Macaca mulata, Homo sapiens*, Henderson *et al.* 1972, 1974; *Zea mays*, Phillips *et al.* 1971, 1974; *Vicia faba*, Scheuermann & Knälmann 1975, Knälmann & Burger 1977, *Hordeum*, Subrahmanyam & Gerlach 1978, Subrahmanyam & Azad 1978; *Triticum*, Flavell & O'Dell 1976; Gerlach *et al.* 1980; Miller *et al.* 1980; Hutchinson & Miller 1982; *Secale*, Miller *et al.* 1980; *Aegilops*, Hutchinson & Miller 1982) by *in situ* DNA-rRNA hybridisation experiments.

Silver-staining methods have been developed for the differential staining of nucleolar-organiser regions (Ag-NORs) both in animal chromosomes (Howell *et al.* 1975; Goodpasture & Bloom 1975; Bloom & Goodpasture 1976) and those of plants (Hizume *et al.* 1980; Sato *et al.* 1980; Lacadena *et al.* 1984a).

It was found that the sites detected by the silver staining corresponded exactly to those obtained by *in situ* hybridisation with rRNA (Miller *et al.* 1976a). Furthermore, it was demonstrated in human cells that the size of the silver-stained regions was positively correlated with the amount of label present after hybridisation *in situ* with rRNA (Warburton & Henderson 1979).

Since the silver-staining reaction of NORs can be considered as an indication of genetic activity, it can be used to analyse gene activity at the rDNA sites with conventional light microscopy. According to several authors (Miller *et al.* 1976a, b; Howell 1977; Schmiady *et al.* 1979) only those NORs that were functionally active during interphase are stained by silver (Ag-NORs) at the next mitotic metaphase. In other words, in this context we will refer to NOR activity as a "potential activity", i.e. the ability of the silver-stained mitotic NOR to be transcribed in the preceding or in the following interphases (see Medina *et al.* 1986).

152   *J.R. Lacadena* et al.

Amphiplasty

Amphiplasty is the term proposed to denote morphological changes which occur in chromosomes following interspecific hybridisation. When the change affects individual chromosomes of the complement it is called differential amphiplasty. This phenomenon was first described by Navashin (1928, 1934) in some interspecific hybrids of *Crepis*: the secondary constriction of the SAT-chromosome of one of the parental species is lacking in the hybrid, the satellite being retracted onto its own chromosome arm and not distinguishable. McClintock (1934) noted that differential amphiplasty was related to the formation of nucleoli in such a way that only chromosomes with a secondary constriction were active in the formation of nucleoli. Differential amphiplasty has been reported in many other interspecific hybrids, both in plants and animals (see Table 1). The NOR competition is cytologically expressed as amphiplasty.

Table 1. Differential amphiplasty reported in interspecific hybrids and amphiploids.

Plants	References
Salix	Wilkinson 1941, 1944
Ribes	Keep 1960, 1962, 1971
Agropyron	Heneen 1962
Solanum	Yeh & Peloquin 1965
Hordeum	Kasha & Sadasivaiah 1971; Lange & Jochemsen 1976; Jessop & Subrahmanyam 1984; Linde-Laursen et al. 1986
Nicotiana	Gerstel et al., 1978
Hordeum x Secale	Ramsay & Dyer 1983
Aegilops x Secale	Cermeño & Lacadena 1985
Lolium x Festuca	Canride et al. 1986
Amphiploids	
Triticum 4x – Secale	Darvey 1974; Thomas & Kaltsikes 1983; Lacadena et al. 1984a
Triticum 6x – Secale	Cermeño et al. 1984a
Aegilops ventricosa – Triticum 4x	Orellana et al. 1984
Aegilops ventricosa – Secale	Orellana et al. 1984
Triticum 6x – Agropyron	Lacadena et al. 1984b
Triticum 4x – Haynaldia (Dasypyrum)	Friebe et al. 1987
Aegilops squarrosa – Secale	Cermeño et al. 1987
Animals	
Xenopus	Honjo & Reeder 1983; Cassidy & Blackler 1974
Drosophila	Durica & Krider 1978
Siamang x Gibbon	Shafer, Howell & Warburton, cited by Howell 1977
Mule	Kopp et al. 1986
Mouse – Human (hybrid cells)	Eliceiri & Green 1969; Bramwell & Handmaker 1971; Marshall et al. 1975; Miller et al. 1976 a, b; Croce et al. 1977

In spite of the apparent suitability of the silver procedures to analyse the "potential activity" of NORs in relation to the phenomenon of differential amphiplasty, this kind of analysis was used in animals only in mouse-human hybrid cells (Miller *et al.* 1976a, b; Croce *et al.* 1977). On the other hand, first attempts in plant material were carried out by Anastassova-Kristeva *et al.* (1979, 1980) and Nicoloff *et al.* (1979), analysing by means of Ag-NORs the phenomenon of "nucleolar dominance" in barley translocation lines. Some years later, Ramsay & Dyer (1983) analysed amphiplasty by silver-staining in barley × rye hybrids.

Nucleolar competition in *Triticeae*

To carry out a complete cytogenetical analysis of the phenomenon of nucleolar competition it is necessary to know how many active (or potentially active) NORs are present in the cell and to identify which SAT-chromosomes carry active NORs. The first question can be answered using the silver-staining technique, while the second could be solved if the chromosome complement of the species analysed shows an appropriate banding pattern which enables us to identify individual chromosomes with accuracy. On the other hand, the occurrence of natural allopolyploidy and the possibility of obtaining artificial hybrids, alloploids and derivatives makes the *Triticeae* a model biological system for the analysis of both natural and artificial nucleolar competition.

Nucleolar activity ("potential activity") was analysed in mitotic metaphase cells using a modified, highly reproducible, silver-staining procedure which we have developed (Lacadena *et al.* 1984a), based on that of Hizume *et al.* (1980). In addition, individual SAT-chromosomes were identified by their C-banding patterns. The C-banding procedure was carried out according to Giráldez *et al.* (1979).

The comparative analysis of somatic metaphase chromosomes by phase contrast, C-banding and Ag-staining was made according to the following rationale: the nucleolar-organiser chromosomes (SAT-chromosomes) were identified both with phase contrast and C-banding from comparisons of the same metaphase cells. On the other hand, silver-stained nucleolar organiser regions (Ag-NORs), i.e. the potentially active NORs, were also identified, corresponding to the secondary constrictions observed previously in the same cells by phase contrast. Obviously, the ideal situation would be if silver-staining and C-banding techniques could be applied to the same cell. However, it has unfortunately not been possible in our material.

The *Triticeae* include diploid (2n=14), allotetraploid (2n=28) and allohexaploid (2n=42) species. It is accepted that the different genomes (x=7) of diploid species may arise from a common ancestral genome, thus the seven chromosomes of each genome have the corresponding homoeologues in any other genome. The homoeologous groups are numbered from 1 to 7. It has been shown that the SAT-chromosomes of diploid species belong to homoeologous groups 1, 5 and/or 6 (see Table 2).

Table 2. Identification of SAT-chromosomes in diploid species of Triticeae.

Species	Genome constitution	SAT chromosomes	References
Triticum monococcum	AA	1A, 5A	Gerlach et al. 1980; Miller et al. 1983
Aegilops speltoides	BB(SS)	1B, 6B	Hutchinson & Miller 1982; Dvorak et al. 1984
Aegilops squarrosa	DD	5D	Hutchinson & Miller 1982
Secale cereale	RR	1R	Miller et al. 1980; Appels et al. 1980
Aegilops umbellulata	UU	1U, 5U	Martini et al. 1982
Agropyron elongatum (Elytrigia elongata)	EE	5E, 6E	Dvorak 1980; Dvorak et al. 1984
Hordeum vulgare	$H^V H^V$	$5H^V$, $6H^V$	Islam 1980, Linde-Laursen et al. 1982
Haynaldia villosa (Dasypyrum villosum)	VV	one (1V?)*	Friebe et al. 1987
Aegilops comosa	MM	two*	Chennaveraiah 1960; Teoh & Hutchinson 1983; Teoh et al. 1983; Cermeño et al. 1984b
Aegilops uniaristata	$M^u M^u$(UnUn)	one*	" "
Aegilops caudata	CC	two*	" "
Aegilops longissima	$S^l S^l$	two*	" "
Aegilops sharonensis	$S^{sh} S^{sh}$	two*	" "
Aegilops bicornis	$S^b S^b$	two*	" "

*Cytologically characterized by their C-banding patterns but not identified in relation to the homoeologous group.

As indicated above, the NOR competition is cytologically expressed as differential amphiplasty. This nucleolar competition can be considered and analyzed as natural amphiplasty in allopolyploid species, or as artificial amphiplasty in interspecific hybrids, artificial amphiploids or their derivatives such as chromosome addition or substitution lines.

1. Natural amphiplasty

Natural amphiplasty was studied with the silver-staining procedure in the genera *Triticum* (Cermeño *et al.* 1984a) and *Aegilops* (Cermeño *et al.* 1984b). The results are in Table 3.

The nucleolar competition found in tetraploid and hexaploid wheats presents

Table 3. Analysis of natural amphiplasty in <u>Triticum</u> spp. and <u>Aegilops</u> spp. using the silver-staining procedure.

Species	Genome constitution	No. of Ag-NORs at metaphase	Active chromosomes	Maximum number of nucleoli at interphase
<u>Triticum turgidum durum</u>	AABB	4	6B>1B	4
T. dicoccoides	AABB	4	6B>1B	4
T. aestivum	AABBDD	4–6	6B>1B>>5D>>1A	6
T. <u>compactum</u>	AABBDD	4	6B>1B>>5D	5
T. spelta	AABBDD	4	6B>1B>>5D	5
<u>Aegilops umbellulata</u>	UU	4	1U>5U	4
Ae. comosa	MM	2–4	E>F	4
Ae. <u>uniaristata</u>	UnUn	2	C	2
Ae. <u>caudata</u>	CC	4	A>C	4
Ae. squarrosa	DD	2	B	2
Ae. speltoides	SS(BB)	2–4	D and E	4
Ae. longissima	$S^l S^l$	4	E and F	4
Ae. sharonensis	$S^{sh} S^{sh}$	4	D and E	4
Ae. bicornis	$S^b S^b$	4	C and E	4
Ae. ovata	$UUM^o M^o$	4	1U>5U	4
Ae. columnaris	$UUM^c M^c$	4–5	$1U>5U>>M^c(E)$	6
Ae. biuncialis	$UUM^b M^b$	4–6	$1U>5U>>M^b(E)$	6
Ae. variabilis	$UUS^v S^v$	4	1U>5U	4
Ae. triuncialis	UUCC	4	1U>>5U	4
Ae. cylindrica	CCDD	4	C(A)>>C(C)	4
Ae. ventricosa	$DDM^v M^v$	2	$M^v(C)$	2
Ae. crassa 4x	$DDM^{cr} M^{cr}$	4	$M^{cr}(C)>>D(B)$	4
Ae. crassa 6x	$DDD^2 D^2 M^{cr} M^{cr}$	6	$M^{cr}(E)>M^{cr}(F)=D^2$	6
Ae. triaristata 6x	$UUM^t M^t M^{t2} M^{t2}$	6	$1U>>5U>>M^t$ or M^{t2}	6
Ae. <u>juvenalis</u>	$DDUUM^j M^j$	4–6	$1U>5U>>M^j(E)$	6

some interesting characteristics. All tetraploid wheats (with the possible exception of *Triticum carthlicum*) show two pairs (1B and 6B) of nucleolar-organiser chromosomes (Hutchinson & Miller 1982; Lacadena *et al.* 1984a; Cermeño *et al.* 1984a). This means that the SAT-chromosomes from the A genome (1A and 5A) have been completely suppressed.

Hexaploid wheat, *Triticum aestivum*, shows four pairs (1B, 6B, 5D, 1A) of nucleolar-organiser chromosomes (Crosby 1957; Bhowal 1972; Flavel & Smith 1974; Flavell & O'Dell 1976). Chromosomes 1B and 6B are considered as "strong" and 1A and 5D as "weak" nucleolar-organisers since the NORs on 6B and 1B possess 90% of the total rRNA gene complement (60% and 30%, respectively) (Flavell & Smith 1974; Flavell & O'Dell 1976).

Analysis of the NOR activity in the hexaploid wheats *T. compactum*, *T. spelta* and *T. aestivum* with the silver-staining technique (Cermeño *et al.* 1984a), provides evidence for only four Ag-NORs, corresponding to chromosomes 1B and 6B. This is certainly a limiting factor of the silver technique: in a few cases the Ag-staining seems to have insufficient resolving power to reveal all the potentially active NORs which are present in the cell. However, this deficiency can be overcome taking into account the maximum number of nucleoli observed at interphase, each nucleolus being organised by an active NOR. From the analysis of the distribution of the number of nucleoli at interphase in different nulli-tetrasomic plants it was concluded that chromosomes 5D and 1A are also active.

It is worth mentioning that NOR activity of chromosome 1A is usually neglected when considering hexaploid wheat. However, it is interesting to point out that the silver-staining observations confirm previous results with other techniques; thus, micronuclei formed at the tetrad stage of meiosis in plants monosomic for chromosome 1A were able to produce a nucleolus. This was observed both with light microscopy (Crosby 1957) and electron microscopy (Medin *et al.* 1980). In summary, the following graded series of activity of nucleolar-organiser chromosomes can be established in *Triticum aestivum*: 6B $>$ 1B $\gg$ 5D $\gg$ 1A.

Some questions remain unanswered: why is the NOR of chromosome 5A completely suppressed both in tetraploid and hexaploid wheats? Why is the NOR of chromosome 1A functional (although weakly) in hexaploid wheat but not in tetraploid wheat?

In relation to the natural amphiplasty, analysis was made in the genus *Aegilops*, and it was concluded that with the exception of *Ae. crassa* (4x), amphiplasty occurs in all tetraploid and hexaploid species analysed. The U genome of *Ae. umbellulata* completely suppresses the NOR activity of the genomes M^o (*Ae. ovata*), S^v (*Ae. variabilis*), D (*Ae. juvenalis*) and C (*Ae. triuncialis*), and that of one pair of the nucleolar-organiser chromosomes of the genomes M^c (*Ae. columnaris*), M^b (*Ae. biuncialis*), M^j(*Ae. juvenalis*) and M^t (*Ae. triaristata*). On the other hand, the nucleolar activity of the D genome is completely suppressed by the genomes U (in *Ae. juvenalis*), C (in *Ae. cylindrica*) and M^v (in *Ae. ventricosa*) (See Fig. 1).

2. Artificial amphiplasty

The *Triticeae* show the possibility of forming many interspecific hybrids, amphidiploids and other genome combinations as well as derivatives such as chromosome addition and substitution lines. The results, shown in Table 4, are based on Lacadena *et al.* (1984a, b), Cermeño *et al.* (1984a, b), Orellana *et al.* (1984), Santos *et al.* (1984), Lacadena & Cermeño (1985), Cermeño & Lacadena (1985), Cermeño *et al.* (1987), Friebe *et al.* (1987), Cermeño (unpublished) and Ramsay & Dyer (1983).

The main conclusions obtained are as follows:

1) The NOR activity of chromosome 1R of rye, *Secale cereale*, is suppressed when in competition with any other nucleolar-organiser chromosomes. This has been observed in a variety of wheat-rye combinations:

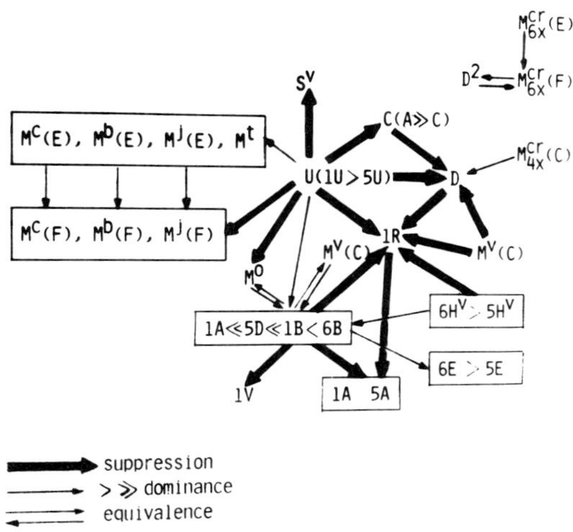

Figure 1. Diagram showing nucleolar suppression and dominance relationships between the different genomes or nucleolar-organiser chromosomes of *Triticeae* species. Letters in brackets represent SAT chromosomes of the corresponding genomes. Numbers stand for homoeologous groups. Based on Lacadena (1985).

i) hexaploid triticale, AABBRR (Lacadena *et al.* 1984a);

ii) hexaploid wheat × rye hybrid, ABDR and octoploid triticale, AABBDDRR (Cermeño *et al.* 1984a);

iii) tetraploid triticales, DDRR and (A/B) (A/B) RR, in the presence of the chromosome pairs 6B and/or 1B (Cermeño *et al.* 1987);

iv) different wheat-rye derivatives with genome constitutions ABRR and ABRRR, and the chromosome addition line AABBDD+1R1R (Cermeño *et al.* 1984a). Not even the three doses of 1R present in ABRRR plants are able to overcome the suppression barrier of the A and B genomes of wheat (it should be noted that in autotetraploid plants of rye, RRRR, there are four active NORs).

v) in the amphidiploid *Aegilops ventricosa-Secale cereale*, DDM^VM^VRR, the NOR activity of 1R is also suppressed (Orellana *et al.* (1984);

vi) in barley (*Hordeum vulgare*) × rye (*S. cereale*) hybrids, genome constitution H^VR, only the barley chromosomes 6H^V and 5H^V show NOR activity (Ramsay & Dyer 1983);

vii) in *Aegilops variabilis* × *S. cereale* (US^VR), *Ae. triuncialis* × *S. cereale* (UCR), *Ae. biuncialis* × *S. cereale* (UM^bR), *Ae. biuncialis* × *S. vavilovii* (UM^bR^V) and *Ae. juvenalis* × *S. cereale* (UM^jDR) hybrids, the 1U and 5U chromosomes show a strong dominance, respectively, over the S^V, C, M^b and M^j nucleolar-organiser chromosomes, while NOR activity of chromosome 1R is completely suppressed (Cermeño & Lacadena, 1985);

viii) exceptionally, NOR activity of chromosome 1R is present in the (A/B) (A/B)RR tetraploid triticale lacking chromosomes 1B and 6B of wheat (Cermeño *et al.* 1987).

Table 4. Analysis of artificial amphiplasty in <u>Triticeae</u> using the silver-staining procedure.

Species and hybrid combination	Genome and chromosome constitution	No. Ag-NORs at metaphase	Active chromosomes	Maximum No. nucleoli at interphase
Rye, <u>Secale cereale</u>	RR	2	1R	2
Hexaploid triticale	AABBRR	4	6B>1B	4
Hexaploid wheat x rye	ABDR	2-3	6B>1B>>5D	3
Octoploid triticale	AABBDDRR	4-6	6B>1B>>5D	6
Tetraploid triticale	(A/B)(A/B)RR			
	1B1B...6B6B...RR	3	6B>1B	4
	1B1B...6A6A...RR	2	1B	2
	1A1A...6A6A...RR	2	1R	2
	$DDR^{C}R^{C}$	2	5D	2
	$DDR^{S}R^{S}$	2	5D	2
Wheat-rye derivatives				
Addition line	AABBDD+1R1R	4	6B>1B>>5D	6
Triticale 6x x rye 2x	ABRR	2	6B>1B>>1A-5A?	3
Triticale 6x x rye 4x	ABRRR	2	6B>1B>>1A-5A?	3
Tetraploid rye	RRRR	4	1R	4
Amphiploids				
Ae. ventricosa-T.turgidum	$AABBDDM^{V}M^{V}$	6	$6B>1B=M^{V}(C)$	6
Ae. ventricosa-T. dicoccum	$AABBDDM^{V}M^{V}$	6	$6B>1B=M^{V}(C)$	6
Ae. ventricosa-T. aethiopicum	$AABBDDM^{V}M^{V}$	6	$6B>1B=M^{V}(C)$	6
Ae. ventricosa-S. cereale	$DDM^{V}M^{V}RR$	2	$M^{V}(C)$	2
Chromosome addition lines				
T. aestivum-Hordeum vulgare	$AABBDD+5H^{V}5H^{V}$	6	$6B>1B=5H^{V}>>5D$	8
	$AABBDD+6H^{V}6H^{V}$	6	$6H^{V}>6B>1B>>5D$	7
(any non SAT)	$AABBDD+ H^{V} H^{V}$	4	6B>1B>>5D	6
Hordeum vulgare	$H^{V}H^{V}$	4	$6H^{V}>5H^{V}$	4
T. aestivum-Agropyron elongatum	AABBDD+5E5E	4	6B>1B>>5E=5D	7
	AABBDD+6E6E	4-6	6B>1B=6E>>5D	7
(any non SAT)	AABBDD+ E E	4	6B>1B>>5D	6
Agropyron elongatum	EE	4	6E>5E	4
Amph. (T. aestivum-A. elongatum	AABBDDEE	4-8	6B>1B=6E>5E>>5D	10
T. aestivum-Ae. umbellulata	AABBDD+1U1U	2-3	1U>6B>1B>>5D	6
	AABBDD+5U5U	2-3	5U>>6B>1B>>5D	7
(any non SAT)	AABBDD+ U U	3-4-6	6B>1B>>5D	5-6
T. aestivum-Ae. ovata	$AABBDD+6M^{O}6M^{O}$		$6B>1B=6M^{O}>>5D$	
T. aestivum-Haynaldia villosa (Dasypyrum villosum)	AABBDD+1V1V	4	6B>1B>>5D	5
(any non SAT)	AABBDD+ V V	4	6B>1B>>5D	4-6
Haynaldia villosa	VV	2	1V	2
Amph. (T. dicoccum-H. villosa)	AABBVV	4	6B>1B	4
Hybrids				
Hordeum vulgare x S. cereale	$H^{V}R$	2	$6H^{V}>5H^{V}$	
Ae. variabilis x S. cereale	$US^{V}R$	2	$1U>5U>>?S^{V}?$	3
Ae. triuncialis x S. cereale	UCR	2	1U>5U>>?C?	3
Ae. biuncialis x S. cereale	$UM^{b}R$	2	$1U>5U>>M^{b}(E)$	3
Ae. biuncialis x S. vavilovii	$UM^{b}R^{V}$	2	$1U>5U>>M^{b}(E)$	3
Ae. juvenalis x S. cereale	$UM^{j}DR$	2-3	$1U=5U>M^{j}(E)$	4

2) The nucleolar activity of U genome (1U > 5U) from *Aegilops umbellulata* seems to be the strongest in the *Triticeae* since, in addition to suppressing NOR activities of nucleolar-organiser chromosomes of the genomes S^V, M^O, M^C, M^b, M^j, C, D in allopolyploid *Aegilops* species (Cermeño *et al.* 1984b), it suppresses the NOR activity of chromosome 1R of rye (Cermeño & Lacadena, 1985) and behaves as a dominant in *Triticum aestivum–Ae. umbellulata* addition lines (Martini *et al.* 1982; Lacadena & Cermeño 1985).

3) The behaviour of nucleolar-organiser chromosomes of tetraploid and hexaploid wheats is variable. They suppress the NOR activity of chromosome 1R of rye (Lacadena *et al.* 1984a; Cermeño *et al.* 1984a) and the SAT-chromosome of *Haynaldia villosa* (*Dasypyrum villosum*) Friebe *et al.* 1987) but behave only as dominants in the presence of *Agropyron elongatum* (*Elytrigia elongata*) SAT-chromosomes (Lacadena *et al.* 1984b). On the other hand, nucleolus-organiser chromosomes of wheat are dominated by the U genome of *Ae. umbellulata* (Martini *et al.* 1982; Lacadena & Cermeño 1985) and are dominated or co-dominated by barley SAT-chromosomes (Santos *et al.* 1984) and, likewise, show co-dominance with SAT-chromosomes of genomes M^V from *Ae. ventricosa* (Orellana *et al.* 1984) and M^O from *Ae. ovata* (Cermeño, unpublished).

4) In spite of the D genome behaving the most weakly in alloploid species of *Aegilops* (*Ae. cylindrica, Ae. ventricosa, Ae. crassa* 4x and 6x, *Ae. juvenalis*), it suppresses the NOR activity of chromosome 1R of rye (*S. cereale* and *S. silvestre*) in tetraploid triticales DDR^CR^C and DDR^SR^S, as indicated above (Cermeño *et al.* 1987).

5) It should be very interesting to analyse the genome combinations AARR and AADD, but the corresponding plants are not available yet. All the results discussed are represented graphically in Fig. 1.

Genetic control of nucleolar competition and mechanisms of nucleolar organiser activity and Ag-NOR staining

Obviously, it would be very interesting to determine how the genetic control is carried out and which chromosomes are involved. There is clear evidence in wheat that the nucleolar-organiser chromosomes themselves are responsible for nucleolar activity, as can be seen in the case of micronuclei in meiotic tetrads, comprising only chromosome 1A, which are able to produce a nucleolus (Crosby 1957; Medina *et al.* 1980). However, the genetic control of nucleolar activity in wheat appears to be rather complex, as was pointed out by several authors (Mohan & Flavell 1974; Viegas & Mello-Sampayo 1975; Flavell & O'Dell 1976, 1979; Martini *et al.* 1982). Although it has been suggested (Flavell & Martini 1982) that the experimental design could be difficult since there are probably many genes involved in the regulation of the total nucleolar activity, the utilisation of appropriate aneuploid combinations might solve the problems. For instance, the expression of NOR activity of the chromosome 1R of rye in tetraploid triticale (A/B) (A/B) RR lacking the wheat chromosomes 1B and 6B (Cermeño *et al.* 1987) could be a good example.

The analysis of molecular mechanisms of nucleolar organiser activity in wheat has been carried out by Flavell and his collaborators. The differential expression of ribosomal RNA genes was analysed by Flavell *et al.* (1986) in wheat, concluding that differences in gene expression are associated with differences in chromatin structure and cytosine methylation.

Short sequences act as enhancers of transcription by binding to specific regulatory proteins which stimulate the attachment of transcriptional complexes, hence these authors conclude that differential expression of arrays of ribosomal RNA genes results when genes have different numbers of enhancer repeats or a higher affinity for the regulatory proteins. They linked the molecular model to the cytological observations as follows: 1) the regulatory protein is induced at telophase and binds to rRNA genes; which genes are bound is determined on a competitive basis by their affinity for the protein; 2) genes associated with protein are able to uncoil, become associated with RNA polymerase I and are responsible for development of the nucleolus; 3) the genes associated with the regulatory protein maintain the conformation after their replication; 4) at prophase the regulatory protein remains attached to a subset of the genes maintaining the differential condensation, thus the subset of genes selected by the regulatory protein is maintained through the cell generations.

As pointed out by Flavell *et al.* (1986), cytological observations indicate that active and inactive rDNA genes are organised in different ways. Inactive genes lie in supercoiled chromatin outside nucleoli whereas active genes are dispersed and gather nucleolar material around themselves. However, Busch *et al.* (1982) observed under electron microscopy that some genes lie in a condensed comformation (the so-called fibrillar centres, Stahl 1982) within the nucleolus and have specific (silver binding) nucleolar proteins bound to them. The genes included in the fibrillar centres are not transcribed but can be rapidly activated, uncoiled and transcribed in contrast to the genes which are condensed in heterochromatin outside the nucleolus. The observations of Appels *et al.* (1986) confirm the above statements: rye RNA genes are preferentially condensed and less dispersed in interphase nucleoli in wheat × rye hybrids in which the rye NOR is suppressed.

In relation to the specific mechanisms of Ag-staining, several hypotheses have been proposed. For example, Swarzacher *et al.* (1978) proposed that Ag-stainable material was a component of ribonucleic protein accumulating around active NORs. In mitosis some of this material remains at the NORs. Their electron micrographs of human cells revealed that the Ag-stainable substance was located on the outside of the NORs or around them but not in the chromosomes themselves. This agrees with cytochemical tests made by Howell (1977) in cricket oocyte chromosomes which revealed that the silver binds neither to the rDNA nor to transcribed rRNA, but rather to proteins which rapidly associate with the freshly transcribed rRNA. On the other hand, Hubbell *et al.* (1979) proposed different roles for the Ag-proteins (see review by Howell 1982).

Medina *et al.* (1983) pointed out that the NOR silver-staining should be associated with rDNA chromatin decondensation rather than with transcriptional activity. This idea was confirmed by the use of inhibitors of transcription which do not abolish silver-staining (Sánchez-Pina *et al.* 1984).

In 1986, Medina and collaborators analysed the pattern of nucleolar silver-staining both in the interphase and mitosis of plant cells. They observed that during interphase the highest density of silver grains occurred in the portions of the dense fibrillar component nearest to the fibrillar centres. At mitosis, not only the NOR was stained but also the pro-nucleolar material in anaphase and telophase. Two different proteins (or sets of proteins) are suggested to be responsible for this staining pattern: one of them is associated with NOR chromatin and the other with early ribosomal precursors. Thus, they conclude that the "Ag-NOR staining reveals either transcriptionally active or potentially active chromatin, characterised by an uncoiled, extended structure, regardless of the state of transcription".

As they indicate, the high rate of transcription of nucleolar DNA could explain the necessity of keeping the functional rDNA chromatin permanently decondensed, whether being transcribed or not, even during mitosis. This would be the structure of inactive, but "potentially active", NOR chromatin. It would represent the ability of the mitotic Ag-NOR to be transcribed in the preceding or following interphase. By contrast, when rDNA genes are fully repressed, their chromatin would be condensed without showing any affinity for silver.

Obviously, a link between the molecular and cytological approaches mentioned above would be welcome.

Acknowledgement

This work has been funded by the CAICYT of Spain (project number 2062/83).

References

Anastassova-Kristeva, M., H. Nicoloff, R. Rieger, G. Künzel and A. Hagberg 1979. Nucleolus organizer activity as affected by chromosome repatterning in barley. *Barley Genet. Newslett.* 9, 9–12.

Anastassova-Kristeva, M., R. Rieger, G. Künzel, H. Nicoloff and A. Hagberg 1980. Further evidence on "nucleolar dominance" in barley. *Barley Genet. Newslett.* 10, 3–6.

Appels, R., W.L. Gerlach, E.S. Dennis, H. Swift and W.J. Peacock 1980. Molecular and chromosomal organization of DNA sequences coding for the ribosomal RNAs in cereals. *Chromosoma* 78, 293–311.

Appels, R., L.B. Moran and J.P. Gustafson 1986. The structure of DNA from the rye (*Secale cereale*) Nor-RI locus and its behaviour in wheat backgrounds. *Can. J. Genet. Cytol.* 28, 665–685.

Bhowal, J.G. 1972. Nucleolar chromosomes in wheat. *Z. Pflanzenzüchtg.* 68, 253–257.

Bloom, S.E. and C. Goodpasture 1976. An improved technique for selective silver-staining of nucleolar regions in human chromosomes. *Hum. Genet.* 34. 199–206.

Bramwell, M.E. and S.D. Handmaker 1971. Ribosomal RNA synthesis in human-mouse hybrid cells. *Biochim. biophys. Acta* 232, 580–583.

Busch, H., M.A. Lischwe, J. Michalik, P.-K. Chan and R.K. Busch 1982. Nucleolar proteins of special interest: silver-staining proteins B23 and C23 and antigens of human tumour nucleoli. In *The nucleolus*, E.G. Jordan & C.A. Cullis, eds, 43–71, Cambridge, UK: Cambridge Univ. Press.

Carnide, V., J. Orellana and M.A.M. do Valle Ribeiro 1986. Nucleolar organiser activity in *Lolium* and *Festuca*. 1. *Lolium multiflorum, Festuca arundinacea* and *Lolium– Festuca* hybrids. *Heredity* 56, 311–317.

Cassidy, D.M. and A.W. Blackler 1974. Repression of nucleolar organiser activity in an interspecific hybrid of the genus *Xenopus*. *Develop. Biol.* 41, 84-96.

Cermeño, M.C., B. Friebe, F.J. Zeller and K.-D. Krolow 1987. Nucleolar competition in different (A/B) (A/B)RR and DDRR tetraploid triticales. *Heredity* 58, 1–4.

Cermeño, M.C. and J.R. Lacadena 1985. Nucleolar organiser competition in *Aegilops–* rye hybrids. *Can. J. Genet. Cytol.* 27, 479–483.

Cermeño, M.C., J. Orellana, J.L. Santos and J.R. Lacadena 1984a. Nucleolar organiser activity in wheat, rye and derivatives analyzed by a silver-staining procedure. *Chromosoma* 89, 370–376.

Cermeño, M.C., J. Orellana, J.L. Santos and J.R. Lacadena 1984b. Nucleolar activity and competition (amphiplasty) in the genus *Aegilops*. *Heredity* 53, 603–611.

Chennaveraiah, M.S. 1960. Karyomorphological and cytotaxonomic studies in *Aegilops*. *Acta Horti. Gotoburgensis* 23, 85–178.

Croce, C.M., A. Talavera, C. Basilico and O.J. Miller 1977. Suppression of production of mouse 28S ribosomal RNA in mouse-human hybrids segregating mouse-chromosomes. *Proc. Nat. Acad. Sci.* 74, 694–697.

Crosby, A.R. 1957. Nucleolar activity of lagging chromosomes in wheat. *Am. J. Bot.* 44, 813–822.

Darvey, N.L. 1974. Nucleolar behaviour in hexaploid triticale. *EWAC Newsletter* 4, 69–70.

Durica, D.A. and H.M. Krider 1978. Studies on the ribosomal RNA cistrons in interspecific *Drosophila* hybrids. II. Heterochromatic regions mediating nucleolar dominance. *Genetics* 89, 37–64.

Dvořak, J. 1980. Homoeology between *Agropyron elongatum* chromosomes and *Triticum aestivum* chromosomes. *Can. J. Genet. Cytol.* 22, 237–259.

Dvořak, J. M.W. Lassner, R.S. Kota and K.C. Chen 1984. The distribution of the ribosomal RNA genes in the *Triticum speltoides* and *Elytrigia elongata* genomes. *Can. J. Genet. Cytol.* 26, 628–632.

Eliceiri, G.L. and H. Green 1969. Ribosomal RNA sythesis in human-mouse hybrid cells. *J. Mol. Biol.* 41, 253–260.

Flavell, R.B. and G. Martini 1982. The genetic control of nucleolus formation with special reference to common bread wheat. In *The nucleolus*. E.G. Jordan and C.A. Callis, eds, 113-128, Cambridge, UK: Cambridge Univ. Press.

Flavell, R.B. and M. O'Dell 1976. Ribosomal RNA genes on homoelogous chromosomes of groups 5 and 6 in hexaploid wheat. *Heredity* 37, 377–385.

Flavell, R.B. and M. O'Dell 1979. The genetic control of nucleolus formation in wheat. *Chromosoma* 71, 135–152.

Flavell, R.B., M. O'Dell, W.F. Thompson, M. Vincentz, R. Sardana and R.F. Barker 1986. The differential expression of ribosomal RNA genes. *Phils. Trans. R. Soc. Lond.* B 314, 385–397.

Flavell, R.B. and D.B. Smith 1974. The role of homoeologous group 1 chromosomes in the control of rRNA genes in wheat. *Biochem. Genet.* 12, 271–279.

Friebe, B., M.C. Cermeño and F.J. Zeller 1987. C-banding polymorphism and the analysis of nucleolar activity in *Dasypyrum villosum* (L.) Candargy, its added chromosomes to hexaploid wheat and the amphiploid *Triticum dicoccum–D. villosum*. *Theor. Appl. Genet.* 73, 337–342.

Gerlach, W.L., T.E. Miller and R.B. Flavell 1980. The nucleolus organisers of diploid wheats revealed by in situ hybridization. *Theor. Appl. Genet.* 58, 97–100.

Gerstel, D.U., J.A. Burns and L.G. Burk 1978. Cytoplasmic male sterility in *Nicotiana*, restoration of fertility, and the nucleolus. *Genetics* 89, 157–169.

Giráldez, R., M.C. Cermeño and J. Orellana 1979. Comparison of C-banding pattern in the chromosomes of inbred lines and open pollinated varieties of rye, *Secale cereale* L. *Z. Pflanzenzüchtg.* 83, 40–48.

Goodpasture, C. and S.E. Bloom 1975. Visualization of nucleolar organiser regions in mammalian chromosomes using silver-staining. *Chromosoma* 53, 37–50.

Henderson, A.S., D. Warburton and K.C. Atwood 1972. Localisation of ribosomal DNA in the human chromosome complement. *Proc. Nat. Acad. Sci.* 69, 3394–3398.

Henderson, A.S., D. Warburton and K.C. Atwood 1974. Localisation of rDNA in the chromosome complement of the Rhesus (*Macaca mulatta*). *Chromosoma* 44, 367–370.

Heneen, W.K. 1962. Karyotype studies in *Agropyron junceum, A. repens* and their spontaneous hybrids. *Hereditas* 48, 471–502.

Hennig, W., P. Vogt, G. Jacob and I. Siegmund 1982. Nucleolus organiser regions in *Drosophila* species of the *repleta* group. *Chromosoma* 87, 279–292.

Hizume, M., S. Sato and A. Tanaka 1980. A highly reproducible method of nucleolus organising regions staining in plants. *Stain Technol.* 55, 87–90.

Honjo, T. and R.H. Reeder 1973. Preferential transcription of *Xenopus laevis* ribosomal RNA in interspecies hybrids between *X. laevis* and *X. mulleri. J. Mol. Biol.* 80, 217–228.

Howell, W.M. 1977. Visualization of ribosomal gene activity: Silver stains proteins associated with rRNA transcribed from oocyte chromosomes. *Chromosoma* 62, 361–367.

Howell, W.M. 1982. Selective staining of nucleolus organizer regions (NORs). *Cell Nucleus* 11, 80–142.

Howell, W.M., T.E. Denton and J.R. Diamond 1975. Differential staining of the satellite regions of human acrocentric chromosomes. *Experientia* 31, 260–262.

Hubbell, H.R., L.I. Rothblum and J.R. Diamond 1979. Identification of a silver binding protein associated with the cytological silver staining of an actively transcribing nucleolar region. *Cell. Biol. Int. Rep.* 3, 615–622.

Hutchinson, J. and T.E. Miller 1982. The nucleolar organisers of tetraploid wheats revealed by in situ hybridization. *Theor. Appl. Genet.* 61, 285–288.

Islam, A.K.M.R. 1980. Identification of wheat-barley addition lines with N-banding of chromosomes. *Chromosoma* 76, 365–376.

Jessop, C.M. and N.C. Subrahmanyam 1984. Nucleolar number variation in *Hordeum* species, their haploids and interspecific hybrids. *Genetica* 64, 93–100.

Kasha, K.J. and R.S. Sadasivaiah 1971. Genome relationships between *Hordeum vulgare* L. and *H. bulbosum* L. *Chromosoma* 35, 264–287.

Keep, E. 1960. Amphiplasty in *Ribes. Nature* 188, 339.

Keep, E. 1962. Satellite and nucleolar number in hybrids between *Ribes nigrum* and *R. grossularia* and in their backcrosses. *Can. J. Genet. Cytol.* 4, 206–218.

Keep, E. 1971. Nucleolar suppression, its inheritance and association with taxonomy and sex in the genus *Ribes. Heredity* 26, 443–451.

Knälmann, M. and E.-C. Burger 1977. Cytologische Lokalisation von 5S und 18/25S RNA Genorten in Mitose-Chromosomen von *Vicia faba. Chromosoma* 61, 177–192.

Kopp, E., B. Mayr and W. Schleger 1986. Species-specific non-expression of ribosomal RNA genes in a mammalian hybrid, the mule. *Chromosoma* 94, 346–352.

Lacadena, J.R. 1985. Comportamiento cromosómico. *Investigacíon y Ciencia* (Spanish version of *Scient. Amer.*) 110, 92–103.

Lacadena, J.R. and M.C. Cermeño 1985. Nucleolus organiser competition in *Triticum aestivum–Aegilops umbellulata* chromosome addition lines. *Theor. Appl. Genet.* 71, 278–283.

Lacadena, J.R., M.C. Cermeño, J. Orellana and J.L. Santos 1984a. Evidence for wheat-rye nucleolar competition (amphiplasty) in triticale by silver-staining procedure. *Theor. Appl. Genet.* 67, 207–213.

Lacadena, J.R., M.C. Cermeño, J. Orellana and J.L. Santos 1984b. Analysis of nucleolar activity in *Agropyron elongatum*, its amphiploid with *Triticum aestivum* and the chromosome addition lines. *Theor. Appl. Genet.* 68, 75–80.

Lange, W. and G. Jochemsen 1976. Karyotypes, nucleoli, and amphiplasty in hybrids between *Hordeum vulgare* L. and *H. bulbosum* L. *Genetica* 46, 217–233.

Linde-Laursen, I., R. von Bothmer and N. Jacobsen 1986. Giemsa C-banded karyotypes of *Hordeum* taxa from North America. *Can. J. Genet. Cytol.* 28, 42–62.

Linde-Laursen, I., H. Doll and G. Nielsen 1982. Giemsa C-banding patterns and some biochemical markers in a pedigree of European barley. *Z. Pflanzenzuchtg.* 88, 191–219.

Marshall, C.J., S.D. Handmaker and M.E. Bramwell 1975. Synthesis of ribosomal RNA in synkaryons and heterokaryons formed between human and rodent cells. *J. Cell Sci.* 17, 307–325.

Martini, G., M. O'Dell and R.B. Flavell 1982. Partial inactivation of wheat nucleolus organisers by the nucleolus organiser chromosomes from *Aegilops umbellata. Chromosoma* 84, 687–700.

McClintock, B. 1934. The relation of a particular chromosomal element to the development of the nucleoli in *Zea mays. Z. Zellforsch. mikrosk. Anat.* 21, 294–328.

Medina, F.J., M.C. Risueño and J.R. Lacadena 1980. Ultrastructural aspects of tetrads with micronuclei in a wheat monosomic. *Cell Biol. Int. Rep.* 4, 569–577.

Medina, F.J., M.C. Risueño, M.A. Sánchez-Pina and M.E. Fernández-Gómez 1983. A study of nucleolar silver-staining in plant cells. The role of argyrophilic proteins in nucleolar physiology. *Chromosoma* 88, 149–155.

Medina, F.J. E.L. Solanilla, M.A. Sánchez-Pina, M.E. Fernández-Gómez and M.C. Risueño 1986. Cytological approach to the nucleolar functions detected by silver-staining. *Chromosoma* 94, 259–266.

Miller, D.A., V.G. Dev, R. Tantravahi and O.J. Miller 1976a. Suppression of human nucleolus organiser activity in mouse-human somatic hybrid cells. *Exp. Cell Res.* 101, 235–243.

Miller, O.J., D.A. Miller, V.G. Dev, R. Tantravahi and C.M. Croce 1976b. Expression of human and suppression of mouse nucleolar organiser activity in mouse-human somatic cell hybrids. *Proc. Nat. Acad. Sci.* 73, 4531–4535.

Miller, T.E., W.L. Gerlach and R.B. Flavell 1980. Nucleolus organiser variation in wheat and rye revealed by in situ hybridisation. *Heredity* 45, 377–382.

Miller, T.E., J. Hutchinson and S.M. Reader 1983. The identification of the nucleolus organiser chromosomes of diploid wheat. *Theor. Appl. Genet.* 65, 145–147.

Mohan, J. and R.B. Flavell 1974. Ribosomal RNA cistron multiplicity and nucleolar organisers in hexaploid wheat. *Genetics* 76, 33–44.

Navashin, M.S. 1928. Amphiplastie — eine neue Karyologische Erscheinung. *Proc. Int. Conf. Genet.* 5, 1148–1152.

Navashin, M.S. 1934. Chromosome alterations caused by hybridisation and their bearing upon certain general genetic problems. *Cytologia* 5, 169–203.

Nicoloff, H., M. Anastassova-Kristeva, R. Rieger and G. Künzel 1979. "Nucleolar dominance" as observed in barley translocation lines with specifically reconstructed SAT chromosomes. *Theor. Appl. Genet.* 55, 247–251.

Orellana, J., J.L. Santos, J.R. Lacadena and M.C. Cermeño 1984. Nucleolar competition analysis in *Aegilops ventricosa* and its amphiploids with tetraploid wheats and diploid rye by the silver-staining procedure. *Can. J. Genet. Cytol.* 26, 34–39.

Phillips, R.L., R.A. Kleese and S.S. Wang 1971. The nucleolus organiser region of maize (*Zea mays* L.): Chromosomal site of DNA complementary to ribosomal RNA. *Chromosoma* 36, 79–88.

Phillips, R.L., R.A. Kleese and S.S. Wang 1974. The nucleolus organiser region of maize (*Zea mays* L.): tests for ribosomal gene compensation on magnification. *Genetics* 77, 285–297.

Ramsay, G. and A.F. Dyer 1983. Nucleolar organiser suppression in barley × rye hybrids. In *Kew Chromosome Conference II*, P.E. Brandham, & M.D. Bennett, eds, London, UK: George Allen & Unwin.

Ritossa, F.M. and S. Spiegelman 1965. Localisation of DNA complementary to ribosomal RNA in the nucleolus organiser region of *Drosophila melanogaster. Proc. Nat. Acad. Sci.* 53, 737–745.

Sánchez-Pina, M.A., F.J. Medina, M.E. Fernández-Gómez and M.C. Risueño 1984. "Ag-NOR" proteins are present when transcription is impaired. *Biol. Cell.* 50, 199–202.

Santos, J.L., J.R. Lacadena, M.C. Cermeño and J. Orellana 1984. Nucleolar organiser activity in wheat-barley addition lines. *Heredity* 52, 425–429.

Sato, S., M. Hizume and S. Kawamura 1980. Relationship between secondary constrictions and nucleolus organising regions in *Allium sativum* chromosomes. *Protoplasma* 105, 77–85.

Scheuerman, W. and M. Knälmann 1975. Localisation of ribosomal cistrons in metaphase chromosomes of *Vicia faba* (L). *Exp. Cell Res.* 90, 463.

Schmiady, H., M. Münke and K. Sperling 1979. Ag-staining of nucleolus organiser regions on human prematurely condensed chromosomes from cells with different ribosomal RNA gene activity. *Exp. Cell Res.* 121, 425–428.

Stahl, A. 1982. The nucleolus and nucleolar chromosomes. In *The nucleolus*, E.G. Jordan and C.A. Cullis, eds, 1-24, Cambridge, UK: Cambridge Univ. Press.

Subrahmanyam, N.C. and A.A. Azad 1978. Trisomic analysis of ribosomal RNA cistron multiplicity in barley (*Hordeum vulgare* L.). *Chromosoma* 69, 225–264.

Subrahmanyam, N.C. and W. Gerlach 1978. Cytological localisation of rRNA cistrons in *Hordeum*: a test for the non-expression of the nucleolar organiser region in interspecific hybrids. *Barley Genet. Newslett.* 8, 99–100.

Swarzacher, H.G., A.V. Mikelsaar and W. Schnedl 1978. The nature of the Ag-staining of nucleolus organiser regions: electron- and light-microscopic studies on human cells in interphase, mitosis and meiosis. *Cytogenet. Cell Genet.* 20, 24–39.

Teoh, S.B. and J. Hutchinson 1983. Interspecific variation in C-banded chromosomes of diploid *Aegilops* species. *Theor. Appl. Genet.* 65, 31–40.

Teoh, S.B., T.E. Miller and S.M. Reader 1983. Interspecific variation in C-banded chromosomes of *Aegilops comosa* and *Ae. speltoides*. *Theor. Appl. Genet.* 65, 343–348.

Thomas, J.B. and P.J. Kaltsikes 1983. Effect of chromosomes 1B and 6B on nucleolus formation in hexaploid triticale. *Can. J. Genet. Cytol.* 25, 292–297.

Viegas, W.S. and T. Mello-Sampayo 1975. Nucleolar organisation in the genus *Triticum*. *Broteria*, 44, 121–133.

Wallace, H. and M.L. Birnstiel 1966. Ribosomal cistrons and the nucleolar organiser. *Biochim. Biophis. Acta*, 144, 295–310.

Warburton, D. and A.S. Henderson 1979. Sequential silver-staining and hybridisation in situ on nucleolus organising regions in human cells. *Cytogenet. Cell Genet.* 24, 168–175.

Wilkinson, J. 1941. The cytology of the cricket bat willow (*Salix alba* var. *caerula*). *Ann. Bot.* 5, 149–165.

Wilkinson, J. 1944. The cytology of *Salix* in relation to its taxonomy. *Ann. Bot.*, 8, 269–284.

Yeh, B.P. and S.J. Peloquin 1965. The nucleolus associated chromosome of *Solanum* species and hybrids. *Amer. J. Bot.*, 52, 626.

Chromosome behaviour in wheat × maize, wheat × sorghum and barley × maize crosses

D.A. Laurie and M.D. Bennett

Institute of Plant Science Research, Maris Lane, Trumpington, Cambridge CB2 2LQ, UK

In a plant breeding context, a wide-cross can be defined as an interspecific or intergeneric hybrid, either between two crop species or between a crop species and one of its wild relatives. Karyotypically stable wide-crosses are valuable because the addition of alien germplasm increases the range of genetic variation which is available for manipulation by plant breeders. Karyotypically unstable wide-crosses which eliminate one of the parental chromosome complements may be useful for the production of haploids (Kasha & Kao 1970). The subject of wide-crossing in relation to plant breeding has recently been reviewed by several authors (e.g. Sharma & Gill 1983; Zeller & Hsam 1983; Riley & Law 1984; Fedak 1985; Mujeeb-Kazi & Kimber 1985; Goodman *et al.* 1987).

Usually, sexual wide-crosses combine species from the same tribe but the crosses considered here span a much greater taxonomic distance. Wheat and barley are members of the Pooideae while maize and sorghum are members of the Panicoideae and wheat × maize, wheat × sorghum and barley × maize crosses thus link two subfamilies of the Gramineae (Hutchinson 1959). Similar crosses were investigated by Zenkteler & Nitzsche (1984), who reported fertilisation in several novel cereal wide-crosses, including wheat × maize, from light microscope studies of thick (10–15μ m) sections cut from ovaries. We have used cytogenetic studies to confirm and extend their observations. This paper summarises observations on early seed development in wheat × maize crosses and compares these to results from wheat × sorghum and barley × maize crosses.

Proof of fertilisation in wheat × maize, wheat × sorghum and barley × maize crosses

To demonstrate fertilisation parents were chosen whose 4C nuclear DNA contents and relative chromosome sizes within the genome were known (Table 1). The hexaploid wheat 'Chinese Spring' (*Triticum aestivum* L., 2n=42) and the diploid barley 'Sultan' (*Hordeum vulgare* L., 2n=14) were used as female parents while the single-cross F_1 hybrid maize 'Seneca 60' (*Zea mays* L., 2n=20) and the grain sorghum line 'S9B' (*Sorghum bicolor* (L.) Moench, 2n=20) were used as male parents. Assuming proportionality of chromosome DNA content and chromosome size, it is clear from Table 1 that hybrid zygotes in all three crosses should show a clear size difference between the chromosomes from the female parent and those from the male. The largest maize chromosome should be

Table 1. Relative genome and chromosome sizes

	4C nuclear DNA content (pg)	Ratio of largest to smallest chromosome	Estimated DNA content of smallest chromosome (pg)
Female parents			
'Chinese Spring' wheat	69.3(*1)	1.44 : 1(*3)	1.38
'Sultan' barley	22.2(*1)	1.28 : 1(*4)	1.33

	4C nuclear DNA content (pg)	Ratio of largest to smallest chromosome	Estimated DNA content of largest chromosome (pg)
Male parents			
'Seneca 60' maize	9.84(*2)	2.00 : 1(*5)	0.70
'S9B' sorghum	3.23(*2)	1.52 : 1(*5)	0.20

*1 Bennett & Smith (1976); *2 Laurie & Bennett (1985); *3 Furuta et al. (1984); *4 Heslop-Harrison & Bennett (1983); *5 Bennett & Smith (unpublished).

about half the size of the smallest wheat or barley chromosome and the largest sorghum chromosome should be only about one seventh the size of the smallest wheat chromosome.

Zygotes at metaphase were obtained as described in Laurie & Bennett (1986, 1987). All but one had the expected F_1 combination, confirming that fertilisation occurred. The 9 zygotes from 'Chinese Spring' wheat × 'Seneca 60' maize crosses had 21 large wheat chromosomes and 10 small maize chromosomes (Fig. 1), 7 of the 8 zygotes from 'Chinese Spring' wheat × 'S9B' sorghum crosses had 21 large wheat chromosomes and 10 very small sorghum chromosomes (Fig. 2), and all 8 zygotes from 'Sultan' barley 'Seneca 60' maize crosses had 7 large barley chromosomes and 10 small maize chromosomes (Fig. 3). The exceptional zygote, from a 'Chinese Spring' wheat × 'S9B' sorghum cross, had 21 wheat chromosomes and 20 sorghum chromosomes, either as a result of fertilisation by an unreduced male gamete or because both sperm nuclei had fused with the egg cell.

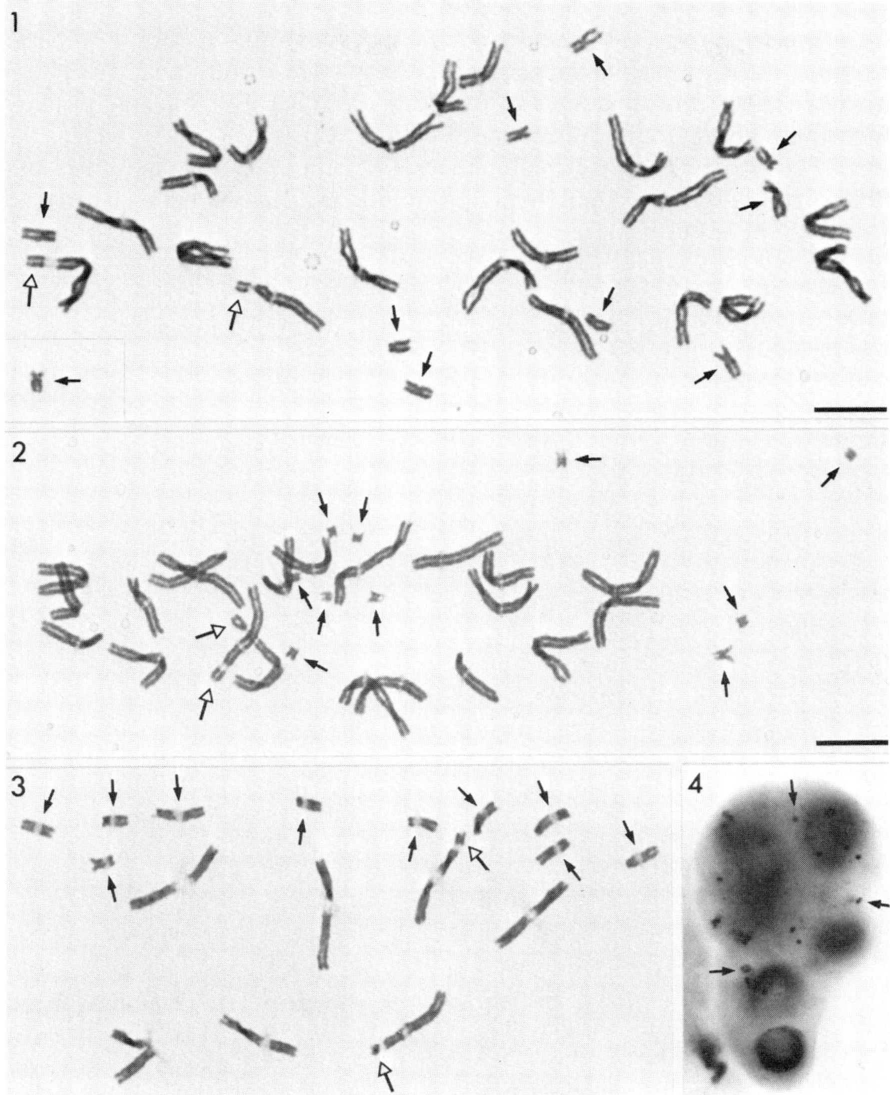

Figures 1–3. Zygotes at metaphase from **1** 'Chinese Spring' wheat × 'Seneca 60' maize (one maize chromosome (insert) lay slightly to one side of the other chromosomes), **2** 'Chinese Spring' wheat × 'S9B' sorghum and **3** 'Sultan' barley × 'Seneca 60' maize. Solid arrows indicate the cere 10 maize chromosomes (Figs 1 and 3) and the 10 sorghum chromosomes (Fig. 2). Open-headed arrows indicate the satellited chromosomes of the female parent. Note the well-defined centromeres on the maize chromosomes in Fig. 3. **4** An embryo from a 'Chinese Spring' wheat × 'N103A' maize cross fixed 48h after pollination. Numerous micronuclei are visible (examples arrowed). All bars represent 10 μm

Endosperm is expected to arise from the fusion of a sperm nucleus with the two polar nuclei of the female parent. In wheat × maize and wheat × sorghum crosses endosperm formation was relatively rare (Table 2), but one primary endosperm mitosis at metaphase was seen in a 'Chinese Spring' × 'Seneca 60' cross. This had the expected combination of 42 large wheat chromosomes and 10 small maize chromosomes. No primary endosperm mitosis was seen in wheat × sorghum or barley × maize crosses.

Table 2. Frequency of embryo and endosperm development in 'Chinese Spring' wheat / 'Seneca 60' maize, 'Chinese Spring' wheat / 'S9B' sorghum and 'Sultan' barley / 'Seneca 60' maize crosses fixed 48h after pollination.

	no. of florets examined	no. with only an embryo	no. with only an endosperm	no. with an embryo and an endosperm	Total fertilized n	%
Chinese Spring wheat (krl, kr2) ×						
Seneca 60 maize	343	80	8	12	100	29.2 *1
Chinese Spring wheat (krl, kr2) ×						
S9B sorghum	100	57	2	10	69	69.0 *2
Sultan barley ×						
Seneca 60 maize	100	4	8	16	28	28.0

*1 Data pooled from Laurie & Bennett (1986, 1987); *2 Data from Laurie & Bennett (in preparation).

The frequency of embryo and endosperm development in florets fixed 48h after pollination

Table 2 shows data from crosses made in a controlled environment cabinet at 20+1°C. The three crosses described above showed fertilisation at remarkably high frequencies in view of the taxonomic distances spanned. In 'Chinese Spring' wheat × 'Seneca 60' maize crosses 23% of florets contained an embryo, 2% contained an endosperm and 3% contained an embryo and an endosperm. Thus 26.8% (92/343) florets contained an embryo. In 'Chinese Spring' wheat × 'S9B' sorghum 57% of the florets contained an embryo, 2% contained an endosperm and 10% contained an embryo and an endosperm so that in this cross about 2 out of 3 (67/100) florets contained an embryo. These two crosses were similar in that most florets in which fertilisation had occurred contained only an embryo, the polar nuclei remaining unfertilised.

'Sultan' barley × 'Seneca 60' maize crosses differed in that the proportion of florets containing an endosperm exceeded the proportion containing an embryo. Thus 4% contained an embryo, 8% contained an endosperm and 16% had both an embryo and an endosperm. In total 20% of 'Sultan' × 'Seneca 60' florets contained an embryo.

The fate of embryos and endosperms in 'Chinese Spring' wheat × 'Seneca 60' maize crosses

Post-fertilisation development was studied in most detail in this cross. In zygotes at metaphase the centromeres of the maize chromosomes were often difficult to see and they showed little evidence of binding spindle microtubules. Delayed movement of the maize daughter chromosomes was seen in the single zygote anaphase found and 41 out of 50 2-celled embryos had one or more micronuclei, indicating that elimination of one or more maize chromosomes could occur as early as the zygotic mitosis.

Interspecific *Hordeum* hybrids which show elimination of one parental genome also show a correlation between elimination and a reduced centromere size, which is thought to reflect a reduced capacity for binding spindle microtubules (Finch & Bennett 1983).

All embryos with 4 or more cells contained micronuclei and no maize chromosomes were observed in metaphase or anaphase figures from embryos with 8 or more cells, indicating that all maize chromosomes were lost in the first three rounds of division. This rapid elimination of maize chromosomes should produce embryos whose cells contain a haploid complement of wheat chromosomes. Such embryos might be expected to be viable since haploid wheat plants can be recovered from 'Chinese Spring' × *H. bulbosum* crosses following loss of the *bulbosum* chromosomes (Barclay 1975). However, although some 'Chinese Spring' × 'Seneca 60' embryos grew to several thousand cells, most showed an early arrest of development and none survived transfer to embryo culture.

The failure of embryos to survive was probably due to the absence, or poor development, of the endosperm. In florets fixed 48h after pollination endosperm was present in only 13% (12/92) of the florets containing an embryo and always consisted of abnormally few nuclei compared with 'Chinese Spring' selfs made under similar conditions (Bennett *et al.* 1973). Furthermore the endosperms had cytological aberrations including the presence of micronuclei, abnormally shaped nuclei and chromatin bridges.

Wheat × maize crosses involving other genotypes

An early experiment showed that 'Chinese Spring' wheat hybridised with the southern Mexican maize race 'Zapalote Chico' (Laurie & Bennett 1986).

Subsequently we observed other genotype combinations to see whether they hybridised and whether any gave superior karyotype stability. This was of interest, since the ratio of hybrid to haploid progeny in *H. vulgare* × *H. bulbosum* crosses is influenced by parental genotype (Simpson *et al.* 1980; Pickering 1983, 1984). All six wheat genotypes used could be fertilised by maize (Table 3), despite 4 of the wheats' carrying dominant alleles at one or both of the "crossability" loci *Kr1* and *Kr2*.

Dominant alleles of *Kr1*, located on the long arm of chromosome 5B (Lange & Riley 1973; Sitch *et al.* 1985) and *Kr2*, located on the long arm of chromosome 5A (Sitch *et al.* 1985), reduce fertilisation frequency in crosses between wheat and several alien species including rye (*S. cereale*) and *H. bulbosum* (Riley & Chapman 1967; Snape *et al.* 1979; Falk & Kasha 1981, 1983; Thomas *et al.* 1981; Sitch *et al.* 1985) by inhibiting alien pollen tube growth at the base of the wheat style and in the transmitting tissue of the wheat ovary (Lange & Wojciechowska 1976; Jalani & Moss 1980).

Wheats carrying *Kr1*, which has the more powerful effect, and/or *Kr2* might therefore be expected to show a reduced frequency of fertilisation in comparison to 'Chinese Spring' when pollinated with maize, but this was not always the case. 'Chinese Spring (Hope 5B)', in which the 5B chromosome of 'Chinese Spring' has been replaced by the 5B chromosome from 'Hope', carries *Kr1* but shows no reduction in fertilisation frequency in comparison to 'Chinese Spring' when pollinated with 'Seneca 60' maize (Laurie & Bennett 1987). 'Highbury', carrying both *Kr1* and *Kr2*, shows a reduction in fertilisation frequency to about half the level seen in 'Chinese Spring' when pollinated with 'Seneca 60' maize (Laurie & Bennett 1987). Since the 'Chinese Spring (Hope 5B)' data indicate that *Kr1* does not prevent fertilisation it might be argued that this is due to the action of *Kr2*, but this is unlikely because 'Chinese Spring (Hope 5A)' did not show a similar reduction in crossability (Table 3).

All the maize genotypes used were able to fertilise wheat, fertilisation being recorded in all but 1 of the 26 cross combinations studied. The one failure was in a cross where only 11 florets were examined. The sample sizes are as yet too small to allow meaningful quantitative comparison, but the data suggest that several other maize genotypes can be as efficient as 'Seneca 60' in effecting fertilisation.

Many maize stocks differ from one another in 4C nuclear DNA content (Laurie & Bennett 1985, Rayburn *et al.* 1985), which is correlated with variation in the amount of heterochromatin (Rayburn *et al.* 1985), but in the maize stocks used here this variation had no obvious effect on karyotype stability or embryo and endosperm development. Maize chromosomes from low DNA content stocks such as 'Seneca 60' and 'Gaspe Flint', or stocks with low amounts of heterochromatin such as 'Tama Flint' and 'Knobless Wilbur's Flint', appeared to be lost as rapidly as those from the high DNA content and high heterochromatin stock 'Zapalote Chico'. Likewise, using the tetraploid maize stocks N103A, N103B

Table 3. Frequency of embryo and endosperm development in other hexaploid wheat × maize crosses.

	no. of florets examined	no. with only an embryo	no. with only an endosperm	no. with an embryo and an endosperm	Total fertilized n	%
Chinese Spring (krl, kr2) crossed with:						
1) Zapalote Chico (Oaxaca 57)	175	25	8	9	42	24.0 *1
2) Ac carrier	40	10	1	0	11	27.5
3) Tama Flint	44	6	2	0	8	18.2
4) Gaspe Flint	26	5	1	0	6	23.1
5) Knobless Wilbur's Flint	25	2	0	0	2	8.0
6) N103A tetraploid	37	2	2	1	5	13.5
7) N103B tetraploid	42	5	2	2	9	21.4
8) N104A tetraploid	33	4	3	1	8	24.2
Total	422	59	19	13	91	21.6
Chinese Spring (Hope 5A) (krl, Kr2) crossed with:						
1) Seneca 60	22	4	2	3	9	40.9
2) Zapalote Chico (Oaxaca 57)	10	2	3	0	5	50.0
3) Ac carrier	25	4	2	0	6	24.0
4) Gaspe Flint	27	3	2	2	7	25.9
5) N103A tetraploid	11	0	0	0	0	0.0
6) N103B tetraploid	14	4	2	0	6	42.9
Total	109	17	11	5	33	30.3
Chinese Spring (Hope 5B) (Krl, kr2) crossed with:						
1) Seneca 60	189	36	7	15	58	30.7 *2
2) Zapalote Chico (Oaxaca 57)	26	4	6	1	11	42.3
3) Ac carrier	27	10	2	0	12	44.4
4) Tama Flint	23	1	0	0	1	4.3
5) Gaspe Flint	20	3	1	1	5	25.0
6) Knobless Wilbur's Flint	26	4	1	0	5	19.2
7) N103A tetraploid	28	0	1	1	2	7.1
8) N103B tetraploid	30	2	1	1	4	13.3
Total	369	60	19	19	98	26.6
Chinese Spring nullisomic for 5B and tetrasomic for 5D (- , kr2) crossed with:						
Seneca 60	27	7	0	0	7	25.9
Highbury (Krl, Kr2) crossed with:						
Seneca 60	194	19	0	9	28	14.4 *2
Hope (Krl, Kr2) crossed with:						
Seneca 60	32	2	3	2	7	21.9
Overall total	1153	164	52	48	264	22.9

*1 Additional data plus data from Laurie & Bennett (1986); *2 Data from Laurie & Bennett (1987).

or N104A as pollen parents did not increase karyotype stability (Fig. 4) or embryo viability.

One haploid wheat plant, from a 'Chinese Spring' × 'Zapalote Chico' cross, was produced from these experiments by culturing an embryo excised from an immature seed containing no endosperm. Although there can be no proof that this embryo arose from a hybrid zygote this result was encouraging, since it showed that a haploid embryo could be capable of developing to a culturable stage in the absence of an endosperm.

Chromosome elimination in 'Chinese Spring' wheat × 'S9B' sorghum crosses

This cross showed fertilisation at more than double the frequency seen in 'Chinese Spring' wheat × 'Seneca 60' maize crosses (69% and 29.2% respectively). Most florets in which fertilisation occurred had an embryo but no endosperm (Table 2).

The 'S9B' chromosomes showed no evidence of attachment to the spindle in zygote metaphases, nor in the single zygote anaphase observed where all 20 daughter chromosomes of the sorghum complement remained near the plane of the metaphase plate. Furthermore, all embryos with 2 or more cells had 1 or more micronuclei, and no sorghum chromosomes were seen in metaphase or anaphase figures except those from zygotes. These observations suggest that elimination of sorghum chromosomes was always rapid and could be complete at the zygotic mitosis.

The endosperms in 'Chinese Spring' × 'S9B' crosses were like those from crosses with 'Seneca 60' maize in having abnormally few nuclei and in showing micronuclei and other aberrations, again suggesting rapid elimination of sorghum chromosomes.

Chromosome elimination in 'Sultan' barley × 'Seneca 60' maize crosses

The fertilisation frequency in 'Sultan' × 'Seneca 60' crosses was similar to that in 'Chinese Spring' wheat × 'Seneca 60' crosses (28% and 29.2% respectively), but the frequency with which the polar nuclei were fertilised was higher (24% and 5.8% respectively, Table 2). Thus, in 'Sultan' × 'Seneca 60' crosses, most embryos (16/20) were accompanied by an endosperm.

Another striking difference was that in all 8 'Sultan' × 'Seneca 60' zygotes seen the maize chromosomes had well-defined centromeres (Fig. 3). Preliminary observations indicate that the maize chromosomes are eventually eliminated, but that at least some are retained for a minimum of four cell cycles.

Chromosome elimination was also observed in endosperms which, as in the crosses discussed above, contained an abnormally low number of nuclei in comparison to barley selfs produced similarly (Bennett *et al.* 1975).

The potential uses of wheat × maize, wheat × sorghum and barley × maize crosses

These crosses provide useful new systems for studying chromosome elimination, but it is of more interest to recover plants from these crosses. Allowing caryopses to develop on plants produced only one wheat haploid from over 1500 florets pollinated with maize, and no plants were recovered from smaller experiments on wheat × sorghum and barley × maize crosses. This low frequency is unsatisfactory and we must either find genotypes which allow embryos to develop to a point where they can be rescued by conventional embryo culture or, preferably, devise culture techniques that enable embryos to develop without an endosperm. Spikelet culture, in which spikelets are transferred to Murashige and Skoog medium two days after pollination and grown for three weeks (Mathias & Boyd pers. comm) may provide such a system, since we have recently used this method to recover 28 haploid wheat plants from 720 ovaries pollinated with maize.

It may therefore be possible to use maize or sorghum as alternatives to *Hordeum bulbosum* for the production of wheat or barley haploids via chromosome elimination. If so, the relative insensitivity of maize to the effect of the wheat crossability loci *Kr1* and *Kr2* would make wheat × maize crosses particularly useful in this context (Laurie & Bennett 1987).

The ability to recover plants would also offer the prospect of gene transfer by, for example, using ionising radiation to induce intergenomic translocations prior to elimination. Particularly exciting is the prospect of transferring active maize transposable elements to wheat or barley which, if successful, might enable genes to be isolated by the technique of 'transposon tagging', a method already successfully exploited in other organisms including maize itself (Fedoroff *et al.* 1984). Data in Table 3 show that wheat was readily hybridised with a maize stock carrying the transposable element Activator (*Ac*).

Gene transfer would be easier if karyotypically stable hybrids could be produced, and the observation that maize chromosomes show well-defined centromeres in barley × maize zygotes and are not so rapidly eliminated as maize or sorghum chromosomes are from wheat indicates that variation in karyotype stability exists between these wide hybrids. In the search for karyotypically stable hybrids it will therefore be of interest to extend these studies to crosses using diploid and tetraploid wheats and their relatives as female parents.

The present results show that fertilisation can occur in wide-crosses which span two subfamilies of the Gramineae and that the frequency of fertilisation is high enough for these crosses to be a worthwhile area for further study. The results also imply that the fertilisation process is sufficiently conserved to allow successful hybridisation between many other Gramineae species previously considered too distantly related to be combined by sexual means.

Acknowledgements

This work was funded by the United Kingdom Overseas Development Administration, project R3797. We thank Miss L. Tiller for technical assistance.

References

Barclay, I.R., 1975. High frequencies of haploid production in wheat (*Triticum aestivum*) by chromosome elimination. *Nature* 256, 410–411.

Bennett, M.D., M.K. Rao, J.B. Smith, and M.W. Bayliss 1973. Cell development in the anther, the ovule, and the young seed of *Triticum aestivum* L. var. Chinese Spring. *Phil. Trans. R. Soc. Lond. B* 266, 39–81.

Bennett, M.D., J.B. Smith and I. Barclay 1975. Early seed development in the Triticeae. *Phil. Trans. R. Soc. Lond. B* 272, 199–227.

Bennett, M.D. and J.B. Smith 1976. Nuclear DNA amounts in angiosperms. *Phil. Trans. R. Soc. Lond. B* 274, 227–274.

Falk, D.E. and K.J. Kasha 1981. Comparison of the crossability of rye (*Secale cereale*) and *Hordeum bulbosum* onto wheat (*Triticum aestivum*). *Can. J. Genet. Cytol.* 23, 81–88.

Falk, D.E. and K.J. Kasha 1983. Genetic studies of the crossability of hexaploid wheat with rye and *Hordeum bulbosum*. *Theor. Appl. Genet.* 64, 303–307.

Fedak, G. 1985. Alien species as sources of physiological traits for wheat improvement. *Euphytica* 34, 673–680.

Fedoroff, N.V., D.B. Furtek and O.E. Nelson 1984. Cloning of the bronze locus in maize by a simple and generalisable procedure using the transposable controlling element Activator (*Ac*). *Proc. Natl. Acad. Sci. U.S.A.* 81, 3825–3829.

Finch, R.A. and M.D. Bennett 1983. The mechanism of somatic chromosome elimination in *Hordeum*. pp. 147–154. In *Kew Chromosome Conference II*, P.E. Brandham and M.D. Bennett, eds, Allen and Unwin, London.

Furuta, Y., K. Nishikawa, T. Makino and Y. Sawai 1984. Variation in DNA content of 21 individual chromosomes among six subspecies in common wheat. *Jap. J. Genet.* 59, 83–90.

Goodman, R.M., H. Hauptli, A. Crossway and V.C. Knauf 1987. Gene transfer in crop improvement. *Science* 236, 48–54.

Heslop-Harrison, J.S. and M.D. Bennett 1983. Prediction and analysis of spatial order in haploid chromosome complements. *Proc. R. Soc. Lond. B* 218, 211–223.

Hutchinson, J. 1959. *The families of flowering plants. Vol. II. Monocotyledons.* Clarendon Press, Oxford.

Jalani, B.S. and J.P. Moss 1980. The site of action of the crossability genes (*Kr1, Kr2*) between *Triticum* and *Secale*. I. Pollen germination, pollen tube growth, and number of pollen tubes. *Euphytica* 29, 571–579.

Kasha, K.J. and K.N. Rao 1970. High frequency haploid production in barley (*Hordeum vulgare* L.). *Nature* 225, 874–876.

Lange, W. and R. Riley 1973. The position on chromosome 5B of wheat of the locus determining crossability with rye. *Genet. Res., Camb.* 22, 143–153.

Lange, W. and B. Wojciechowska 1976. The crossing of common wheat (*Triticum aestivum* L.) with cultivated rye (*Secale cereale* L.). I. Crossability, pollen grain germination and pollen tube growth. *Euphytica* 25, 609–620.

Laurie, D.A., and M.D. Bennett 1985. Nuclear DNA content in the genera *Zea* and *Sorghum*. Intergeneric, interspecific and intraspecific variation. *Heredity* 55, 307–313.

Laurie, D.A., and M.D. Bennett 1986. Wheat × maize hybridisation. *Can. J. Genet. Cytol.* 28, 313–316.

Laurie, D.A. and M.D. Bennett 1987. The effect of the crossability loci *Kr1* and *Kr2* on fertilisation frequency in hexaploid wheat × maize crosses. *Theor. Appl. Genet.* 73, 403–409.

Mujeeb-Kazi, A. and G. Kimber 1985. The production, cytology and practicality of wide hybrids in the Triticeae. *Cereal Res. Commun.* 13, 111–124.

Pickering, R.A., 1983. The influence of genotype on doubled haploid barley production. *Euphytia* 32, 863–876

Pickering, R.A., 1984. The influence of genotype and environment on chromosome elimination in crosses between *Hordeum vulgare* L. × *Hordeum bulbosum* L. *Plant Sci. Lett.* 34, 153–164.

Rayburn, A.L., H.J. Price, J.D. Smith and J.R. Gold 1985. C-band heterochromatin and DNA content in *Zea mays. Amer. J. Bot.* 72, 1610–1617.

Riley, R. and V. Chapman 1967. The inheritance in wheat of crossability with rye. *Genet. Res., Camb.* 9, 259–267.

Riley, R. and C.N. Law 1984. Chromosome manipulation in plant breeding: progress and prospects. In *Gene manipulation in plant improvement*, J.P. Gustafson, ed. 16th Stadler Genetics Symp. pp. 301–322.

Sharma, H.C. and B.S. Gill 1983. Current status of wide hybridisation in wheat. *Euphytica* 32, 17–31.

Simpson, E., J.W. Snape and R.A. Finch 1980. Variation between *Hordeum bulbosum* genotypes in their ability to produce haploids of barley, *Hordeum vulgare. Z. Planzenzuchtg.* 85, 205–211.

Sitch, L.A., J.W. Snape and S.J. Firman 1985. Intrachromosomal mapping of crossability genes in wheat (*Triticum aestivum*). *Theor. Appl. Genet.* 70, 309–314.

Snape, J.W., V. Chapman, J. Moss, C.E. Blanchard and T.E. Miller 1979. The crossabilities of wheat varieties with *Hordeum bulbosum. Heredity* 42, 291–298.

Thomas, J.B., P.J. Kaltsikes and R.G. Anderson 1981. Relation between wheat-rye crossability and seed set of common wheat after pollination with other species in the Hordeae. *Euphytica* 30, 121–127.

Zeller, F.J. and S.L.K. Hsam 1983. Broadening the genetic variability of cultivated wheat by utilising rye chromatin. *Proc. 6th International wheat Genetics Symposium, Kyoto, Japan.* 161–173.

Zenkteler, M. and W. Nitzsche 1984. Wide hybridisation experiments in cereals. *Theor. Appl. Genet.* 68, 311–315.

Consistent chromosome changes in radiation-induced murine leukaemias

G. Breckon, A.R.J. Silver and R. Cox

MRC Radiobiology Unit, Chilton, Didcot, Oxon. OX11 ORD, UK

Ionising radiation is known to induce gross structural changes in mammalian chromosomes and recent advances in molecular genetics have linked specific chromosomal changes with the activation of proto-oncogenes (Taub *et al.* 1982; Shtivelman *et al.* 1985). The aim of our cytogenetic studies on radiation-induced murine leukaemias is to establish correlations between specific chromosome events and structural changes affecting the activities of cellular proto-oncogenes or other genes which may be involved in oncogenesis. Hayata (1983) has described a strong correlation between rearrangements of murine chromosome (Ch)2 and radiation-induced acute myeloid leukaemia in RFM mice. We extend these observations on Ch2 changes to leukaemias arising in irradiated CBA/H mice and discuss data which argue against the involvement of the Ch2 encoded proto-oncogene *c-abl* in these chromosome rearrangements.

Material and method

Induction and passage of leukaemias

CBA/H murine leukaemias were induced by either the bone-seeking radionuclide ^{224}Ra or by X-rays (Major & Mole 1978; Humphreys 1985). In unirradiated mice the spontaneous frequency of myeloid leukaemias was extremely low ($<10^{-3}$) while that for lymphoid leukaemias was estimated to be ca. 1%. Leukaemias were classified as Acute Myeloid Leukaemia (AML) or Acute Lymphoid Leukaemia (ALL) using the following diagnostic criteria.

AML — replacement of sternal bone marrow with abnormal myeloid cells of varying maturity, enlarged spleen of largely myeloid cells and terminal white blood count 10^4–10^5 cells/ml.

ALL — high percentage of malignant lymphoid cells, infiltration of spleen by lymphoid type cells, considerable involvement of peripheral lymph nodes and terminal white blood count 50–500 cells/ml.

A positive confirmation of leukaemia was obtained by passage of spleen cells into syngeneic mice giving rise to leukaemia of the same karyotype. Leukaemias were maintained by *in vivo* passage of 1.3 × 10^3 cells (by intra-peritoneal injection) in 12-week male CBA/H mice. Passage intervals were usually 7 ± 1 days. Karyotypic data were obtained from both primary and early passage material where appropriate.

Chromosome preparation and analysis
Direct *in vivo* metaphases were prepared from haemopoietic tissues and short-term cultures. G bands were obtained using combined ASC and trypsin followed by Giemsa staining. 50–100 G banded metaphases/tissue/sample were analysed for clonal populations. The banding nomenclature of Nesbitt and Franke (1973) was used, but a revised ideogram for the CBA/H was also developed (Breckon, in preparation). 10–25 metaphases were photographed and karyotyped. The criteria of Rowley & Potter (1978) were used for the identification of abnormal clones. Briefly, the observation of at least two 'pseudodiploid' (e.g. translocations, deletions, inversions) or hyperdiploid cells, or three hypodiploid or monosomic cells each sharing the same abnormality, was considered evidence for the presence of an abnormal clone.

Results

Induced leukaemias
In each of the 16 induced AMLs, karyotypic analysis of both direct and short-term culture cells revealed the presence of a dominant aneuploid cell clone. A rearrangement of Ch2 was found in 15/16 cases, three representative examples being shown in Fig. 1. Deletions (8/15) were the most frequent change along with translocations (7/15). Trisomy of Ch6 (4/16), trisomy of Ch15 (4/16) and numerical changes in Y were also observed (Fig. 2). The G-banding system showed that most changes to Ch2 (6/8 deletions; 3/7 translocations) were centred on a segment between the bands C_2 and E_5. The other 4 translocations involve the G to H_4 segment and the 2 deletions involved the C_2 to H_4 segment (Fig. 3). Of the 15 Ch2 events recorded, the 8 deletions were predominantly interstitial and the 7 translocations were often incomplete. Exchange between the two homologues (t2;2) was the most common translocation (3/7). Others observed were t2;3, t2;15, t;y, and an unknown chromosome marker.

Karyotypes of cells from ALL (Fig. 2) showed only 6 events involving Ch2 (two trisomies, 3 translocations and one deletion). The greater number of non-random chromosome rearrangements in AMLs (15/16) distinguished them from ALLs (4/12). Karyotypes of AMLs also showed a high frequency of both trisomies and monosomies when compared to ALLs (Fig. 2).

Figure 1. 'G' banded karyotypes; N CBA/H Male, and three representative primary clones of acute myeloid leukaemia. Deletion of Ch2, N452, and translocations N14 (t2; to mar), N36 (t2; 2) are in double boxes. Trisomies CH1,6,15 are in single boxes, as are numerical changes in Ch Y.

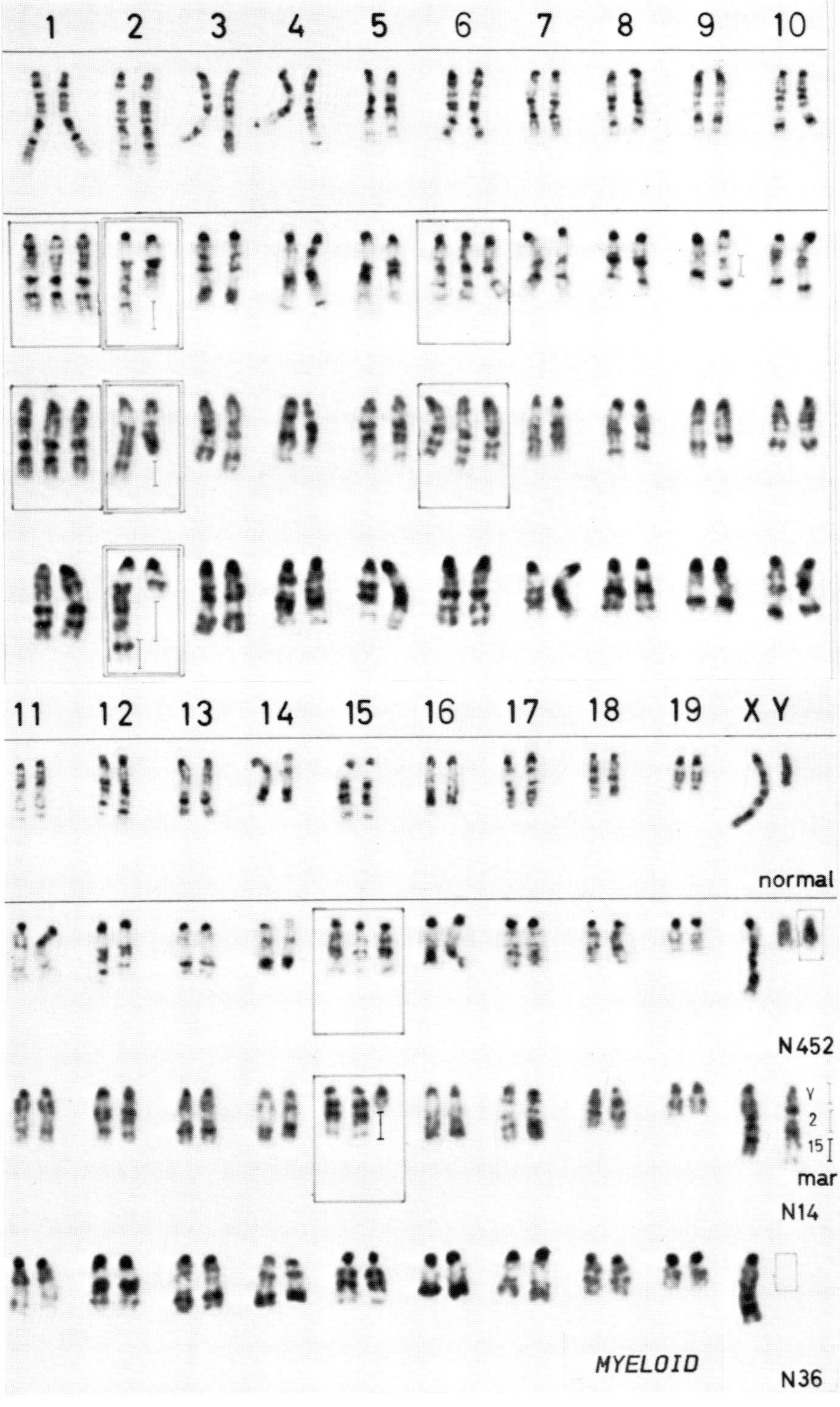

1 2 3 4 5 6 7 8 9 10

11 12 13 14 15 16 17 18 19 X Y

normal

N452

Y
2
15
mar

N14

MYELOID

N36

Discussion

Cytogenetic studies with human leukaemias continue to provide crucial information on the involvement of gross genetic changes in leukaemogenesis and important clues as to the genes involved in oncogenic transformation of haemopoietic cells (Sherr 1987). Animal models of haemopoietic neoplasias have a complementary role to play in directly approaching the inductive and developmental processes that underlie leukaemogenesis.

CHROMOSOME NUMBER		GAIN	LOSS
C	1	TsTsTs	T T Rb.
H	2		DDDDDDDD, TTTTTTT, Ms Ms
R	3	Is	
O	4		Ms Ms T
M	5	TsTsTsTs	
O	6	TsTsTsTsTs	Rb
S	7		Ms
O	8		
M	9	Is	Ms T
E	10		Ms Ms
	11		D
N	12		Ms
U	13		Ms T
M	14		D
B	15	TsTsTsTsTs	T
E	16		
R	17		Rb
	18		
	19		
	X		
	Y	DpDpDpDp	T DDD
	UI	† † † † † †	

Figure 2. Representative changes (gain/loss etc.) observed in each chromosome. In the AML all except one were aneuploids; two had a hyperdiploid clone around the 40% level. Of the ALL, 7 were aneuploid and 5 were 2n = 40 with an apparently normal karyotype, D = deletion, Dp = duplication, In = inversion, Is = insertion, MS = monosomy, Rb = Robertsonian translocation, Tr = translocation, Ts = trisomy, UI = unidentified.

A murine model of AML has been used by Hayata (1983) and in this laboratory to show that abnormalities centred on an interstitial region of Ch2 are clearly involved in the genesis of myeloid leukaemia. Our observation of such events in a proportion of murine lymphoid leukaemias further suggests that these abnormalities are not restricted to cells of the myeloid lineage and therefore may have their origin in relatively uncommitted haemopoietic stem or early progenitor cells.

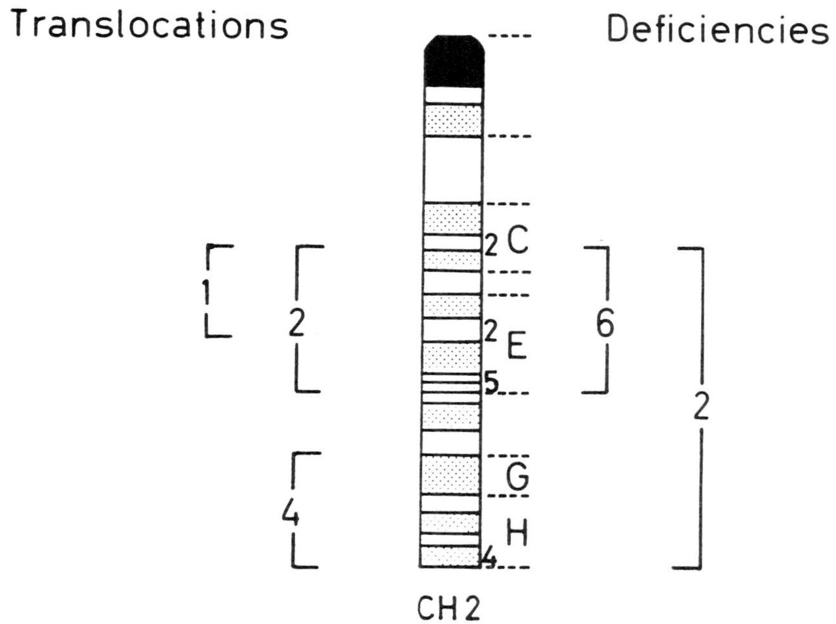

Figure 3. Ideogram of Ch2 showing the segments involved and the frequencies of translocations and deficiencies (deletions) observed in acute myeloid leukaemia.

A major feature of the Ch2 translocations that we have observed in murine leukaemias is their polarity, i.e. there is invariably loss of DNA sequences from one Ch2 copy. In addition to this, the majority of the translocations were incomplete and appeared to involve loss of Ch2 sequence from the genome. This observation, together with the relative lack of specificity of the translocation events, indicates that the Ch2 translocations in murine AML do not parallel the 9: 22 translocation (Ph chromosome) of human chronic myeloid leukaemias (CML) which results in specific activation of the Ch9 encoded human *c-abl* proto-oncogene by a gene fusion mechanism (Shtivelman *et al.* 1985). Furthermore, molecular studies on murine *c-abl* DNA mRNA and protein in AMLs show no evidence of *abl* involvement (Silver *et al.* 1987). Therefore, although murine *c-abl* is thought to map in the region of the C_2 breakpoint in Ch2 it does not appear to be activationally changed by the rearrangements that we have characterised. It is necessary, therefore, to consider an alternative mechanism possibly involving chromosome deletion. Such chromosome deletions characterise a number of human neoplasms including a myeloid disorder (Jacobs 1987; Heubner *et al.* 1985; Le Beau 1986) and current studies in our laboratory are focussed on this problem. The possible association between chromosome deletion and neoplasia is at present the subject of considerable speculation and the murine AML model may well provide a valuable experimental system for investigating the molecular basis and phenotypic consequence of carincogen-induced chromosome DNA sequence loss at specific chromosomal sites.

Acknowledgements

We thank Drs. E. Humphreys and R.H. Mole for supplying CBA/H meyloid leukaemias and D. Malowany for passage and preparation of leukaemic cells. We also thank the animal staff of this laboratory for their help and co-operation and our colleagues for useful discussion during the project. This work was partly funded by C.E.C. contract B16-064 and by H.S.E. support (No. 286) to A.R.J.S.

References

Hayata, I. 1983. Partial deletion of chromosome 2 in radiation induced myeloid leukaemia in mice. In *Radiation-induced chromosome damage in man*, p. 227, ed. Ishihara, T. and Sasaki, M.S., Liss: New York.

Humphreys, E.R., J.R. Loutit, I.R. Major, and V.A. Stones 1985. The induction by ^{224}Ra of myeloid leukaemia and osteosarcoma in male CBA mice. *Int. J. Radiat. Biol.* 47, 239–247.

Heubner, K., M. Isobe, C.M. Croce, D.W. Golde, S.E. Kaufman and J.C. Gasson 1985. The human gene encoding GM-CSF is at 5q21–q32, the chromosome region deleted in the 5q$^-$ anomaly. *Science* 230, 1282–1285.

Jacobs, A. 1987. Human preleukaemia: Do we have a model? *Br. J. Cancer* 53, 1–5.

Le Beau, M.H., C.A. Westbrook, M.O. Diaz, R.A. Larson, J.D. Rowley, J.C. Gasson, D.W. Golde, and C.J. Sherr 1986. Evidence for the involvement of GM-CSF and FMS in the deletion (5q) in myeloid disorders. *Science* 231, 984–987.

Major, I.R. and R.H. Mole 1978. Myeloid leukaemia in X-ray irradiated CBA mice. *Nature* 272, 455–456.

Nesbitt, M.N. and V. Franke 1973. A system of nomenclature of band patterns of mouse chromosomes. *Chromosoma* (Berl.) 41, 145–158.

Rowley, J.D. and D. Potter 1976. Chromosomal banding patterns in acute leukaemia, *Blood*, 705–722.

Sherr, C.J. 1987. Leukaemia and lymphoma. *Cell* 48, 727–729.

Shtivelman, E., B. Lifshitz, R.P. Gale, and E. Canaani 1985. Fused transcript of abl and bcr genes in chronic myelogenous leukaemia. *Nature* 315, 550–a554.

Silver, A.R.J., G. Breckon, W. Masson, D. Malonwany and R. Cox 1987. Studies on radiation leukaemogensis in the mouse. In *Radiation Research: Proceedings of the 8th ICRR.* (E.M. Fielden *et al*) Vol. 2, London: Taylor and Francis (in the press).

Taub, R., I. Kirsch, C. Morton, G. Lenoir, D. Suran, S. Tranick, S. Aaronson and P. Leder 1982. Translocation of the c-myc gene into the immunoglobulin heavy chain locus in human Burkitt lymphoma and murine plasmocytoma cells. *Proc. Natl. Acad. Sci. (USA)* 79, 7839–7841.

Origins and causes of chromosome instability in plant tissue culture and regeneration

A. Karp

Biochemistry Department, Rothamsted Experimental Station, Harpenden, Herts. AL5 2JQ, UK

Techniques are now available for the regeneration of whole plants from cultured cell suspensions, segments, organs, explants and, in an increasing number of species, from isolated protoplasts. These regeneration systems are essentially extensions of vegetative propagation and as such would be expected to give rise to clonal uniformity. However, it has rapidly become apparent that regeneration through a callus phase (a stage of disorganised cell growth) can be associated with remarkable instability. This phenomenon is widespread among plant species and has been given the name of somaclonal variation (Larkin & Scowcroft 1981). In this paper one aspect of the variability will be addressed, namely chromosome instability. The nature and frequency of the chromosome changes will be illustrated by our studies in potato and cereals, factors affecting the variation will be identified, and finally, three possible origins of the variation will be discussed.

Potato

In potato (*Solanum tuberosum* L.) regeneration can be achieved routinely from isolated protoplasts in a wide range of cultivars (Shepard *et al.* 1980; Thomas *et al.* 1982; Sree Ramulu *et al.* 1983; Creissen & Karp 1985). Potato plants regenerated in this way show considerable phenotypic variability and cytological studies have shown that some of this variation relates to changes in the chromosomes.

On average, 50–70% of protoplast-derived potato plants retain their tetraploid complement of 48 chromosomes (Fig. 1a), but between 10–20% are aneuploid at the tetraploid level (Fig. 1b) and 20–30% are octoploid or aneuploid at the octoploid level (Fig. 1c) (Karp 1986). Structural rearrangements, including deletions and translocations have also been observed (Creissen & Karp 1985; Fish & Karp 1987).

At least five factors are known to affect the degree and nature of the chromosome variation in potato; ploidy, genotype, regeneration procedure, tissue culture source and media components. Different cultivars behave differently in tissue culture, in terms of morphogenetic response and stability, and in potato there are numerous examples illustrating both of these aspects (e.g. Tempelaar *et al.* 1985; Wheeler *et al.* 1985).

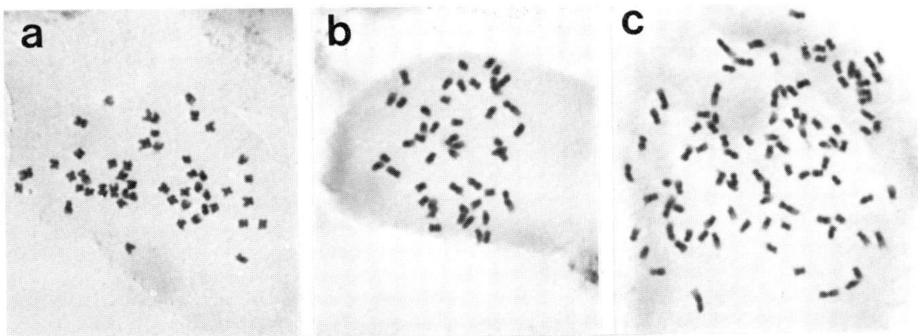

Figure 1. Chromosome variation in protoplast-derived potato plants: **a** normal regenerant (2n=4x=48), **b** aneuploid at the 4x level (2n=47), **c** octoploid (2n=8x=96).

That ploidy is another factor affecting the frequency and nature of the chromosome variation can be illustrated by a study in *Solanum brevidens*. Plants were regenerated from this diploid wild relative of potato using the same procedure as was used for the tetraploid. Instead of 20–30% of regenerated plants doubling in chromosome number, over 75% of the regenerants were found to be tetraploid or aneuploid at the tetraploid level. The remaining 25% were stable diploids (Nelson *et al.* 1986).

Regeneration procedure also affects chromosome stability. Leaving hormones as a separate topic, this includes the choice of regeneration system (i.e. explants vs. protoplasts) and the length of time in culture. Broadly speaking, the longer the time in culture the greater the chance of instability (McCoy *et al.* 1982), but long periods of time (over a year in culture) may be required before such an effect is manifested (Karp & Maddock 1984).

Regeneration from protoplasts requires extensive application of hormones and is a lengthy procedure. It may take 4–6 months from protoplast to plant. In contrast, regeneration from cultured explants is technically simpler and is a much more rapid process. A survey of plants regenerated from leaf explants of the cultivars Champion, Myatts, Ashleaf and Desirée indicated that explant regeneration is associated with greater stability. Over 90% of the regenerants tested were tetraploid (Wheeler *et al.* 1985). However, ploidy is also of importance here, and plants regenerated from monohaploid (2n=x=12) leaf pieces were all found to have doubled to the doubled monohaploid level (2n=2x=24). Interestingly, if leaves are taken from the doubled monohaploid and passed through a second cycle of leaf regeneration, about 50–60% of the erated plants are found to have doubled again to the homozygous tetraploid level (Karp *et al.* 1984). Regeneration from dihaploid (2n=2x=24) leaf pieces also gives 50–60% chromosome doubling, although the exact frequency for both monohaploids and dihaploids depends upon the genotype of the line (Jacobsen 1978; Karp *et al.* 1984).

Tissue culture source is another factor which influences the nature of the chromosome variation. Although only a small sample was analysed, regeneration from tuber pieces gave rise to more aneuploidy than regeneration from leaf stem or rachis explants (Wheeler *et al.* 1985). Similarly, regeneration from tuber protoplasts is associated with a frequency of polyploidy higher than regeneration from mesophyll protoplasts (M.G.K. Jones & A. Karp, unpublished results).

The final factor which affects chromosome stability is possibly the most complex and contradictory. Our interest in hormone effects relates to attempts to improve the efficiency of regenerating tetraploid potato plants from protoplasts, rather than studies of hormonal influences *per se*. The effects of changing the hormone composition at different stages in the regeneration procedure have been studied in the cultivar Maris Bard. When the initial culture medium was varied, no significant differences in chromosome stability were present between plants regenerated from the different treatments (Fish & Karp 1987). However, when the hormone content of the regeneration medium (from callus growth to shoot initiation) was varied, there were highly significant differences in chromosome stability between plants regenerated from the four treatments (Karp & Fish 1987). These data suggest that the callus phase is the sensitive phase, at which hormone manipulations affect the stability of the regenerated plants. The callus is a heterogeneous mixture from which only a few cells give rise to shoots. There is always less chromosome variation in the regenerated plants compared with the callus, suggesting that some form of selection is operating at this stage (Orton 1980; Ogihara 1981). It now remains to understand how such a selection operates and how it could be biased in favour of stable euploid plants.

Cereals

Regeneration of whole plants from protoplasts has been achieved in rice (Fujimara *et al.* 1985) and other tropical cereals, but to date it has not been possible to regenerate from protoplasts of wheat and barley. In these cereals regeneration from cultured immature embryos is the favoured explant system.

Chromosome variation has been observed amongst regenerants from cultured immature embryos of numerous cereals, including wheat (Karp & Maddock 1984), oats (McCoy *et al.* 1982), maize (McCoy & Phillips 1982) and triticale (Armstrong *et al.* 1984). The same five factors described earlier in potato affect the nature and frequency of chromosome variation in the cereals. However, the cereals offer different opportunities for studying the causes and origins of the instability, and it is the different approach adopted in the cereals that will be emphasised, rather than complete treatment of all five factors.

The cereals are not as amenable to tissue culture as potato, but more knowledge in cytogenetics and molecular biology is available, and cereals generally have the added advantage of larger, more easily-handled chromosomes. In the cereals emphasis has therefore been placed on identifying components of the genotype that are factors important in determining chromosome stability.

Ploidy appears to be important in determining the frequency and nature of the chromosome variation observed in cereal regenerants. Regeneration from cultured immature embryos of four cultivars of hexaploid bread wheat (*Triticum aestivum*) was associated with 30% aneuploidy and a high frequency of structural changes (Fig. 2a, b) (Karp & Maddock 1984), but when plants were regenerated from cultured immature embryos of barley (*Hordeum vulgare*, 2n=2x=14), no aneuploidy or polyploidy was observed and only one incidence of chromosome breakage was seen (Karp *et al.* 1987a). Other diploids, in which chromosome stability has been described, include *H. spontaneum* (Breiman *et al.* 1987) (Fig. 2c) and pearl millet (*Pennisetum americanum*) (Swedlund & Vasil 1985).

In contrast, chromosome variation has been reported in regenerated plants of maize (Lee & Phillips 1986) and rye (Linacero & Vazquez 1986). One possible explanation for this is that these diploid cereals contain appreciable amounts of heterochromatin. Evidence relating chromosome breakage with hetero-chromatin has been described by Lee & Phillips (1986) and in wheat-rye hybrids the positions of break-points in structural rearrangements were found to be nearly all associated with heterochromatin (Lapitan *et al.* 1984).

In collaboration with the Athens School of Agricultural Sciences, an experiment has recently been initiated in which plants have been regenerated from cultured immature embryos of two pairs of sister lines of triticale (× *Triticosecale*) and of five 'sister' lines of rye (*Secale cereale*). The triticale lines differ in the heterochromatic content of a particular rye chromosome (6R or 7R) whilst the rye lines (kindly supplied by Dr A. Lukazewski of Missouri-Columbia) differ in only one heterochromatic band. Clear differences in morphogenetic response have been observed between the lines (Bebeli *et al.* 1987). Chromosomes are currently being analysed in the regenerated plants to establish whether instability is related to the presence of the telomeric heterochromatin bands and to morphogenetic response.

Another example where chromosome constitution appears to be of importance was revealed during attempts to achieve regeneration from protoplasts of wheat. Protoplasts isolated directly from wheat tissue appear to have little or no capacity for division. In an attempt to circumvent this block, cell suspensions were established from immature embryos and protoplasts isolated from the cell suspensions. One of these cell suspensions (CB2d) gave protoplasts which divided at low (<1%) frequency (Maddock 1987). A selection-recycling scheme was devised to improve the division frequency in which protoplasts were isolated and cultured and the fastest-growing colonies selected to make new cell suspensions. This recycling was repeated three times and the division frequency increased 10-fold (Wu *et al.* 1987). Chromosomes were examined in the cell suspensions and dividing protoplast lines.

Considerable chromosome loss was found to have occurred in all the lines. Out of a total of 350 cells (50 per line) none was found to contain the normal wheat complement (2n=6x=42). Moreover, numerous structural changes were present. The interesting feature was that in all the lines the most frequently-observed chromosome numbers were around 2n=31 (Fig. 2d). When protoplasts were isolated and recycled from C82d there was some chromosome doubling

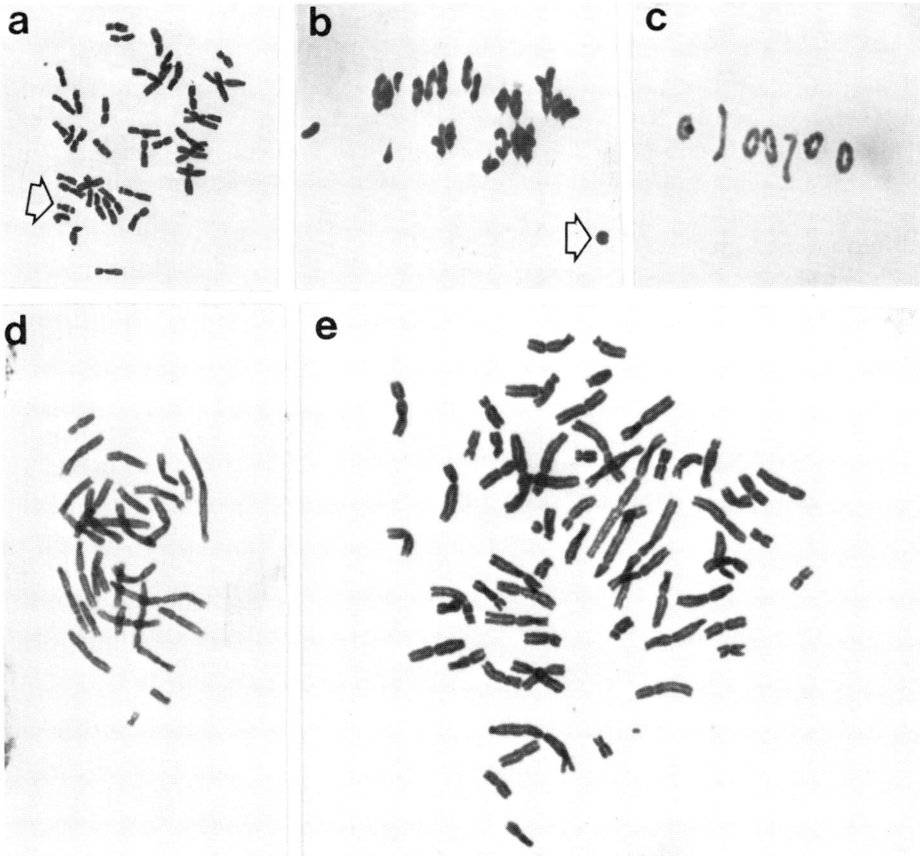

Figure 2. Chromosome variation in cereal cell lines and plants regenerated from cultured immature embryos **a** euploid (2n=6x=42) heterozygous for a deletion (arrow), **b** MI in an aneuploid (2n=43) with an iso-chromosome (arrow), **c** normal MI in *H. spontaneum* regenerant, **d** abnormal wheat karyotype (2n=31) in C82d protoplast line, **e** a protoplast-derived cell of C82d in which chromosome doubling and karyotype restructuring has occurred.

and an increase in structural changes (Fig. 2c). With each successive recycling the frequency of doubled cells and structural changes increased but the most frequently-observed chromosome number remained at 2n=31 (Karp *et al.* 1987b). These data suggest that, in selecting for increased capacity for division, there was loss of whole chromosomes, chromosome arms and chromosome segments from the normal wheat complement, and that this stabilised at around 2n=31. It would seem unlikely that plant regeneration would ever be achieved from such a line. We are presently analysing the lines in greater detail using banding techniques to establish what it is that has been lost from the wheat complement.

The origins of chromosome instability in plant tissue culture and regeneration

The origin of the chromosome instability in plant tissue culture and regeneration is largely unknown. Nevertheless, the identification of factors affecting the variation has led to the recognition of three possible origins.

Although somatic growth is the result of mitotic divisions, variations could accumulate in plant tissues, either as a result of somatic mutation or as part of the differentiation process. The data described earlier, indicating that some tissues give rise to more variation than others, clearly argue that some of the variation is already present in the plant cells, but how much originates in this way is simply not known.

Another possible origin is that the hormones in the medium act (directly or indirectly) as mutagens. The evidence for and against this has been dealt with in another review (Karp & Bright 1985), and earlier on, experiments designed to investigate hormone effects have been described. Overall, it would seem that hormones influence chromosome instability, but are not necessarily the origin of the variation.

Throughout this paper the importance of the callus phase in relation to variability has been stressed. Plant tissue culture systems, such as micro-propagation and meristem tip culture, which do not go through a callus phase, are largely free of somaclonal variation. The third possible origin of chromosome instability is therefore that it results from the 'stress' conditions of the callus phase. This could be viewed in two ways. It could be that the genome is not as stable as might have been supposed originally, but that normally most changes are selected against. During the callus phase such constraints are removed and the variation is allowed to accumulate. Conversely, it could be that the genome is normally stable, but the stress induced by the callus phase results in the activation of otherwise latent mechanisms of change, such as chromosome breakage, sequence amplification and transposition.

The third possible origin of chromosome instability is in accordance with observations that different genotypes behave differently in culture. The occurrence of somaclonal variation therefore provides an opportunity for looking at mechanisms of change in chromosomes of different plant species. However, the other two origins should not be disregarded, and in this respect it is hoped that studies of chromosome instability in culture will lead to a better understanding of the changes that occur during differentiation and the factors affecting the fidelity of mitotic divisions.

Acknowledgements

The author thanks all those members of the Biochemistry Department who regenerated the material described in this article, in particular Dr M.G.K. Jones, Dr Q.S. Wu, Dr G.P. Creissen, Dr R.S. Nelson, Dr S.E. Maddock, Mr N. Fish, Miss P. Bebeli and Miss S.H. Steele.

References

Armstrong, K.C., C. Nakamura and K. Keller 1984. Karyotype instability in tissue culture regenerants of Triticale (× *Triticosecale* Wittmack) cv. Welsh from 6-month-old callus cultures. *Z. Pflanzenzucht.* 91, 233–245.

Bebeli, P., A. Karp and P.J. Kaltsikes 1987. Plant regeneration and somaclonal variation from cultured immature embryos of isogenic lines of rye and triticale differing in their content of heterochromatin. 1. Morphogenetic Response. *Theor. Appl. Genet.* (Submitted).

Breiman, A., D. Rotem, A. Karp and H. Shaskin 1987. Heritable somaclonal variation in wild barley (*Hordeum spontaneum*). *Theor. Appl. Genet.* 74, 104–112.

Creissen, G.P. and A. Karp 1985. Karyotypic changes in potato plants regenerated from protoplasts. *Plant Cell Tissue Organ Culture* 4, 171–182.

Fish, N. and A. Karp 1986. Improvements in regeneration from protoplasts of potato and studies on chromosome stability. 1. The effect of initial culture media. *Theor. Appl. Genet.* 72, 405–412.

Fujimara, T., M. Sakurai, H. Akagi, T. Negishi and A. Hirose 1985. Regeneration of rice plants from protoplasts. *Plant Tissue Culture Letts.* 2, 74–75.

Jacobsen, E. 1982. Polyploidization in leaf callus tissue and in regenerated plants of dihaploid potato. *Plant Cell Tissue Organ Culture* 1, 77–84.

Karp, A. 1986. Chromosome variation in regenerated plants. In *Genetic manipulation in plant breeding*. W. Horn, C.J. Jensen, W. Odenbach and O. Schieder, eds, 547–554. Walter de Gruyter & Co. Berlin.

Karp, A. and S.W.J. Bright 1985. On the causes and origins of somaclonal variation. *Oxford Surveys of Plant Molecular and Cell Biology* Vol 2, B.J. Miflin, ed. 199–234. Oxford University Press.

Karp, A. and N. Fish 1987. Improvements in regeneration from protoplasts of potato and studies on chromosome stability. 2. The effect of regeneration media. *Theor. Appl. Genet.* (Submitted).

Karp, A., R. Rissiott, M.G.K. Jones and S.W.J. Bright 1984. Chromosome doubling in monohaploid and dihaploid potatoes by regeneration from cultured leaf explants. *Plant Cell Tissue Organ Culture* 3, 363–373.

Karp. A., S.H. Steele, S. Oarmar, M.G.K. Jones, P.R. Shewry and A. Breiman 1987a. Relative stability among barley plants regenerated from cultured immature embryos. *Genome* 29, 405–412.

Karp, A., Q.S. Wu, S.H. Steele and M.G.K. Jones 1987b. Chromosome variation in dividing protoplasts and cell suspensions of wheat. *Theor. Appl. Genet.* 74, 140–146.

Larkin, P.J. and W.R. Scowcroft 1981. Somaclonal variation — a novel source of variability from cell cultures for plant improvement. *Theor. Appl. Genet.* 60, 197–214.

Lee M. and R.L. Phillips 1987. Genomic rearrangements in maize induced by tissue culture. *Genome* 29, 122–128.

Linacero, R. and A.M. Vazquez 1986. Somaclonal variation in plants regenerated from embryo calluses in rye (*Secale cereale* L.) In *Genetic Manipulation in Plant Breeding*. W. Horn, C.J. Jensen, W. Odenvbach and O. Schieder, eds, 479–481. Walter de Gruyter & Co, Berlin.

Maddock, S.E. 1987. Suspension and protoplast culture of hexaploid wheat (*Triticum aestivum* L.) *Plant Cell Rep.* 6, 23–26.

McCoy, T.J., R.L. Phillips and H.W. Rines 1982. Cytogenetic analysis of plants regenerated from oat (*Avena sativa*) tissue cultures. High frequency of partial chromosome loss. *Can. J. Genet. Cytol.* 24, 37–50.

McCoy, T.J. and R.L. Phillips 1982. Chromosome stability in maize (*Zea mays*) tissue cultures and sectoring in some regenerated plants. *Can. J. Genet. Cytol.* 24, 559–563.

Nelson, R.S., A. Karp and S.W.J. Bright 1986. Ploidy variation in *Solanum brevidens* plants regenerated from protoplasts using an improved culture system. *J. Exp. Bot.* 37, 253–261.

Ogihara, Y. 1981. Tissue culture in *Haworthia* - Part 4. Genetic characterisation of plants regenerated from callus. *Theor. Appl. Genet.* 60, 353–363.

Orton, T.J. 1980. Chromosome variability in tissue cultures and regenerated plants of *Hordeum*. *Theor. Appl. Genet.* 56, 101–112.

Shepard, J.F., D. Bidney and E. Shahin 1980. Potato protoplasts in crop improvements. *Science* 208, 17-24.

Sree Ramulu, K., P. Dijkhuis and S. Roest 1983. Phenotypic variation and ploidy level of plants regenerated from protoplasts of tetraploid potato (*Solanum tuberosum* L. cv 'Bintje'). *Theor. Appl. Genet.* 65, 329–338.

Swedlund, B. and I.K. Vasil 1985. Cytogenetic characterization of embryogenic callus and regenerated plants of *Pennisetum americanum* (L.) K. Schum. I. In cultured immature embryos. *Bot. Gaz.* 143, 454–465.

Tempelaar, M.J., E. Jacobsen, M.A. Ferwerda and M. Hartogh 1985. Changes of ploidy level by *in vitro* culture of monohaploid and polyploid clones of potato. *Z. Pflanzenzucht.* 95, 193–200.

Thomas, E. 1981. Plant regeneration from shoot-culture-derived protoplasts of tetraploid potato (*Solanum tuberosum* cv. Maris Bard). *Plant Sci. Lett.* 23, 84–88.

Wheeler, V.A., N.E. Evans, D. Foulger, K.J. Webb, A. Karp, J. Franklin and S.W.J. Bright 1984. Plant regeneration from explant cultures of fourteen potato cultivars and study of cytology and morphology of regenerated plants. *Ann. Bot.* 55, 309–320.

Wu, Q.S., M.T.J. de Both and M.G.K. Jones 1986. Improvement in the division frequency of wheat (*Triticum aestivum*) suspension culture protoplasts by recycling. (Submitted).

CHROMOSOME DISPOSITION

Parental genome separation in F1 hybrids between grass species

M.D. Bennett

Plant Breeding Institute, Maris Lane, Trumpington, Cambridge CB2 2LQ, UK

(*Present address: Jodrell Laboratory, Royal Botanic Gardens, Kew, Richmond, Surrey TW9 3DS, UK*)

The fusion of parental genomes (male and female) occurs during fertilisation. Given the mechanics of fertilisation, especially the fact that parental genomes are contained in separate nuclei prior to fusion, a marked tendency for chromosomes from each parental genome to remain spatially separated at zygotic mitosis might be expected, and this tendency has been noted in many materials. For example, Figure 3 in the article by Laurie & Bennett, elsewhere in this volume, shows a clear tendency for a side-by-side separation of the seven large chromosomes from barley, and the ten much smaller chromosomes from maize, in a barley × maize zygotic metaphase.

Spatial separation of parental genomes is also seen in somatic hybrid cells formed after fusion of protoplasts, and subsequently of nuclei, from different species. For example, Constabel *et al.* (1975) show in their Figure 5 a clear tendency for the large chromosomes from *Pisum sativum* to be separated side-by-side from the small chromosomes from *Glycine max* in a metaphase cell of the non-sexual, somatic hybrid. Their Figure 6 shows a similar separation at interphase between a dense parental genome from barley (with $4C$ = about 22 pg of DNA) and the markedly less dense genome from carrot (with $4C$ = only about 4 pg of DNA). It also illustrates an important general point, namely that all the available evidence shows that gross features in the relative spatial disposition of chromosomes at metaphase (such as parental genome separation) clearly reflect their spatial disposition at interphase in undistorted meristematic cells. Figures 1–4 from Constabel *et al.* (1977), which show a clear tendency towards parental genome separation at both interphase and metaphase in somatic cell hybrids of *Glycine max* and *Vicia hajastana*, support this conclusion. Interestingly, their plate also shows variation in the form of genome separation — large chromosomes from *Vicia* tending to separate from small chromosomes from *Glycine* side-by-side in Figure 3, but concentrically in Figures 1 & 4. The cells shown in these figures had divided five or six times during four weeks in culture, suggesting that parental genome separation may persist long after nuclear fusion. Further proof that this is so after normal fertilisation is provided by a phenomenon recognised since the work of Haecker (1895) as 'gonomery'. This describes the more or less separate grouping of maternal and paternal chromosomes which persists during cleavage stages of some animals after the fusion of the male and female pronucleus.

Gonomery occurs in some plants also. For example, Fig. 1 shows prometaphase from an 8-celled proembryo of an F1 hybrid between *Hordeum vulgare* and *H. bulbosum*. The seven chromosomes from *H. vulgare*, whose large centromeres are circled, typically still form a group, separate from the *H. bulbosum* chromosomes with small or indistinguishable centromeres.

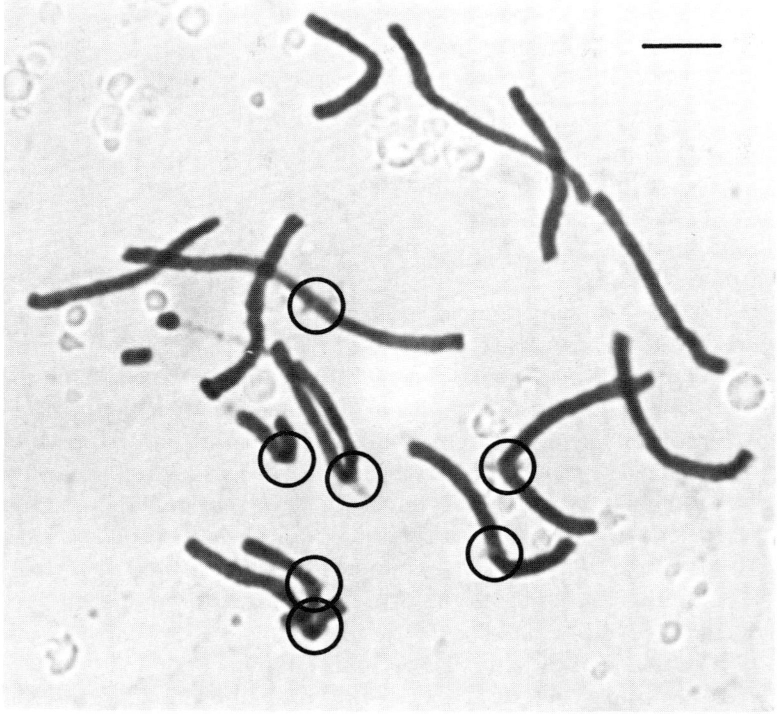

Figure 1.   Prometaphase from an 8-celled proembryo of *Hordeum vulgare* cv. Tuleen 346 × *H. bulbosum* clone L6 showing a tendency for parental genomes to remain separate. Centromeres of *vulgare* chromosomes visible in the light microscope are circled while those of *bulbosum* chromosomes which were less clear or indistinguishable are unmarked. Bar = 5 μm.

It is well known that the relative positions of chromosomes in a nucleus are often reproduced to a marked degree after mitosis in its daughter nuclei. Thus, once established in the zygote, a detectable tendency towards parental genome separation might be expected to persist for several subsequent cell cycles in early embryo development. However, without some positive control to maintain it, spatial separation of parental genomes would gradually disappear due to movements which alter and progressively randomise the relative positions of chromosomes during successive mitotic cycles. Such a control was unknown. Thus, it has been generally assumed that after fertilisation parental genomes progressively intermingle, so that chromosomes become randomly arranged within the nucleus early in the development of multicellular higher organisms.

A main purpose of this paper is to show that this view is incorrect, at least for interspecific diploid hybrids of grasses.

Parental genome separation in root-tip metaphases of *Hordeum vulgare* × *Secale africanum*

The extensive work on genome separation described below began in 1980 when Dr R. A. Finch and I first noted a clear tendency for the parental genomes to be separate in squashed root-tip metaphases of F1 hybrid plants between barley (*Hordeum vulgare* cv. Sultan) and a wild rye (*Secale africanum*). The parental sets are easily distinguished by size, the seven chromosomes from *Secale* all being larger than those from *Hordeum*.

Typically the set from barley tended to be grouped towards the centre of the squash while the set from *Secale* tended to be outside and around it. The tendency was highly significant, both in many individual cells (Finch *et al.* 1981) and overall.

The tendency for parental genome separation was highly significant in squashed metaphase cells, but it was important to discover whether it was a technical artefact perhaps produced by squashing, or whether it indicated a normal arrangement *in vivo*. This question was answered using the EM serial thin-section reconstruction technique which allows accurate determination of both the identities and the undistorted 3-dimensional placement of chromosomes in a single cell as *in vivo* (Finch *et al.* 1981; Bennett 1984a; Bennett *et al.* 1986). As in light microscope studies, chromosomes within a reconstructed cell were all easily distinguished as to parental origin (those from *Secale* all had larger volumes than those from *Hordeum*).

Fig. 2 shows a polar view of the positions as *in vivo* of the chromosomes on a metaphase plate of a reconstructed cold-treated root-tip mitosis, and illustrates the basis of two tests for a tendency towards concentric genome separation. The first compares the mean distance from the 3-dimensional cell mid-centromere point (marked as a cross in Fig. 2) of centromeres from each parental genome. The second tests how often it is possible to draw a circle which includes all the centromeres from one parent but excludes all those from the other parent. (N.B. For simplicity, only the positions of the centromeres are shown in Fig. 3 for replicate cells.) For example, the cell in Fig. 2 is drawn in panel one of Fig. 3 which shows polar views of the positions of centromeres on congressed metaphase plates of all the nine root-tip metaphase cells of this hybrid which were reconstructed. These cells show striking evidence of a trend towards concentric parental genome separation with the centromeres of the larger *Secale*

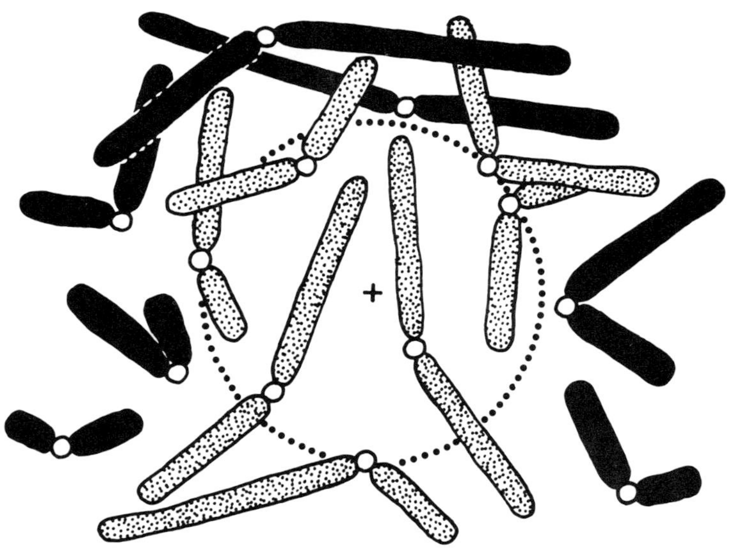

Figure 2.   A polar view of a reconstructed metaphase plate in a root tip cell of *Hordeum vulgare* cv. Sultan (stippled chromosomes) × *Secale africanum* (solid chromosomes). Centromeres are shown as small circles, and the cell mean centromere point is marked by a cross.

chromosomes mostly outside and around the centromeres of the smaller *Hordeum* chromosomes. In three of the nine cells (i.e. 33%) all seven *Hordeum* centromeres lay inside a central circle which excludes all seven *Secale* centromeres, a distribution expected by chance in not more than about 1.23% of cells. This excess frequency of occurrence over expectation was highly significant. Comparison of the mean distance of *Hordeum* and *Secale* centromeres from the centre of the metaphase plate in the nine reconstructed cells showed that *Hordeum* centromeres were on average much closer to the mid-point in every cell, and significantly so in six out of the nine individual cells (Finch & Bennett 1981). However, these tests can become increasingly artificial as cells, and the shapes of their constrained metaphase plates, become more elongated (Schwarzacher-Robinson *et al.* 1987). Consequently, another test is being developed which compares the number of centromeres from one parent contained in the polygon of minimum perimeter containing all the centromeres of the other parent (Heslop-Harrison & Bennett — unpublished). Thus in a computer simulation where two sets of seven points were placed at random on an approximately circular metaphase plate, and polygons of minimum perimeter drawn around each, repeated 10,000 times, the mean number of points from one set included in the polygon for the other set was 2.1932. Using this test, genome separation is indicated by the number of centromeres of either parent contained in polygons for the other parent deviating significantly from expectation, (i.e. from 2.1932 when 2n = 14). Scores tending significantly to zero for both parents indicate a significant tendency towards a side-by-side arrangement of parental genomes, while scores tending significantly towards zero for one parent but towards seven for the other parent, indicate a significant tendency towards concentric parental genome separation. In the nine cells of *Hordeum vulgare* × *Secale africanum* shown in Fig. 3 the total number of centromeres from *Secale* contained in polygons for *Hordeum* was three, while the total number of *Hordeum* centromeres contained in polygons for *Secale* (omitted from Fig. 3 for clarity) was 41. Compared with random expectation of 19.74, these

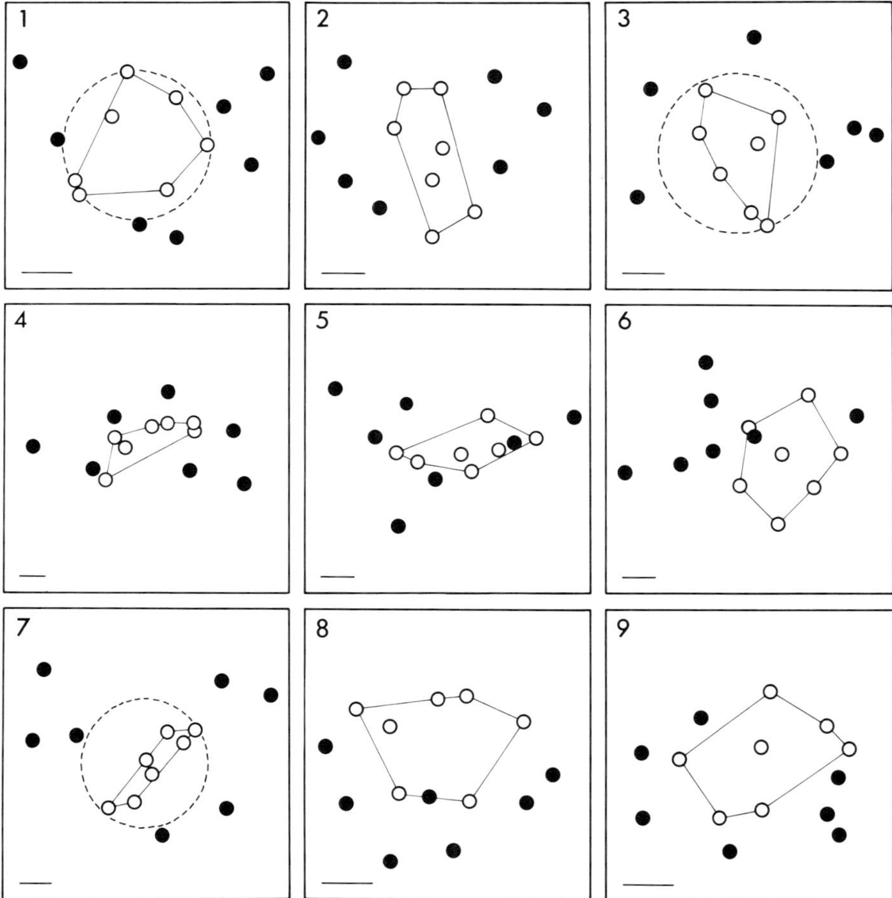

Figure 3. Polar views of positions of *Hordeum* (○) and *Secale* (●) centromeres on metaphase plates, and polygons of least perimeter including all *Hordeum* centromeres, in nine serially sectioned reconstructed root tip metaphase cells of *H. vulgare* × *S. africanum*. Reproduced from Schwarzacher-Robinson *et al.* (1987). Bar = 2 µm.

numbers were respectively very highly significantly lower and higher. Clearly, all three tests agree in showing a significant tendency for, or towards, concentric parental genome separation.

These results were obtained for reconstructed cells using chromosome and centromere coordinates as *in vivo*. Thus, it can be safely concluded that this example of parental genome separation is not a technical artefact, caused for example by squashing, but that it is real in the living metaphase cells of this hybrid.

How common is parental genome separation?

This above conclusion immediately posed another question. Is *Hordeum vulgare* × *Secale africanum* unusual, or is parental genome separation common in somatic cells of interspecific hybrids? Following the original study, a similar significant tendency towards concentric separation of parental genomes was also found in 45 reconstructed metaphase cells of all the other intergeneric or interspecific F1 hybrids in which quantitative studies have been made, namely in: *Hordeum vulgare* × *Secale cereale* (20 × R) (Finch & Bennett 1981); *H. chilense* × *S. africanum* (Schwarzacher-Robinson *et al.* 1987); *H. vulgare* × *H. bulbosum*; *H. marinum* × *H. vulgare*; and *Aegilops squarrosa* × *Secale cereale* (mostly unpublished, but see Bennett 1984b). Moreover, even a cursory examination shows that the tendency towards concentric parental genome separation clearly exists in reconstructed root-tip metaphases of other interspecific hybrids such as *H. chilense* × *S. montanum* which await quantitative analysis.

Linde-Laursen & von Bothmer (1984), studying C-banded root tip metaphase squashes by light microscopy noted, but did not quantify, a clear tendency for parental genome separation in the diploid F1 hybrid *H. vulgare* × *Psathyrostachys fragilis*. Moreover, using similar methods, Linde-Laursen & Jensen (1984) showed quantitatively in four polyploid hybrids (namely *H. roshevitzii* × *H. vulgare*, *H. jubatum* × *H. vulgare*, *H. vulgare* × *Triticum aestivum* and *H. vulgare* × *H. brevisubulatum* ssp. *turkestanicum*) that "*H. vulgare* chromosomes were located nearer the centre of the metaphase plate than the non-*H. vulgare* chromosomes". Moreover, other quantitative studies (Gleba *et al.* 1983; 1987) have demonstrated a significant tendency for parental genome separation at metaphase in mature somatic hybrids of *Nicotiana chinensis* + *Atropa belladona*, and *Nicotiana plumbaginifolia* + *N. sylvestris*, formed by protoplast fusion.

Clearly, therefore, parental genome separation is not a rarity in somatic cells among wide hybrids of plants. Indeed it is common, having been found, for example, in at least ten interspecific *Hordeum* hybrid combinations involving four genomes, and also in dicotyledon and animal examples.

Does parental genome separation persist throughout the cell cycle?

So far it has been shown that parental genome separation, of one form or another, occurs in many wide hybrids at mitotic metaphase. However, it has been asked whether the condition is transiently established, perhaps as the result of congressional movements, only at mitotic metaphase. A clear answer to this question has come from comparisons of reconstructions of ten male archesporial nuclei at various stages of pre-meiotic mitosis from the hybrid *Hordeum vulgare* cv. Sultan × *Secale africanum* (Bennett & Smith — unpublished). Reconstruction showed that in these pre-meiotic nuclei at prophase, the chromosomes are all attached to the still intact nuclear membrane, at or close to their telomere(s), and congressional movements have not yet begun. Although more tedious than at metaphase, it is usually possible to follow each chromosome within a reconstructed prophase nucleus and to identify it as to parental origin by volume just as at metaphase, chromosomes from *Secale* being all larger than any from *Hordeum*. Fig. 4 shows polar views of the positions of centromeres at the centromeric pole of all the four diploid pre-meiotic nuclei which were reconstructed in this hybrid. All show parental genome separation for centromeres of one form or the other. Thus cell 175 in Fig. 4 has perfect

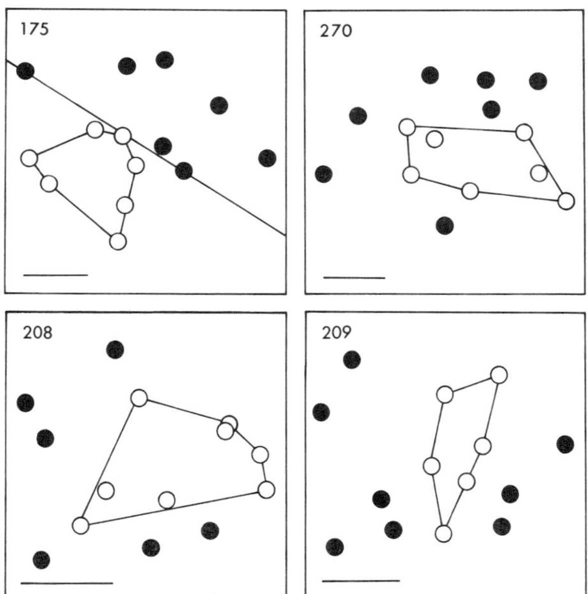

Figure 4. Polar views of positions at the polfeld of *Hordeum* (○) and *Secale* (●) centromeres at prophase, and polygons of least perimeter including all *Hordeum* centromeres, in four reconstructed male archesporial cells at pre-meiotic mitosis in F1 *H. vulgare* × *S. africanum* (2n = 14). Centromeres from parental genomes separate either side of a line in cell 175. Bar = 2 μm.

side-by-side separation of parental genomes (expected by chance in only about 0.5% of cells), while the other three cells all show a tendency towards concentric separation of parental genomes — neatly confirming that both forms of genome separation are indeed real! As all the chromosomes are attached to the nuclear membrane at their telomere(s) this observed placement of genomes at prophase almost certainly reflects their relative positions at G2 of the previous interphase also.

Two of the four sectioned anthers also contained archesporial sectors which had spontaneously doubled their chromosome numbers and four recently doubled cells, now with 26 or 27 chromosomes, were reconstructed including those at prophase, metaphase and telophase of pre-meiotic mitosis illustrated in Fig. 5. At each stage of mitosis the centromeres of *Secale* chromosomes tend to

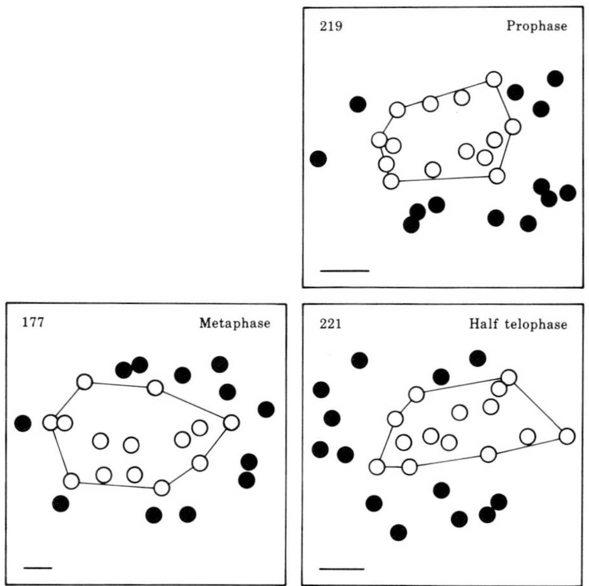

Figure 5. Polar views of the positions of *Hordeum* (○) and *Secale* (●) centromeres, and polygons of least perimeter including all *Hordeum* centromeres in three reconstructed spontaneously doubled (2n = 26 or 27) male archesporial cells at three stages of pre-meiotic mitosis in F1 *H. vulgare* × *S. africanum*. Bar = 2 μm.

surround those of *Hordeum* chromosomes. Despite the higher number of chromosomes in each cell, drawing the polygon of minimum perimeter for the *Hordeum* genomes shows no *Secale* centromere contained in any *Hordeum* polygon. Clearly, therefore, parental genome separation is not just transiently present at metaphase. It persists throughout mitosis and (almost certainly) throughout interphase too. Indeed, rather than being transiently established at metaphase due to congression, such movements may rather tend to reduce the tendency towards parental genome separation seen at this stage below that occurring previously at prophase and interphase.

Does parental genome separation persist during development?

The above observations on archesporial cells also help to answer another question. Thus, given that genome separation was real in root-tip cells, it was important to find out if this was unusual, or whether the phenomenon was widespread in time and throughout the plant. Detailed observations of *Hordeum vulgare* × *Secale africanum* and *H. vulgare* × *H. bulbosum* combined to show that genome separation occurs in roots of various sizes, from plants of various ages, and in different years and environments. Moreover, it was found in root, anther, embryo and endosperm nuclei.

Together, these observations show that genome separation is not just transiently present at one stage or in one tissue, but that it can persist for years and throughout the development of different tissues and organs. Thus, all the indications are that genome separation is the norm, rather than the exception, in somatic cells of these hybrids.

How is parental genome separation determined?

Given that genome separation is real and widespread, another important question concerns how it is determined. Is it under genetic control, or is it a packing phenomenon due to some mechanical constraint that preferentially brings smaller chromosomes near the centre of the metaphase plate? Many examples have been noted where small chromosomes are preferentially located at the centre of the metaphase plate, for example in chicken × Coturnix quail hybrids (Bammi *et al.* 1966). Indeed, several workers have claimed this tendency to be a general rule in chromosome disposition (e.g. Therman & Denniston 1984).

Perhaps the concentric genome separation in our original hybrid (*Hordeum vulgare* × *Secale africanum*) is simply due to chromosome size. Perhaps small chromosomes tend to lie nearer to the mean centromere point than larger chromosomes, so that *Hordeum* chromosomes are nearer the centre of the plate simply because they are smaller than those of *Secale*.

Conclusive evidence that this is not so came first from another intergeneric hybrid; between *Hordeum chilense* and the same male parent as before — *Secale africanum*. Again all seven *Hordeum* chromosomes were smaller than any *Secale* chromosome. Again, as the polar views of all nine root-tip metaphase plates reconstructed show (Fig. 6), there was a highly significant tendency towards concentric separation of the parental genomes. However, in this hybrid the

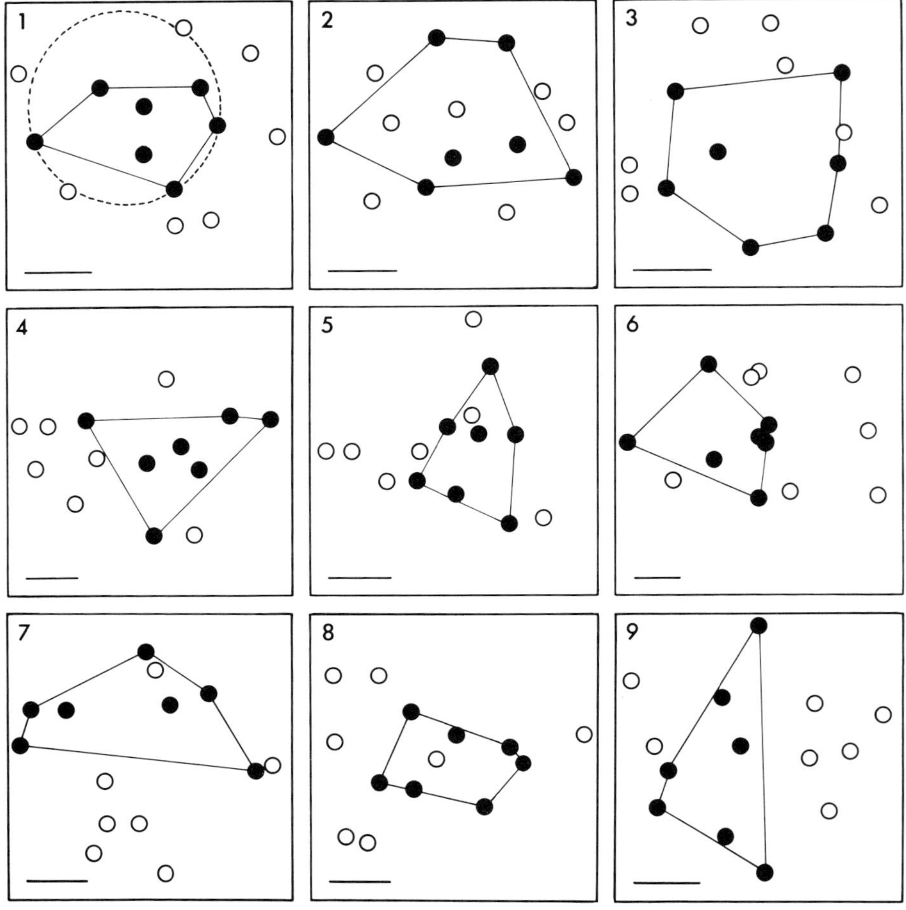

Figure 6. Polar views of positions of *Hordeum* (○) and *Secale* (●) centromeres on metaphase plates, and polygons of least perimeter including all *Secale* centromeres, in nine serially sectioned reconstructed cells of *H. chilense* × *S. africanum*. Reproduced from Schwarzacher-Robinson *et al.* (1987). Bar = 2 μm.

polarity of this separation was the reverse of that orginally found in *H. vulgare* × *S. africanum*, so that here the centromeres of the large chromosomes from *Secale* showed a very highly significant tendency to be grouped centrally and surrounded by centromeres of the smaller chromosomes from *Hordeum* (Schwarzacher-Robinson *et al.* 1987).

Analysis showed that the ratio of the mean distance of centromeres from *Hordeum* and *Secale* to the mean centromere point in these 9 cells was 1:1.28, and this difference was very highly significant. Drawing polygons of minimum perimeter for parental genomes showed that the total number of *Hordeum* centromeres contained in polygons for *Secale* in these nine reconstructed cells was 5, while the total number of *Secale* centromeres contained in polygons for *Hordeum* (omitted from Fig. 6 for clarity) was 34. Compared with random expectation (19.74) these numbers were, respectively, very significantly lower and higher (Schwarzacher-Robinson *et al.* 1987)).

Later, other hybrids were also found where in reconstructed root-tip cells the centromeres of the seven markedly smaller chromosomes were on average significantly farther from the cell mean centromere point than centromeres of the seven markedly larger chromosomes (e.g. *Aegilops squarrosa* × *Secale cereale* (Bennett 1984b) and *H. chilense* × *A. montanum* (Bennett — unpublished). At the same time, several other examples were found where, as in the original hybrid *H. vulgare* × *S. africanum,* the centromeres of the seven markedly larger chromosomes were significantly farther from the cell mean centromere point than those of the seven markedly smaller chromosomes (e.g. *H. vulgare* × *Secale cereale* — Finch & Bennett 1981).

As the centromeres of the parental genome with the larger chromosomes can be either to the inside or the outside of centromeres from the parental genome with the smaller chromosomes, then clearly concentric parental genome separation is not a packing phenomenon determined by relative chromosome size *per se,* but it is genetically determined (Schwarzacher-Robinson *et al.* 1987).

This conclusion is also self-evident *firstly* when two parental genomes, one with large chromosomes and one with all smaller chromosomes, are separate side-by-side yet about equidistant from the cell mean centromere point (as in cell 175 in Fig. 4); and *secondly* in hybrids showing a strong tendency towards a concentric separation of parental genomes even though the mean and range of chromosome sizes in the two parents are very similar, as in barley × bread wheat (Linde-Lauresen & Jensen 1984). Interestingly Gleba *et al.* (1987) who recently reported spatial separation of parental genomes in both cultured hybrid cells and plants derived from interspecific protoplast fusion, also concluded that this arrangement is "epigenetically controlled".

Can parental genome separation change in form during development?

Using light microscopy, Gleba *et al.* (1987) analysed chromosome disposition in metaphase squashes in *Nicotiana chinensis* + *Atropa belladona*, and in *Nicotiana plumbaginifolia* + *N. sylvestris*. These somatic cell hybrids are both relatively stable, and all the chromosomes can be easily distinguished as to parental origin by size alone. They showed that the form of genome separation changed from a side-by-side arrangement (which they termed 'segmental') found in the first divisions of protoplast fusion products, to a concentric form (which they termed 'radial') found in long-term cultured somatic hybrids.

A similar result has been found in the sexual hybrid *Hordeum marinum* ssp. *marinum* × *H. vulgare* cv. Tuleen 346 (Bennett & Finch — unpublished). This hybrid showed parental genome separation in each tissue and stage which we examined. However, Fig. 7 summarises the remarkable changes in the topology

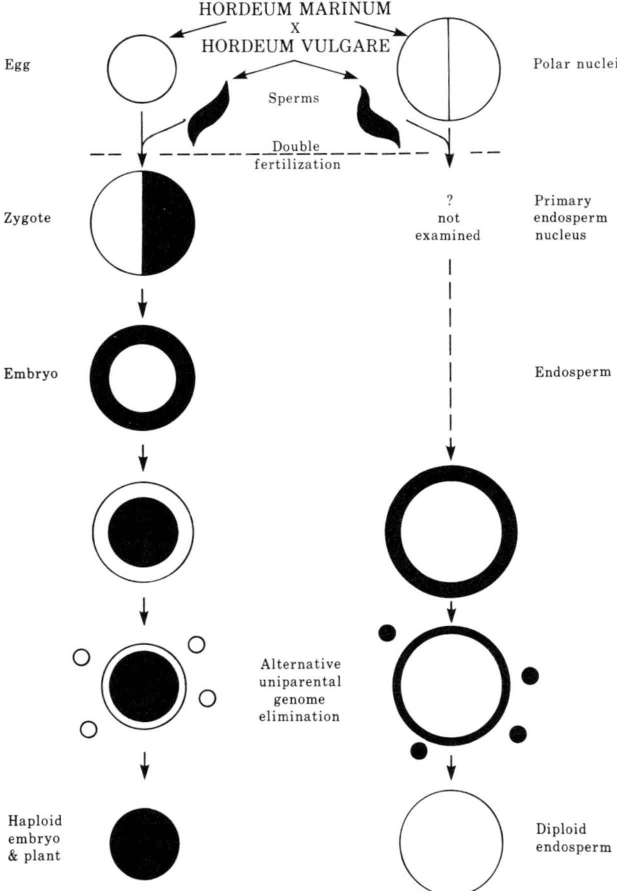

Figure 7. A diagrammatic illustration of developmental variation in the relative positions of domains containing centromeres from *Hordeum marinum* (in outline) and *H. vulgare* (shaded) on metaphase plates of mitosis from early embryo and endosperm stages in F1 *H. marinum* ssp. *marinum* × *H. vulgare* cv. Tuleen 346, during the first 5 days after pollination. This material displays alternative genome elimination (Finch 1983), eliminating *marinum* chromosomes from embryo but *vulgare* chromosomes from endosperm nuclei.

of genome separation at metaphase noted in this material. After fertilisation of the egg, there was a significant tendency towards a side-by-side separation of parental genomes in the diploid zygote. Within a day this changed to a concentric arrangement in embryo cells, with *vulgare* chromosomes around *marinum* ones. It soon reversed in polarity, so that *marinum* chromosomes were now *peripheral* to *vulgare* ones. Subsequently, while retaining this arrangement, elimination of seven *marinum* chromosomes occurred, producing haploid nuclei with only a *vulgare* genome, which after embryo culture yielded a plant with the haploid complement of Tuleen 346 barley (Finch 1983). The triploid primary endosperm mitosis was not examined. Presumably it had the usual side-by-side parental genome separation. Be that as it may, early endosperm metaphase was seen and it had a highly significant tendency to a concentric form of genome separation with *vulgare* centromeres peripheral to *marinum* ones. Unlike in the embryo, this pattern did not subsequently reverse before chromosome elimination began. However, a haploid set of *vulgare* chromosomes was soon subsequently eliminated at successive mitoses, producing diploid endosperm nuclei with 14 *marinum* chromosomes. Such observations provide further evidence showing that genome separation is under genotypic control.

Together, the above results for the relatively stable interspecific cell hybrids formed by protoplast fusion, and for the unstable sexual hybrid between *H. marinum* and *H. vulgare*, allow another important conclusion to be firmly drawn. This is that while it may persist throughout the life cycle, parental genome separation is nevertheless not immutable, as it can vary in form during development, both within and between tissues.

References

Bammi, R.K., R.N. Shoffner and G.J. Heiden 1966. Non random association of somatic chromosomes in the chicken-Coturnix quail hybrid and the parental species. *Can. J. Genet. Cytol.* 8, 537–543.

Bennett, M.D. 1984a. Towards a general model for spatial law and order in nuclear and karyotypic architecture. *Chromosomes Today* 8. 190–202.

Bennett, M.D. 1984b. Nuclear architecture and its manipulation. In *Gene Manipulation in Plant Improvement*, J.P. Gustafson, ed. 469–502. New York: Plenum.

Bennett, M.D., J.B. Smith and A.G. Seal 1986. The karyotype of the grass *Zingeria biebersteiniana* (2n = 4) by light and electron microscopy. *Can. J. Genet. Cytol.* 28. 554–562.

Constabel, F., D. Dudits, O.L. Gamborg and K.N. Kao 1975. Nuclear fusion in intergeneric heterokaryons. A note. *Can. J. Bot.* 53. 2092–2095.

Constabel, F., G. Weber and J.W. Kirkpatrick 1977. Sur la compatibilité des chromosomes dans les hybrides intergénériques de cellules de *Glycine max* × *Vicia hajastana*. *Compte Rendus Acad. Sci. Paris* 285, D 319–322.

Finch, R.A. 1983. Tissue-specific elimination of alternative whole parental genomes in one barley hybrid. *Chromosoma* 88, 386–393.

Finch, R.A. and M.D. Bennett 1981. Spatial separation of mitotic parental genomes in *Hordeum* × *Secale* hybrids. In *Barley Genetics IV*, M.J.C. Asher, R.P. Ellis, A.M. Hayter and R.N.H. Whitehouse, eds, 746–750. Edinburgh: Edinburgh University Press.

Finch, R.A., J.B. Smith and M.D. Bennett 1981. *Hordeum* and *Secale* mitotic genomes lie apart in a hybrid. *J. Cell Sci.* 52, 391–403.

Gleba, Y.Y., V.P. Momot, A.N. Okolot, N.N. Cherup, M.V. Skarshynskaya and V. Kotov 1983. Genetic processes in intergeneric cell hybrids *Atropa* + *Nicotiana* 1. Genetic constitution of cells of different clonal origin. growing *in vitro. Theor. Appl. Genet.* 65, 269–276.

Gleba, Y.Y., A. Parokonny, V. Kotov, I. Negrutiu and V. Momot 1987. Spatial separation of parental genomes in hybrids of somatic plant cells. *Proc. Natl. Acad. Sci. U.S.A.* 84, 3709–3713.

Haecker, V. 1895. Über die Selbständigkeit der vater- und mötterlichen Kernsubstanz wahrend der Embryonalentwicklung von *Cyclops. Arch. mikr. Anat.* 46, 579.

Laurie, D.A. and M.D.Bennett 1988. Chromosome behaviour in wheat × maize, wheat × sorghum and barley × maize crosses. In this volume.

Linde-Laursen, I. and J. Jensen 1984. Separate location of parental chromosomes in squashed metaphases of hybrids between *Hordeum vulgare* L. and four polyploid, alien species. *Hereditas* 100, 67–73.

Linde-Laursen, I. and R. von Bothmer 1984. Somatic cell cytology of the chromosome eliminating, intergeneric hybrid *Hordeum vulgare* × *Psathyrostachys fragilis. Can. J. Genet. Cytol.* 26, 436-444.

Schwarzacher-Robinson, R., R.A. Finch, J.B. Smith and M.D. Bennett 1987. Genotypic control of centromere positions of parental genomes in *Hordeum* × *Secale* hybrid metaphases. *J. Cell Sci.* 87. 291–304.

Therman, E. and C. Denniston 1984. Random arrangement of chromosomes in *Uvularia* (Liliaceae). *Pl. Syst. Evol.* 147, 289–297.

Chromatin and centromeric structures in interphase nuclei

J.S. Heslop-Harrison, M. Huelskamp, S. Wendroth, M.D. Atkinson, A.R. Leitch and M.D. Bennett

Plant Breeding Institute, Trumpington, Cambridge, CB2 2LQ, UK

At the last Kew Chromosome Conference, held in 1982, there were several papers on metaphase chromosome disposition. Since then, evidence for non-random spatial organisation within the metaphase nucleus, including the non-random distribution of chromosomes on the plate, has continued to increase (e.g. Mathog *et al.* 1984; Rickards 1986; Murray 1986), although both the findings and analyses have remained controversial.

The metaphase chromosome is not highly active in transcription, DNA replication or gene control. Thus, if chromosome disposition is important, as the results from metaphase studies indicate (e.g. Bennett 1988), it is also necessary to describe the positioning of interphase chromosomes and their decondensed chromatin. For example, spatial constraints on gene activity might apply only at interphase stages of the cell cycle. This paper presents some preliminary data from our morphological studies of electron-dense chromatin and centromere disposition in interphase nuclei using serial section reconstructions. It does not attempt to review the biochemical and light microscopic data recently presented in an excellent review by Hubert & Bourgeois (1986).

Techniques

In any ideal study of chromosome disposition at metaphase as *in vivo*, the position and the identity of every chromosome must be known (Heslop-Harrison & Bennett 1984); the EM serial thin section reconstruction technique allows this ideal to be achieved in some metaphases. We have reported various results on the association of chromosomes in genome groups (Bennett 1988), lack of association of homologous pairs of chromosomes (Heslop-Harrison *et al.* 1988) and the ordering of chromosomes within haploid sets (Heslop-Harrison & Bennett 1984).

Serial section reconstructions of consecutive sections through interphase nuclei were made to try to find positions and ultimately identities of chromatin segments. Fig. 1 shows photographs of four individual sections through an interphase nucleus. Some of the major features visible in electron micrographs of interphase nuclei are indicated, including the areas of chromatin, the centromeric structures and nucleoli, which were clearly seen.

In the present work, data from two types of dividing nuclei are presented. Root tips from the grass *Aegilops umbellulata* Zhuk. (2n=2x=14) were used as an example of a somatic tissue (Heslop-Harrison (1983) discussed metaphase chromosome disposition in this material). The two root tip nuclei referred to in the present work were designated 67B and 67C, and were adjacent to each other in one root tip. The second cell type was from the male archesporium of the intergeneric F1 sexual hybrid *Hordeum vulgare* L. cv. Sultan × *Secale africanum* Stapf, which was near the last mitotic division before the cells entered meiosis. The hybrid was made and treated as described by Finch *et al.* (1981). One or two chromosomes were sometimes eliminated, and some cells became tetraploid, but the normal number was 2n=14. The line has been maintained vegetatively for a number of years.

For electron microscopy, fresh tissues were fixed in buffered glutaraldehyde, embedded in Spurr's resin and serially sectioned at 0.1 μm thickness before photographing in a transmission electron microscope (Bennett *et al.* 1979). Since the nuclei were typically 5 to 10 μm in diameter, 50 to 100 sections per nucleus were required for each complete series. Areas of the nucleus, chromatin and centromeric structures were quantified in each section by tracing on a digitising tablet connected to a microcomputer, or by using a video image digitiser and a microcomputer-based image analysis system. To obtain volumes, the areas were summed and multiplied by the section thickness.

Evidence that our fixation technique does not alter chromatin distribution comes from Gregoire *et al.* (1984), who examined chromatin distribution in living Rat Kangaroo ovary cells using non-toxic concentrations of the DNA fluorescent label Hoechst 33342. They found a chromatin network in interphase nuclei, with some degree of bending and grouping of chromocentres around the nucleolar regions. They concluded that their results "supported the concept that, in living cells, the chromatin network was similar to that observed after fixation and usual pretreatments of the cells" for electron microscopy.

Nuclear and chromatin volumes

The volumes of the nuclei and chromatin are given in Table 1. Within both materials, interphase chromatin volumes (at least in the very small number of cells measured so far) ranged from well under half to slightly less than the average metaphase chromatin volumes.

It is worth noting that most of the DNA in the chromatin volume being measured is non-coding, since probably less than 1% of the total DNA is required to encode gene products. Although most of the electron-dense chromatin within a nucleus is visibly connected throughout its volume, a few small pieces of chromatin were not connected by noticeably electron-dense pieces. Almost certainly, connections of more diffuse or de-condensed chromatin are present.

The overall electron density of the chromatin in the interphase nuclei is very similar to that in metaphases, but the average size of pieces seen in the sections is

Table 1. Volumes of nuclei and chromatin (μm^3) and numbers of centromeric structures in various nuclei.

Nucleus type and code	Nuclear volume	Chromatin volume	Number of centromeric structures
Hordeum vulgare × Secale africanum hybrid			
Archesporial interphases			
212	159	–	–
200	174	42	–
199	189	49	–
211	223	–	9
207	226	–	13
210*	310	114	–
213	362	113	7
Root-tip metaphases[+]			
EMI	No membrane	116	14
EM4	"	134	14
EM2	"	135	14
EM3[#]	"	140	14
Aegilops umbellulata			
Root-tip interphases			
67B	304	62	12
67C	330	117	10
Root-tip metaphases			
Mean of 10	No membrane	128 ±20	14

* including micronucleus   [+] from Finch et al. 1981

[#] 13 chromosome aneuploid   – unmeasured value

much smaller in the former. Consequently, any errors from measuring thin-cut pieces which are not the full section thickness, and any inaccuracy in finding the edges of chromatin pieces, will cause much greater errors in estimating volumes at interphase than at metaphase. However, it is reasonable to assume that volumes from interphase nuclei in Table 1 correlate with stage in the cell cycle, and that the smaller nuclear and chromatin volumes came from nuclei at G1 or early S, containing largely unreplicated chromatin. Those with the larger chromatin volumes, similar to metaphases, were from nuclei which were in late S or G2. Measurements of the total chromatin volumes per cell have also been made at prophase and telophase and are respectively about the same as, or roughly one half that of metaphase volumes within the test materials which included pre-meiotic archesporial cells of *H. vulgare* × *S. africanum*.

If chromatin volumes measured in G2 interphase nuclei are similar to those at metaphase, there are implications for chromosome despiralisation at telophase.

Some models assume that metaphase chromosomes 'uncoil' or 'despiral' to the level of chromonemata at interphase (e.g. Comings 1968). Other authors have reported that metaphase chromosomes never decondense to the point that they lose their three-dimensional integrity, but remain in distinct domains throughout interphase (Stack *et al.* 1977). Sparrow (1942) reported that the prophase to metaphase chromatid contraction ratio is not less than six to one, which agrees with our results from serially-sectioned interphases. However, this result may apply only to chromatin segments including much of the non-coding DNA, and not to the despiralisation of the small proportion of chromatin including the transcriptionally or otherwise active DNA in the nucleus.

Centromeric structures

Centromere distribution and Rabl-configuration

In nuclei of cells from rapidly-dividing tissues or cultures, there is a general view that chromosomes are arranged in a Rabl-type configuration at interphase (Rabl 1885), with the centromeres clustered at one end of the nucleus in the chromocentre or *Polfeld,* while the telomeres are clustered at the opposite end. Both centromeres and telomeres are often attached to the nuclear envelope (e.g. Fussell 1977; Ashley 1979; Tanaka 1981). Other, more differentiated cell types do not behave in the same way; in particular, interphase cells of the nervous system show a different arrangement of centromeres (Manuelidis 1985; Boni & Mintz 1986; Hadlaczky *et al.* 1986).

Fig. 2 shows two computer rotations of nucleus 213, and the grouping of the seven centromeric structures is clearly visible. The rotations also show that centromeric structures tend to be adjacent to the nuclear envelope in these cell types.

Figure 1. Electron micrographs showing four individual sections through an interphase nucleus (code 213) of the hybrid *Hordeum vulgare* (barley) × *Secale africanum* (a wild rye). The whole nucleus comprises eighty 0.1 μm sections and the section numbers are given in the lower right-hand corner of each part of the figure. **A** two centromeric structures (arrowed); **B** nucleolus (N) and chromatin running into the nucleolus (arrowed); **C** & **D** sections taken 0.5 μm from the first and last section through the nucleus, showing the large variation in percentage of the nuclear volume accounted for by chromatin in different regions of the nucleus.

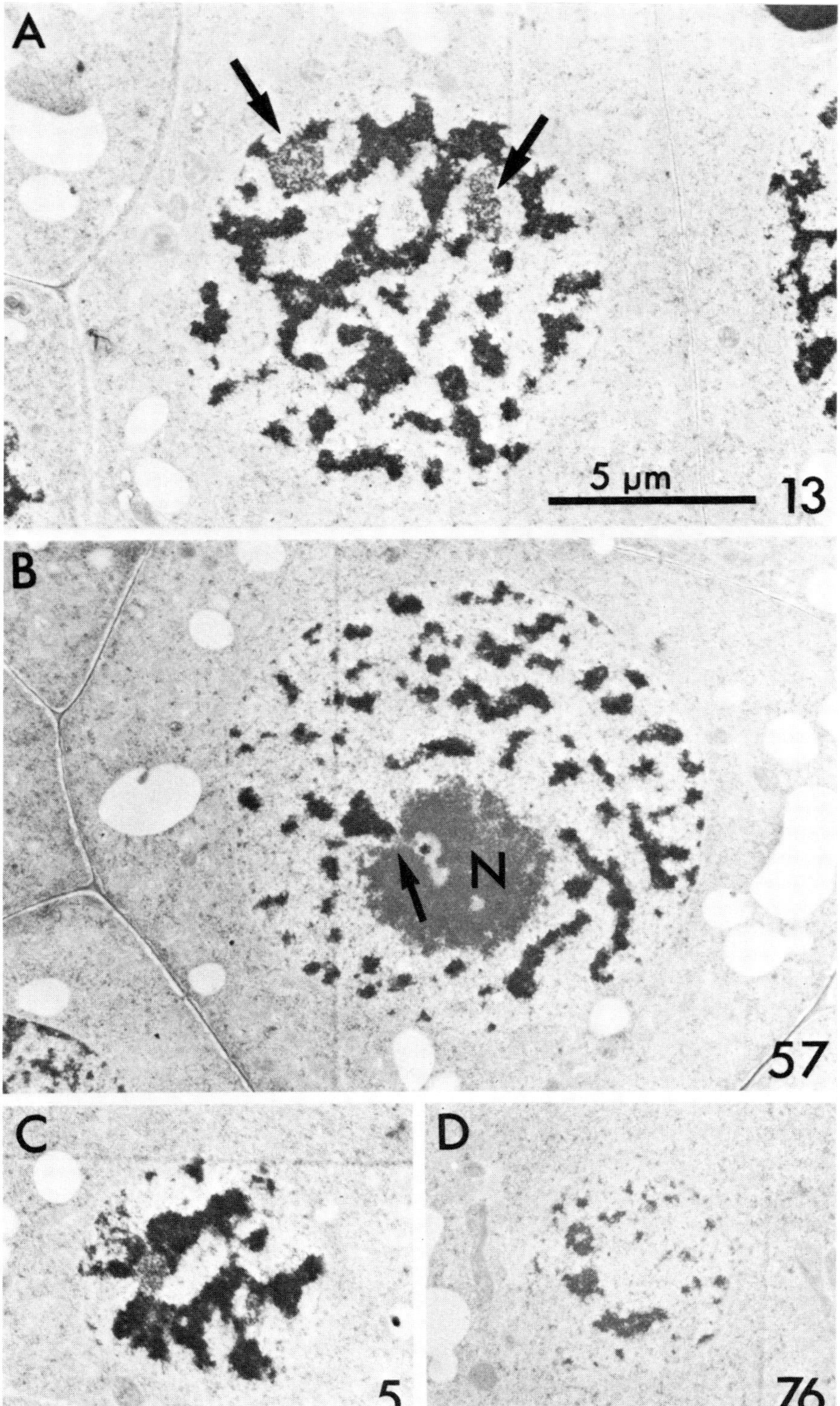

A
5 µm
13
B
N
57
C
5
D
76

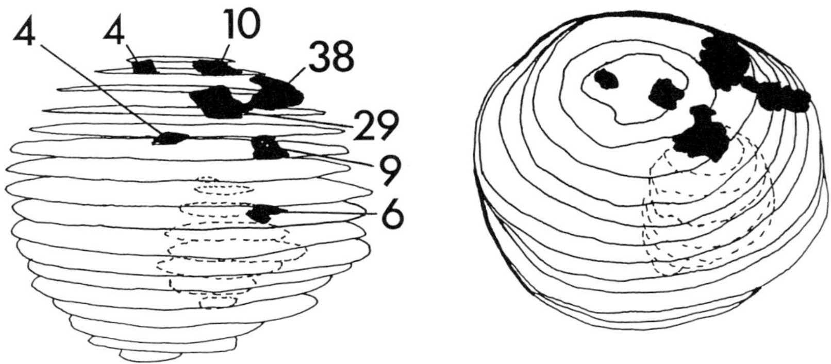

Figure 2. Two rotated projections of the nuclear envelope (unbroken lines), nucleolus (broken line) and centromeric structures (solid areas) in the nucleus shown in Fig. 1. The seven centromeric structures are mainly clustered near the surface of the nucleus. Volumes of each centromeric structure as a percentage of their total volume are given on the left-hand projection.

Centromere fusion

In reconstructions of metaphases from serial sections, the centromeres were discrete and were always separated by a minimum distance (Heslop-Harrison & Bennett 1983). Indeed, when compared with a random separation of points distributed over a simulated metaphase plate, there were significant differences, with many more centromeres lying close to each other in the simulations than in the actual reconstructions. In all four reconstructed interphases analysed (Table 1), the number of centromeric structures which could be counted in the reconstructions was less than the probable number of chromosomes present in the nucleus. This might have been because some of the centromeric structures were not visible as lighter grey areas in the sections because of their small size (particularly in the wide hybrid; Schwarzacher-Robinson *et al.* 1987), or, probably more often, because centromeres sometimes fused at interphase and then separated shortly before the next prophase (as reported in *Allium* by Moens & Church 1977). Volumes of centromeric structures within each nucleus varied widely, possibly reflecting (1) differences in chromosome volume which correlate with centromere volume (Bennett *et al.* 1981), (2) fusion of centromeres at interphase and (3) differential electron-density of centromeres from the two genomes in the hybrid nuclei. The relative contributions of the three and perhaps other factors is unknown.

Proportion of chromatin in the nucleus

The mean proportion of the intranuclear space occupied by electron-dense chromatin varies substantially in different parts of the interphase nucleus. For example, Fig. 1C and 1D compares sections of hybrid nucleus 213 where the chromatin occupies about 55% and 15% of the nuclear volume (in the 0.1 μm thick sections) respectively. Fig. 3 gives the percentage of chromatin in each section of nucleus 213 and shows the gradient through the nucleus. The

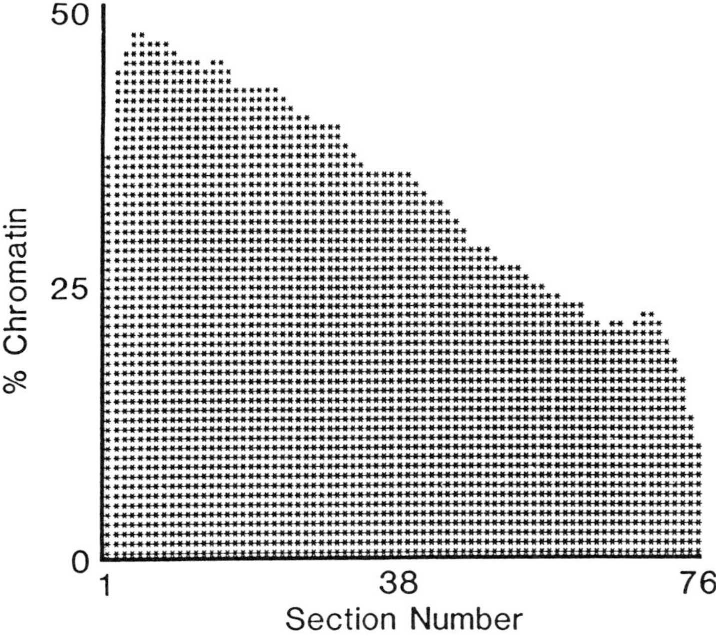

Figure 3. Graph showing the percentage of the total area of each section occupied by electron-dense chromatin (as the rolling mean of five sections) in nucleus 213 of *H. vulgare* × *S. africanum*. The area of the nucleolus was excluded from the measurements. This result was obtained because, by chance, the nucleus was sectioned approximately at right angles to the axis running between centres of the centromere and telomeric clusters at its opposite poles.

proportion of chromatin is highest in the region near the centromere *Polfeld*. At least in bread wheat, recombination sometimes occurs more frequently in regions distal to the centromere (Snape *et al.* 1985). If this phenomenon is general, it might relate to the difference in proportions of chromatin between the centromere and telomere regions of the interphase nucleus. In the hybrid nucleus 213, the sector of the nucleus furthest away from the centromeres had a much lower proportion of chromatin than any other region. Evidence from prophase reconstructions in several materials and the positions of telomeric heterochromatic chromosome segments in interphases of rye and triticale indicates that this zone, away from the centromeric structures, is occupied by the telomeres.

Conclusions

This paper has reported preliminary results from reconstructions of interphase nuclei from two different sources. There is considerable non-randomness in the interphase nucleus, with respect to both chromatin and centromere distribution. In the future, it will probably be possible to observe the positions of transcriptionally active gene segments and controlling regions within the interphase nucleus (Mathog *et al.* 1984). It is possible that the spatial inter-relationships of chromosome segments and genes may be important for the control of gene activitiy (Heslop-Harrison & Bennett 1984; Hubert & Bourgeois 1987).

Acknowledgement

We thank BP Venture Research Unit for support of this research.

References

Ashley, T. 1979. Specific end-to-end attachment of chromosomes in *Ornithogalum virens. J. Cell Sci.* 38, 357–367.

Bennett, M.D. 1988. Parental genome separation in F1 hybrids between grass species. *This volume.*

Bennett, M.D., J.B. Smith, S. Simpson and B. Wells 1979. Intranuclear fibrillar material in cereal pollen mother cells. *Chromosoma* 71, 289–332.

Bennett, M.D., J.B. Smith, J. Ward, and G Jenkins. 1981. The relationship between nuclear DNA content and centromere volume in higher plants. *J. Cell Sci.* 47, 97–115.

Boni, U. De and A.H. Mintz. 1986. Curvilinear, three-dimensional motion of chromatin domains and nucleoli in neuronal interphase nuclei. *Science* 234, 863–866.

Comings, D.E. 1968. Arrangement of chromatin in the nucleus. *Hum. Genet.* 53, 131–143.

Finch, R.A., J.B. Smith and M.D. Bennett. 1981. *Hordeum* and *Secale* mitotic genomes lie apart in a hybrid. *J. Cell Sci.* 52, 391–403.

Fussell, C.P. 1977. Telomere associations in interphase nuclei of *Allium cepa* demonstrated by C-banding. *Expl. Cell Res.* 110, 111–117.

Gregoire, M., D. Harnandez-Verdon and M. Bouteille. 1984. Visualization of chromatin distribution in living PTO cells by Hoechst 33342 fluorescent staining. *Expl. Cell Res.* 152, 38–46.

Hadlaczky, Gy., M. Went and N.R. Ringertz. 1986. Direct evidence for the non-random localization of mammalian chromosomes in the interphase nucleus. *Expl. Cell Res.* 167, 1–5.

Heslop-Harrison, J.S. 1983. The spatial disposition of chromosomes in *Aegilops umbellulata.* In *Kew Chromosome Conference II*, P.E. Brandham and M.D. Bennett, eds, 63–70.

Heslop-Harrison J.S. and M.D. Bennett. 1983. The positions of centromeres on the somatic metaphase plate of grasses. *J. Cell Sci.* 64, 163–177.

Heslop-Harrison J.S. and M.D. Bennett. 1984. Chromosome order — possible implications for development, *J. Embryol. exp. Morph.* 83, Suppl., 51–73.

Heslop-Harrison, J.S., J.B. Smith and M.D. Bennett. 1988. The absence of the somatic association of centromeres of homologous chromosomes in grass mitotic metaphases. *Chromosoma* (in press).

Hubert, J. and C.A. Bourgeois. 1986. The nuclear skeleton and the spatial arrangement of chromosomes in the interphase nucleus of vertebrate somatic cells. *Hum. Genet.* 74, 1–15.

Manuelidis, L. 1985. Indications of centromere movement during interphase and differentiation. *Ann. N.Y. Acad. Sci.* 450, 205–221.

Mathog, D., M. Hochstrasser, Y. Gruenbaum, H. Saumweber and J. Sedat. 1984. Characteristic folding patterns in *Drosophila* salivary gland chromosomes. *Nature* 308, 414–421.

Moens, P.B. and K. Church. 1977. Centromere size, positions and movements in the interphase nucleus. *Chromosoma* 61, 41–48.

Murray, B.G. 1986. Interchange quadrivalents and chromosome order at meiotic metaphase I in *Briza* L. (Gramineae). *Chromosoma* 94, 293–296.

Rabl, C. 1885. Ueber zelltheilung. *Morphologisches Jahrb.* 10, 214–330.

Rickards, G.K. 1986. Position and orientation in the metaphase equator of an interchange quadrivalent of hybrid rye. *Chromosoma* 94, 249–252.

Schwarzacher-Robinson, T., R.A. Finch, J.B. Smith and M.D. Bennett. 1987. Genotypic control of centromere positions of parental genomes in *Hordeum* × *Secale* hybrid metaphases. *J. Cell Sci.* 87, 291–304.

Sparrow, A.H. 1942. The structure and development of the chromosome spirals in microspores of *Trillium. Can. J. Res.* 20 sect. C, 257–266.

Snape, J.W., R.B. Flavell, M. O'Dell, W.G. Hughes and P.I. Payne. 1985. Intrachromosomal mapping of the nucleolar organiser region relative to three marker loci on chromosome 1B of wheat (*Triticum aestivum*). *Theor. Appl. Genet.* 69, 263–270.

Stack, S.M., D.B. Brown and W.C. Dewey. 1977. Visualization of interphase chromosomes. *J. Cell Sci.* 26, 281–299.

Tanaka, N. 1981. Studies on chromosome arrangement in some higher plants. I. Interphase chromosomes in three liliaceous plants. *Cytologia* 46, 343–357.

The spatial distribution of chromosomes in the pre-meiotic nucleus as indicated by asynapsis and chromosome stretching in an interchange quadrivalent

G.K. Rickards

Botany Department, Victoria University, Wellington, New Zealand

A variety of techniques have been used over several decades in studies on the topography of chromosomes in interphase and dividing cells. These techniques include the analysis of chromosome distributions in metaphase cells after spreading or after optical or physical reconstruction of intact cells; of the distribution of induced and spontaneous rearrangements in chromosome complements; and of the position of cloned DNA probes specific for certain chromosome segments after *in situ* hybridisation. These studies have suggested that, in different cell types and different organisms, some chromosomes occupy specific domains (territories) in the interphase nucleus. Controversy exists as to how widespread this phenomenon is and as to what its functional significance might be. The significance of some current data has also been called into question (see Comings 1980, Bennett 1983, Rappold *et al.* 1984, Libbus 1985, Hubert & Bourgeois 1986 for reviews). The present paper introduces a novel approach to the identification of non-random positioning. It uses asynapsis and chromosome "stretching" in an interchange quadrivalent to study the relative placement of two pairs of homologous chromosomes in early prophase I stages of meiosis in *Allium triquetrum* (Liliaceae).

The interchange heterozygote in question has been studied previously (see Rickards 1983 for review), and somatic chromosome studies have shown that the interchange involved the shorter arm of a large metacentric chromosome (number 4) and the short arm of a small acrocentric chromosome (number 6), with both break points being close to the centomeres (Rickards 1977 and Fig. 1A here). This makes the interchange especially interesting because homologous centromeres are carried on chromosome pairs in which one member is large and metacentric, while the other is small and acrocentric.

Assuming full homologous pairing, the expected pachytene configuration of the quadrivalent is as shown in Fig. 1B. The special point of note here is that the centromeres are located in the long, "horizontal" axis, which comprises the non-interchanged chromosome arms; while the non-centromeric, interchanged segments are in the short "vertical" axis. The present study concerns the quadrivalent at early and middle stages of prophase I in pollen mother cells

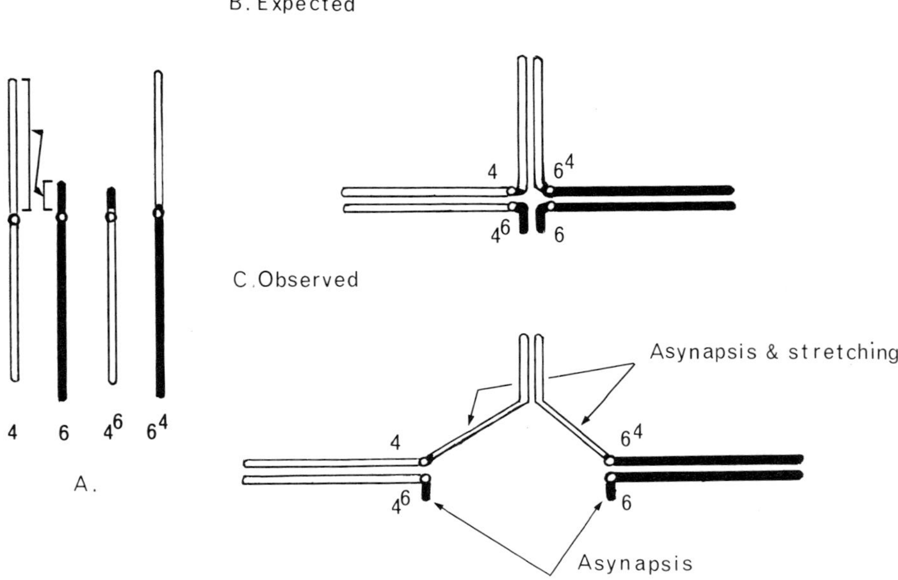

Figure 1. Diagram of the *Allium triquetrum* 4–6 interchange showing **a** the exchange, involving the shorter arm of the metacentric chromosome 4 and the short arm of the acrocentric chromosome 6, with both break points close to the centromeres; **b** the pachytene configuration expected in the heterozygote, assuming full homologous synapsis; and **c** the observed configuration illustrating asynapsis and stretching of the interchanged segments but not the non-interchanged segments.

fixed, stained and prepared as described by Rickards (1977). The configuration observed differs from expectation in two ways, namely (i) localised asynapsis and therefore reduced chiasma frequency and (ii) stretching of unpaired homologous segments, as summarised diagrammatically in Fig. 1C. These two features are interpreted as being consequences of the regular wide separation of the two pairs of centromeres of the quadrivalent, as described below.

Asynapsis

Figures 2–4 are typical examples of pachytene or early diplotene cells. The short arms of chromosomes 4⁶ and 6 (tailed arrows) are *never* synapsed. These unsynapsed arms are always approximately the same lengths relative to each other and relative to the synapsed portions of the horizontal axis. It follows, therefore, that the *long* arms of the acrocentric chromosomes (4⁶ and 6) are always fully synapsed with the homologous portions of chromosomes 4 and 6⁴.

The long arms of chromosomes 4 and 6⁴ always show *some* pairing. This pairing, however, is never complete and is always terminal, with the proximal

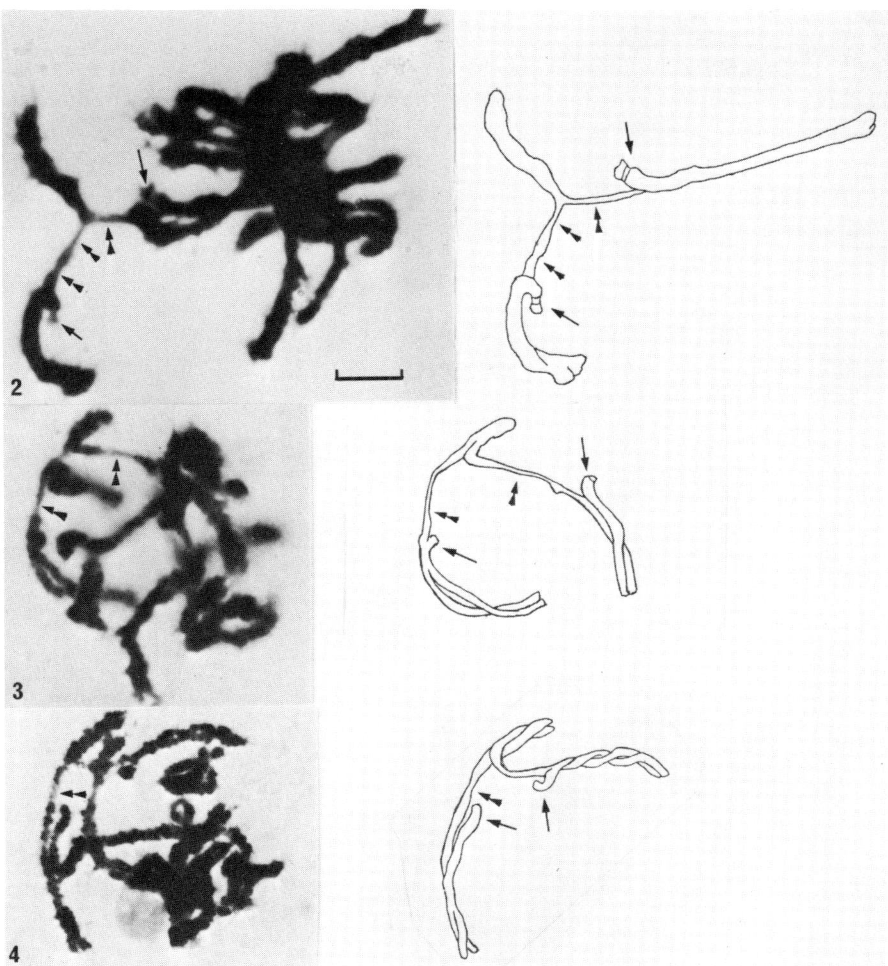

Figures 2–4. Typical examples of the interchange quadrivalent at pachytene-early diplotene, showing asynapsis in short arms of chromosomes 4^6 and 6 (tailed arrows) and asynapsis and stretching in proximal regions of the interchanged arms of chromosomes 4 and 6^4 (double headed arrows), but an absence of these features in the non-interchanged arms. Line diagrams of the quadrivalent are given at the right of each photgraph (as also in Figs 6–9). The bar in Fig. 2 (= 10 μm) applies also to Figs 5–9.

segments showing extensive, continuous asynapsis. Thus while the horizontal axis (which includes the centromere pairs) shows full synapsis, the vertical axis (containing the relatively interchanged segments) shows consistent asynapsis proximal to the break points.

Chiasmata

The maximum configuration observed for this interchange is a *chain* quadrivalent (Fig. 5A); ring quadrivalents have never been observed, neither at diplotene nor at metaphase I. Moreover, while a single chiasma (only) usually forms towards the end of the long arm pair of chromosomes 4 and 6[4], in about 4% of cells unequal bivalents of just one type are produced instead of a quadrivalent, as a result of chiasma failure in these arm pairs (Fig. 5B).

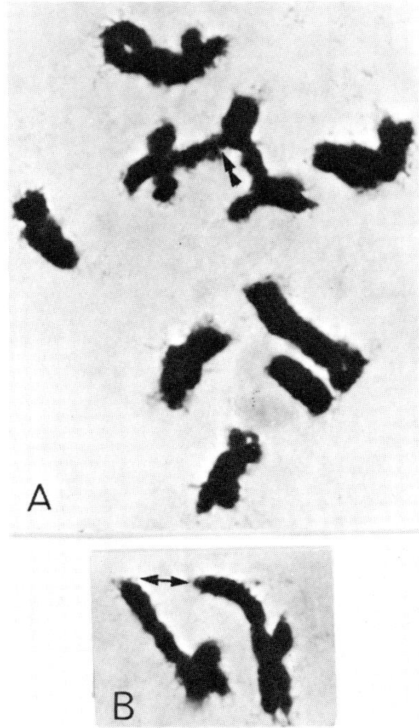

Figure 5. The interchange quadrivalent at late diplotene. **a** The maximum configuration of a chain quadrivalent in which a trace of stretching remains (double-headed arrow); and **b** the minimum configuration of a pair of unequal bivalents formed by chiasma failure in the interchanged arms of chromosomes 4 and 6[4] (arrow).

In *Allium triquetrum*, as in other species of the genus, chiasma frequency is proportional to chromosome length (Levan 1934, Rickards 1970, pp. 93–100). The mean number of chiasmata in the 4/6 quadrivalent, therefore, should be approximately 4.3, this figure being based on the summed lengths of chromosomes 4 and 6 (0.22) relative to the total length of the chromosome complement and on a mean chiasma frequency per cell of 19.6 (data as averages from 10 somatic cells and 25 diplotene cells of normal material obtained from the same population from which the interchange was derived, respectively). The *observed* mean chiasma frequency in the quadrivalent in 27 diplotene cells was 3.5. The difference between observed and expected (0.8, i.e. approaching 1 per quadrivalent) is highly significant (S.E. = 0.112, t = 7.14, degrees of freedom = 26, p≪0.001). Also, based on the behaviour of comparable acrocentric chromosomes in normal material, chiasma formation in the short arms of chromosomes 6 and 4^6 would be expected to produce ring quadrivalents in about 20% of cells, compared to zero observed.

These altered chiasma conditions are consistent with, and presumably a consequence of, the asynapsis described above for the quadrivalent.

Attenuation by stretching

The unpaired segments of chromosomes 4 and 6^4 typically are "stretched" (Figs 2–4). Stretching is indicated by weaker staining, reduced chromosome thickness and increased chromosome length. Stretching is typically uneven and unequal. Regularly these unpaired segments are longer than predicted in relation to other segments of the quadrivalent, and are usually of unequal lengths, despite their homology, because of their unequal stretching.

In later diplotene cells (Figs 6–8) stretching is very commonly seen in chromosome segments equivalent to those that show stretching in pachytene, and is similarly uneven and unequal, giving these segments an irregular and distorted appearance. Especially in diplotene cells, stretching sometimes extends "into" one or more of the three proximal chiasmata of the quadrivalent (Figs 6, 7). Stretching, however, is not found in regions distal to these three chiasmata; and is never found in the short arms of chromosomes 6 and 4^6. It has only rarely been noted in normal bivalents. Stretching can be seen in cells that have not been squashed by thumb pressure during slide preparation (Fig. 9) and is thus not an artifact. Indeed, successive degrees of squashing on the same cell indicate that chromosomes break (not necessarily at stretched regions) during excessive flattening, rather than becoming attenuated. Towards the end of diplotene only traces of stretching remain (Fig. 5A) and by metaphase I virtually all signs of the phenomenon have been lost.

Interpretation

Asynapsis round the break points of interchange configurations at pachytene is well-known (e.g. Burnham 1962), and stretching (and subsequent retraction) is known to occur in chromosomes subject to spindle and other natural and artificially applied forces (Rickards 1975 and references therein). In the present work, however, it is the consistency and the coincidence of the two phenomena (i.e. asynapsis and stretching) coupled with their confinement to specific segments of the interchange quadrivalent that is novel (though see the work of

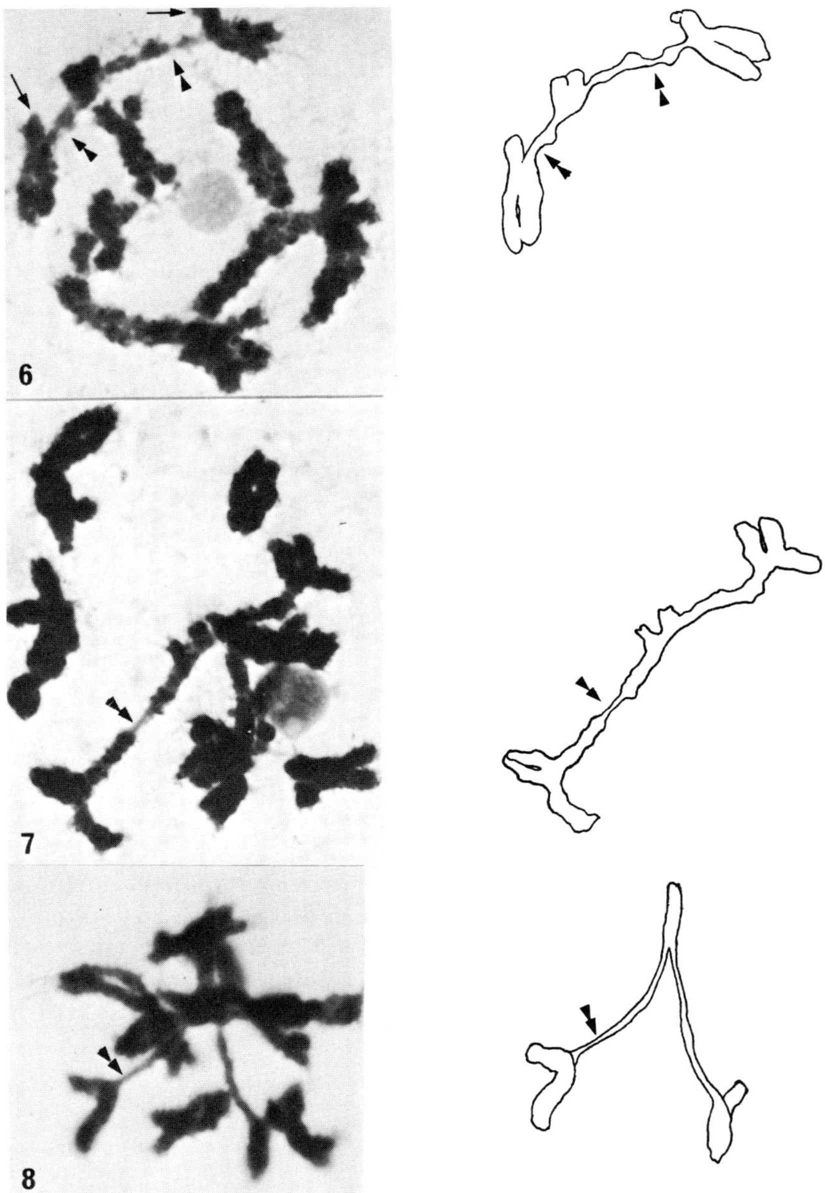

Figures 6–8. Typical examples of the interchange quadrivalent at middle diplotene showing stretching (double headed arrows) of the interchanged arms of chromosomes 4 and 6^4 up to the three most proximal chiasmata.

Hinton 1946) and requires explanation. The evidence presented suggests the following hypothesis.

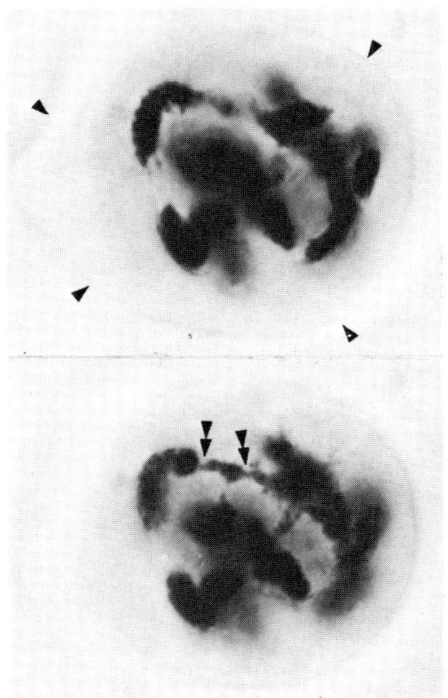
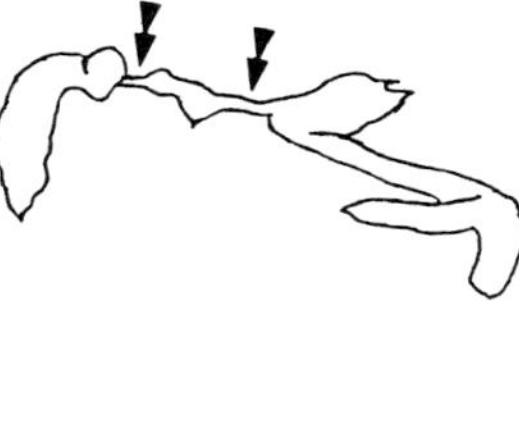

Figure 9. Photographs taken at two focal levels of an unsquashed pollen mother cell showing stretching (double headed arrows) in an undisturbed quadrivalent. The single-headed arrows locate the intact cell wall.

In normal cells, centromere pairs 4 and 6 especially, and probably also the rest of the chromosomes concerned, occupy different and somewhat widely separated domains within the nucleus in the approach to meiotic pairing (Fig. 10A). In the 4/6 interchange cells, this separation is preserved for the centromeres and the non-interchanged arms (the horizontal axis of Figs 1B, C), but is necessarily somewhat compromised for the long, interchanged arm pairs (Fig. 10B). As a consequence of this chromosome topography in interchange cells, homologous centromeres 4 and 4^6, 6 and 6^4, and their non-interchanged chromosome arms synapse readily and fully, because they are close to each other, as they would be in normal cells. This applies also to the *distal* portions of the interchanged arms of chromosomes 4 and 6^4, because of the remoteness of these chromosome ends from their widely separated centromeres. The proximal, interchanged regions of chromosomes 4 and 6^4, however, *cannot* pair, nor can the short arms of chromosomes 6 and 4^6. Forces attempting to pair and/or contract (coil) the unpaired regions of chromosomes 6 and 4^6 cause these regions to stretch, the stretching being expressed because of the distal pairing of the terminal portions of these chromosomes. Stretched regions retract into the rest of the chromosome when pairing and/or contraction forces cease or when the centromeres and chromosome ends are released from attachment to the inside of the nuclear membrane.

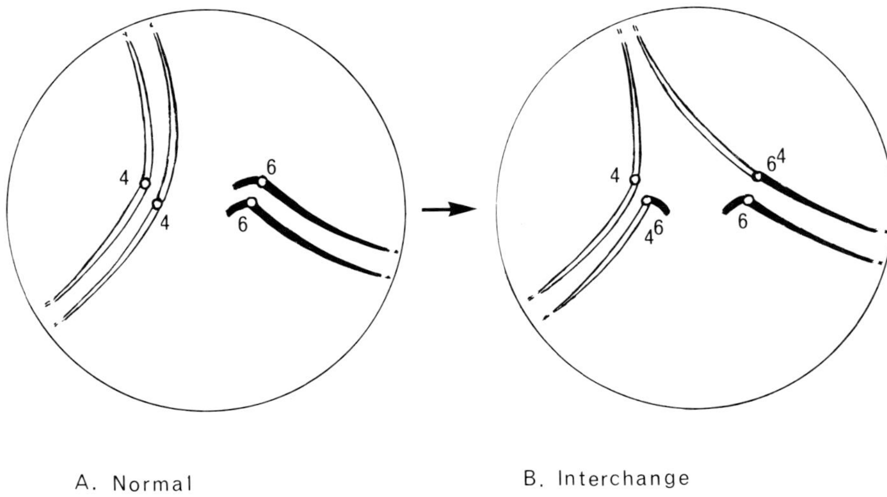

Figure 10. Predicted topography of chromosomes 4 and 6 prior to synapsis in **a** a normal cell and **b** in the 4–6 interchange cell. The view is from the cell "pole", with the centromere pairs at the pole and relatively widely separated. The chromosome arms follow the inside of the nuclear envelope towards the "anti-pole".

The observations reported here are not easily explained if chromosomes were *randomly* positioned in the nucleus. If this were so, some cells would be expected to achieve complete or nearly complete pairing in chromosomes of the quadrivalent; or to produce synapsis and asynapsis, and stretching and non-stretching, in combinations other than the one particular pattern that is consistently seen. Also, specific location of chromosome ends alone does not satisfactorily explain this particular pattern, which predicts that homologous centromeres and ends of long chromosome arms occupy set positions within the nucleus, these positions being perhaps established and maintained by attachment to specific sites on the nuclear envelope, as suggested by Comings (1980) and others.

The wide separation of the two "halves" of the quadrivalent, as represented by the two pairs of homologous centromeres (Figs 2–4 and 6–9), will form the basis of future work aimed at establishing further details of this topography, through synaptonemal complex spreading and/or physical or optical reconstruction of intact cells.

References

Bennett, M.D.1983. The spatial distribution of chromosomes. In *Kew Chromosome Conference II*, P.E. Brandham and M.D. Bennett, eds, 71–79. George Allen & Unwin.

Burnham, C.R. 1962. *Discussions in Cytogenetics*. Burgess, Minneapolis.

Comings, D.E. 1980. Arrangement of chromatin in the nucleus. *Hum. Genet.* 53, 131–143.

Hinton, T. 1946. The physical forces involved in somatic pairing in the Diptera. *J. Exp. Zool.* 102, 237–245.

Hubert, J. & C.A. Bourgeois. 1986. The nuclear skeleton and the spatial arrangement of chromosomes in the interphase nucleus of vertebrate somatic cells. *Hum. Genet.* 74, 1–15.

Levan, A. 1934. Cytological studies in *Allium*. V. *Allium macranthum*. *Hereditas* 18, 349–359.

Libbus, B.I. 1985. The ordered arrangement of chromosomes in the Chinese hamster spermatocyte nucleus. *Hum. Genet.* 70, 130–135.

Rappold, G.A., T. Cremer, H.D. Hager, K.E. Davies, C.R. Miller, & T. Yang. 1984. Sex chromosome positions in human interphase nuclei as studied by *in situ* hybridization with chromosome specific DNA probes. *Hum. Genet.* 67, 317–325.

Rickards, G.K. 1970. *Cytological and cytogenetical studies on normal and interchange Allium triquetrum.* Ph.D. thesis. Victoria University, Wellington.

Rickards, G.K. 1975. Prophase chromosome movements in living house cricket spermatocytes and their relationship to prometaphase, anaphase and granule movements. *Chromosoma* 49, 407–455.

Rickards, G.K. 1977. Prometaphase I and anaphase I in an interchange heterozygote of *Allium triquetrum* (Liliaceae). *Chromosoma* 64, 1–23.

Rickards, G.K. 1983. Orientation behaviour of chromosome multiples of interchange (reciprocal translocation) heterozygoes. *Ann. Rev. Genet.* 17, 443–498.

CHROMOSOME PAIRING AT MEIOSIS AND ITS CONTROL IN DIPLOIDS AND POLYPLOIDS

Chromosome constraints: chiasma frequency and genome size

H. Rees and R.K.J. Narayan

Department of Agricultural Botany, U.C.W., Aberystwyth, Dyfed,Wales, UK

The divergence and evolution of species, be it in form or function, reflect directly upon changes to be found in the genetic material of their chromosomes. In contrast, the divergence and evolution of the chromosomes and of the chromosome complements exhibit patterns of change which would appear to be largely unrelated to and independent of changes in the form and function of the organism. A striking consequence and illustration of such independence is the massive variation in genome size exhibited by closely-related species within many genera of plants and animals. These seemingly unrelated changes in genome size are not, however, achieved in random fashion. Many are strictly constrained, exhibiting well-defined patterns suggesting that the evolution of the chromosome complement has a momentum and direction of its own. This paper deals with some aspects of such contraints and patterns.

Patterns of change

Investigations by Narayan and his colleagues (Rees & Narayan 1977; Narayan 1982; Narayan & Durrant 1983) on diploid species of the genus *Lathyrus* provide information showing how changes in genome size conform to remarkably rigid patterns. The following are three examples.

1. The distribution of DNA amounts within the genus is discontinuous. The species fall into groups with 2C DNA amounts at intervals of about 4 picograms.
2. With increase in DNA amount the "extra" DNA is distributed equally among all seven chromosomes of the haploid complement independently of their size such that the complements become more "symmetrical" with increase in genome size.
3. As the DNA amount increases there is an increment of approximately 4 picograms of repetitive DNA relative to 1 picogram of non-repetitive DNA.

These patterns of change are remarkably well-defined. There are clearly constraints upon the way in which the complements diverge and evolve. It is when we seek to "explain" these constraints that we face problems. It could be argued that the constraints are imposed by selection, implying that the "fitness" of the chromosome complements is restricted to particular amounts, configurations and composition. Alternatively, it could be argued that the patterns displayed reflect the causal mechanisms by which the patterns arise. For example, the ratio of repetitive to non-repetitive DNA in the *Lathyrus* species may simply reflect the rate of amplification of base sequences relative to the rate of "decay" of repetitive sequences to non-repetitive DNA by base sequence divergence (Thompson & Murray 1980). There are good grounds to sustain both arguments. To date, however, there is no compelling evidence for deciding between the one and the other. What is worth emphasis, however, is that insofar as the patterns may bear upon function and fitness the patterns must impose upon the organism their own limitations or "opportunities" for further adaptive change.

The changes in genome size referred to above have been described in detail elsewhere. The main substance of this paper is concerned with the relationship between genome size and recombination at meiosis.

Recombination and genome size

The question posed is this. DNA molecules within the chromosomes have the capacity to recombine, by chiasma formation, with homologous molecules to generate new combinations of genetic material. As the DNA molecules and the chromosomes increase in size with increase in genome size, does it follow that there is a corresponding increase in the amount of recombination, an increase in the chiasma frequency?

Within complements the relationship between chromosome size and chiasma frequency is well-established, in particular by the work of Mather (1937). Leaving aside for the moment the issue of very small chromosomes, he showed that the chiasma frequency per bivalent was directly proportional to the chromosome length (we would now say the DNA content). The question therefore is whether, as the chromosomal DNA content increases with increase in genome size, there is a corresponding and proportionate increase in the chiasma frequency per bivalent?

Before describing the results of our comparison between the complements of different species it is necessary to recall Mather's findings in respect of complements containing very short chromosomes. These form one chiasma independently of length (and DNA content). On the basis of this information Mather went on to argue that the "first" chiasma formed in each and every chromosome was independent of its length. This argument has stood the test of time and it is of relevance not only when comparisons are made between the chiasma frequencies of large and very small chromosomes within complements but, also, for comparisons between species with different numbers of chromosomes. In the genera considered below, namely *Lathyrus* and *Lolium*, there are no "very short" chromosomes. Moreover, the species within each genus have the same basic chromosome number (each has n=7) so that the problems of comparing the chiasma frequencies of chromosomes between the species within each genus are confounded neither by the presence of very short chromosomes nor by different chromosome numbers (cf. Rees & Durrant 1986).

In both genera there is a considerable variation in genome size between species, with a corresponding variation in the average DNA content per chromosome. We can first test whether, as Mather found, the chiasma frequency of the chromosomes within species is proportional to their DNA content. Second, from comparisons between species, we can find out whether the increase in the DNA content of the chromosomes with increase in genome size is accompanied by a corresponding and proportionate increase in the chiasma frequency per bivalent.

a. Lathyrus

Fig. 1 shows the mean chiasma frequency per pollen mother cell plotted against the 2C nuclear DNA amount in 12 *Lathyrus* species. There is no indication whatsoever that increase in genome size, and increase in the DNA content per chromosome, affect the chiasma frequency. The impression is more apparent than real.

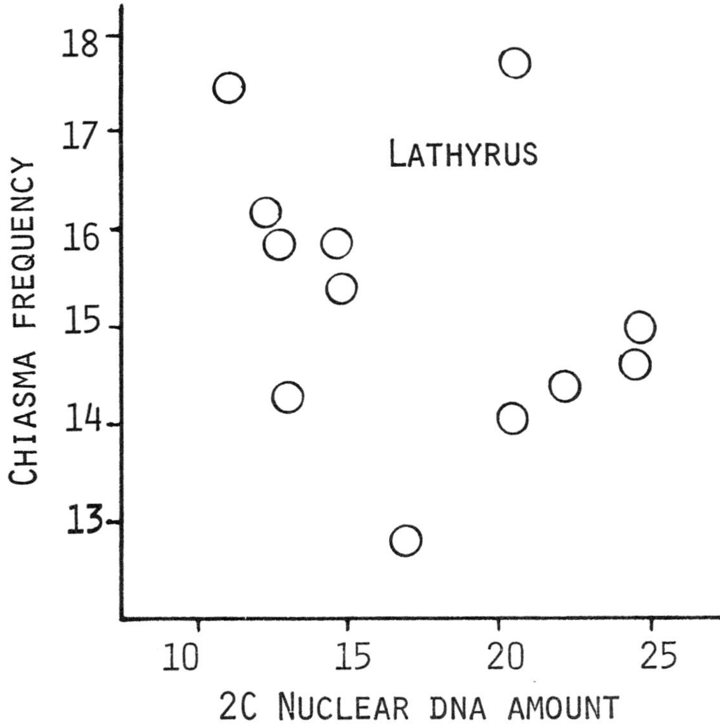

Figure 1. The mean chiasma frequency in pollen mother cells of *Lathyrus* species plotted against the 2C nuclear DNA amount. Data from Rees & Durrant (1986).

In Fig. 2 the average chiasma frequencies of individual bivalents are plotted against the DNA amount of their constituent chromosomes in four of the

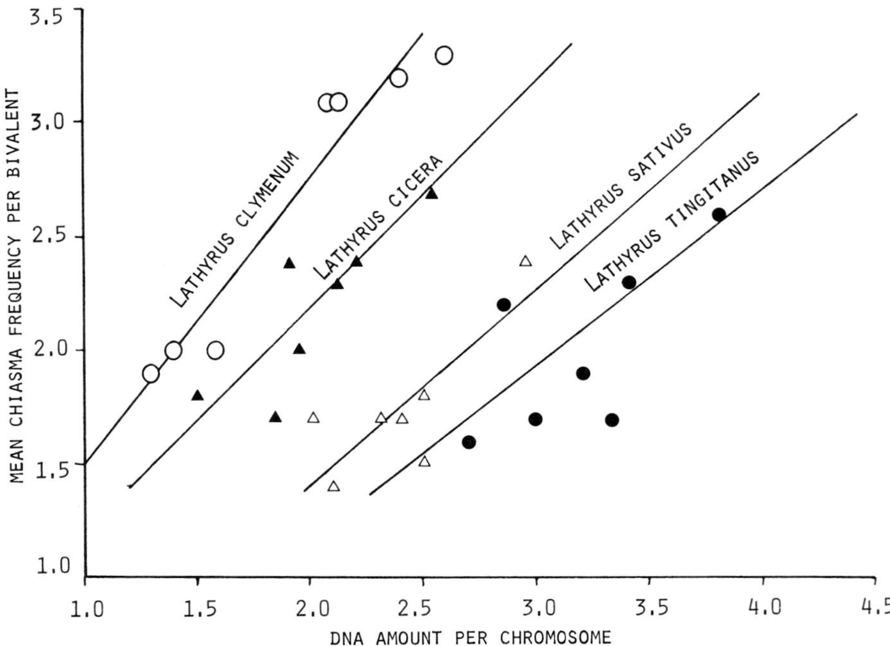

Figure 2. The mean chiasma frequencies of each of the seven bivalents plotted against the DNA amount per chromosome in each of four *Lathyrus* species. DNA amounts, from measurements of individual bivalents by microdensitometry, and their chiasma frequencies are from 20 pollen mother cells in each species. Data kindly provided by Fiona McIntyre.

Lathyrus species with genome sizes ranging from a 2C value of 13.43pg to 22.18pg. Within each species it will be seen that the chiasma frequency increases, as expected, in direct proportion to the DNA amount. A joint regression analysis of variance shows that the regressions are significant (P= <0.001), also that the regression slopes are not significantly heterogeneous. The differences between the means of the regressions, however, are highly significant (P= <0.001). From the latter it follows that the chiasma frequency differs between bivalents with similar DNA content in the different species. The difference follows a regular pattern and is related to genome size. To illustrate the point, in *L. clymenum*, which has a total nuclear 2C DNA content of 13.43 picograms, bivalents with two chiasmata have 1.4 picograms of DNA. With increase in genome size, bivalents with two chiasmata in *L. cicera* (2C=14.64) have 1.8 picograms of DNA; in *L. sativus* (2C=16.78), 2.7 picograms, and in *L. tingitanus* (2C=22.08), 3.1 picograms.

In short, the capacity for chiasma formation per unit of DNA decreases with increase in genome size. Fig. 3 also shows the same trend for chiasma frequencies of 1.5 and 3.0 chiasmata.

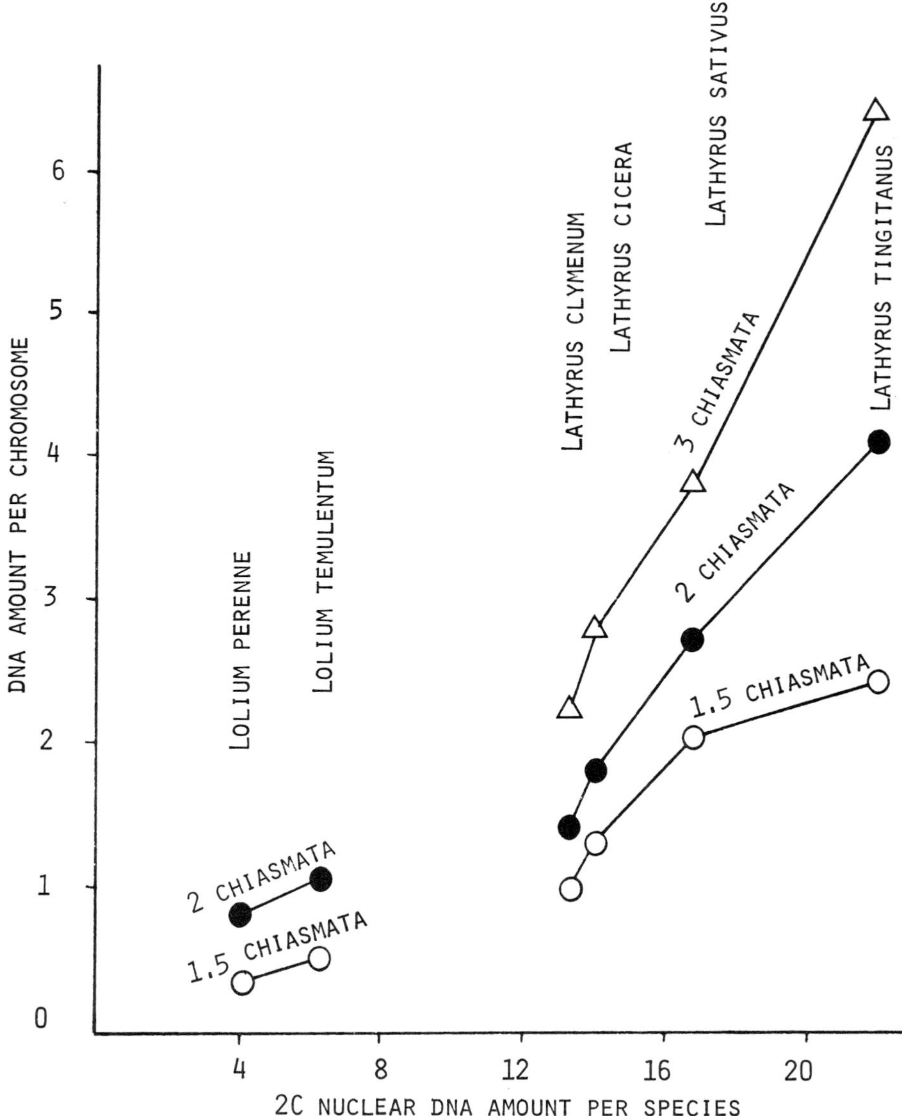

Figure 3. The DNA amount per chromosome associated with varying chiasma frequencies plotted against the 2C nuclear DNA amounts in *Lolium* and *Lathyrus* species.

b. *Lolium*

In Fig. 4 the pollen mother cell chiasma frequencies for 6 *Lolium* species are plotted against the 2C nuclear DNA amounts. In contrast to *Lathyrus* there is a convincing and significant relationship between chiasma frequency and genome size. They are positively related. The immediate impression is that the relation between chiasma frequency and genome size in *Lolium* bears no similarity with that in *Lathyrus*. The impression is misleading.

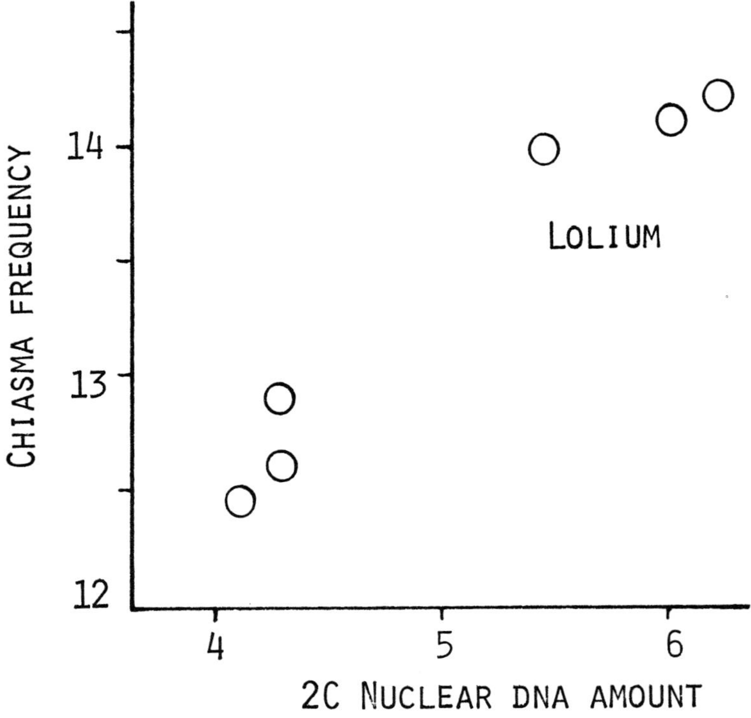

Figure 4. The mean chiasma frequency in pollen mother cells of 6 *Lolium* species plotted against the 2C nuclear DNA amounts. Data from Rees & Durrant (1986).

Fig. 5 shows the mean chiasma frequency in bivalents of *L. perenne* and *L. temulentum* plotted against the mean DNA content of their constituent chromosomes (data in Table 1). In both species the data for small and large classes of chromosomes as distinct from individual chromosomes are presented.

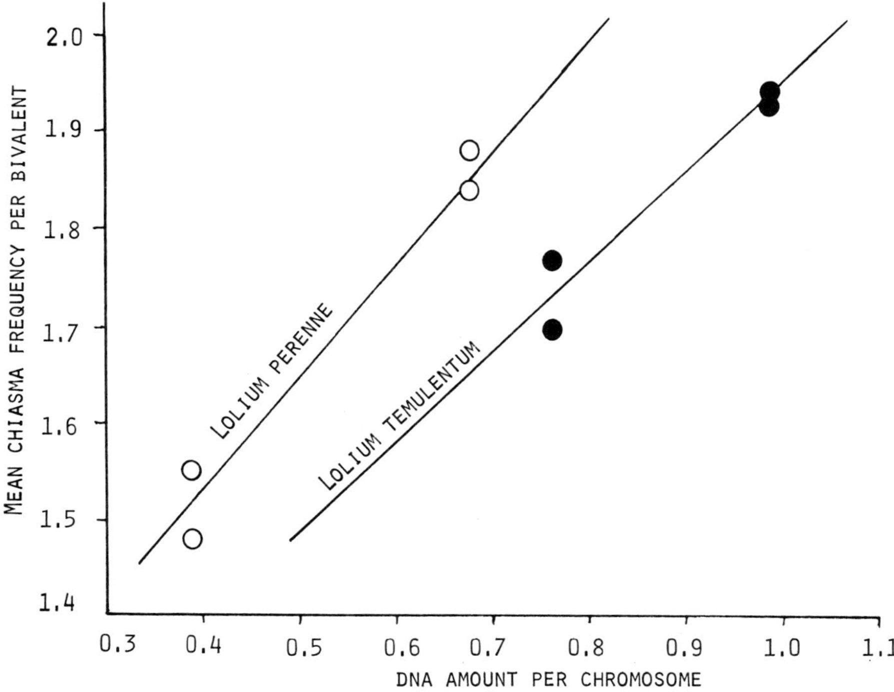

Figure 5. The mean chiasma frequencies of the large and small bivalents in the *Lolium* species plotted against the mean DNA amount of the large and small chromosomes.

Table 1. The mean chiasma frequencies in the five large and two smaller bivalents in L. perenne and in the four large and three smaller bivalents in L. temulentum. Data from 20 pollen mother cells in each of two plants in each species. Also, the mean DNA content in picograms of the chromosomes of the large and small bivalents.

	Chiasma frequency per bivalent		Mean DNA per chromosome	
	Large	Small	Large	Small
L. perenne (1)	1.84	1.48	0.68	0.39
(2)	1.88	1.55		
L. temulentum (1)	1.94	1.70	0.99	0.76
(2)	1.93	1.77		

The justification for this is that the two smaller bivalents in *L. perenne* and the three smaller bivalents in *L. temulentum* are readily identified at first metaphase. As the total nuclear DNA amounts are known, the average DNA amounts for the larger and smaller chromosomes are readily estimated since the DNA contents of the chromosomes within a complement are proportional to their lengths. These were measured in mitotic metaphases. The total 2C nuclear DNA amounts, 4.16 for *L. perenne* and 6.23 for *L. temulentum*, were given by Hutchinson, Rees & Seal (1979).

Within both species, as in *Lathyrus,* the chiasma frequency is proportional to the DNA amount. The joint regression is significant (P = <0.001) and there is no heterogeneity of slopes. As in *Lathyrus* the regression means are significantly different (P = <0.001). It follows that the chiasma frequency differs between chromosomes of similar DNA content in the two species. In *L. perenne*, which has the smaller genome (2C=4.16 picograms), the chromosomes of bivalents with 1.5 chiasmata contain 0.36 picograms, in *L. temulentum* (genome size, 2C=6.23), 0.51 picograms. As in *Lathyrus* the chiasma frequency per unit of DNA decreases with increase in genome size (Fig. 3).

Conclusion

To what extent the pattern of decreasing chiasma frequency per unit of DNA with increasing genome size applies to other genera remains to be seen. Even if it turns out to be a general rule there will certainly be exceptions, imposed by genotypic control and as consequences of chromosome structural change (Rees & Durrant 1986).

Turning to the question of the causal mechanism which underlies the pattern displayed a possibility to consider is one related to the changing composition of the chromosomal DNA with change in genome size. In both *Lathyrus* and *Lolium* increase in genome size is accompanied by a disproportionate increase in the amount of repetitive DNA sequences (Hutchinson, Narayan & Rees 1980). Certainly in chromosome segments comprising very highly repetitive sequences, as in heterochromatin, chiasma formation is precluded. It could well be that a substantial component of the "extra", repetitive DNA fraction is not involved in chiasma formation, such that the number and extent of the "effective" pairing segments at meiosis do not increase in proportion to the change in the total DNA amount.

Whatever the causal mechanism there remains the question as to whether this pattern of change in chiasma frequency is imposed by selection or is an inevitable consequence of the amplification phenomenon which gives rise to variation in genome size. The same question was posed earlier with respect to the distribution of quantitative DNA change, among species, among chromosomes within complements, between repetitive and non-repetitive components. A strong case could be made for either but the evidence to hand is inconclusive. What can be said, and it is worth repeating and emphasising, is that the patterns displayed impose their own constraints upon future change. In the context of recombination the genome size may well determine the potential and the capacity for change and adjustment in the chiasma frequency and, in turn, the potential and the capacity for generating new gene combinations, new genotypes for exposure to selection.

References

Hutchinson, J., R.K.J. Narayan and H. Rees 1980. Constraints upon the composition of supplementary DNA. *Chromosoma* 78, 137–146.

Hutchinson, J., H. Rees and A.G. Seal 1979. An assay of the activity of supplementary DNA in *Lolium*. *Heredity* 43, 411–421.

Mather, K. 1937. The determination of position in crossing-over. 2. The chromosome length-chiasma frequency relationship. *Cytologia, Fujii Jubilee Vol.*, 514–526.

Narayan, R.K.J. 1982. Discontinuous DNA variation in the evolution of plant species: the genus *Lathyrus*. *Evolution* 36 (5), 877–891.

Narayan, R.K.J. and A. Durrant 1983. DNA distribution in chromosomes of *Lathyrus* species. *Genetica* 61, 47–53.

Rees, H. and A. Durrant 1986. Recombination and genome size. *Theor. Appl. Genet.* 73, 72–76.

Rees, H. and R.K.J. Narayan 1977. Evolutionary DNA variation in *Lathyrus*. In *Chromosomes Today*: Vol. 6. A. de la Chappelle and M. Sorsa, eds, Elsevier/North Holland Biomedical Press, Amsterdam.

Thompson, W.F. and M.G. Murray 1980. Sequence organisation in pea and Mung bean DNA and a model for genome evolution. In: *Plant genome*: D.R. Davies and D.A. Hopwood, eds, The John Innes Charity, England.

Pairing autonomy and chromosome size

Y. Hamey, M.T. Abberton, A.J. Wallace and R.S. Callow

Department of Cell and Structural Biology, The University, Manchester, M13 9PL, UK

A century ago, Flemming (1887) observed what Weismann (1887) had predicted as the necessary sexual counterpart to fertilisation and the essential features of meiotic pairing soon began to emerge (Winiwarter 1900; Grégoire 1910; Wilson 1912). The potential of autopolyploidy for investigations of the mechanics of synapsis was fully appreciated by Darlington (1929a). Synapsis being pairwise, multiple associations involve separate regions of one chromosome associating with different partners. The minimum number of such pairing regions must be one more than the maximum number of changes in partner. In autopolyploids of *Tulipa* and *Hyacinthus*, the number of pairing partner exchanges was found to increase with chromosome size to a maximum of six (Newton & Darlington 1929; Darlington 1929b).

Despite occasional observations of triple synapsis in vertebrates (Comings & Okada 1971; Wallace & Hultén 1983; Dollin & Murray 1984), the general pairwise pattern of association in polyploids has been confirmed (Moens 1969; Loidl & Jones 1986; Gillies, Kuspira & Bhambhani 1987). The question of the number of synaptic sites has proved more controversial. Klingstedt (1937) first pointed out that random association at two independent sets of pairing initiators should give rise to a quadrivalent frequency of two-thirds in autotetraploids. Because autotetraploids rarely exceed this value, many organisms were assumed to have no more than two pairing initiators, possibly at the telomeres (Sved 1966). Larger numbers of independent synaptic sites, associating at random, would produce higher multivalent frequencies (Callow & Gladwell 1984). Several lines of evidence argue against the case for only two pairing initiators. Firstly, in addition to Darlington's observations, as many as eight partner exchanges have been observed in triploid *Allium* (Loidl & Jones 1986), five in tetraploid *Schistocerca* (John & Henderson 1962) and four in autotetraploid *Triticum* (Gillies, Kuspira & Bhambhani 1987). Secondly, even homologous chromosomes in autotetraploids can show differential affinity (Giraldez & Santos 1981) so that only isogenic tetraploids are certain to display random association (Callow, Hamey & Pattrick 1984). Large numbers of synaptic sites can therefore be compatible with low multivalent frequencies. Thirdly, there is an impressive body of molecular (Hotta, Tabata & Stern 1984) and ultrastructural (Loidl & Jones 1986) evidence indicating that the number of synaptic sites is far greater than might be predicted even from the highest known frequencies of pairing partner exchange. The inter-relations of number of synaptic sites, pairing partner exchange and multivalent frequency will be examined in this paper and we shall emphasise the responses of such relations to increases in chromosome size.

Variation in the incidence of pairing partner exchange

Statistical studies of variation in the incidence of pairing partner exchange (r) have been possible only since the surface spreading technique has been extended to autopolyploid plant PMCs. Three species have been examined in detail: triploid *Allium sphaerocephalon* (Loidl & Jones 1986), tetraploid *A. vineale* (Loidl 1986) and tetraploid *Triticum monococcum* (Gillies, Kuspira & Bhambhani 1987). With the data provided, we can examine the question of whether the frequency distribution of pairing partner exchange is binomial in form, following the action of a small number of widely spaced synaptic sites, or whether it approximates to a Poisson distribution, where the number of sites is very large but the distances between successive sites, the synaptic intervals, are very small. In both *Allium* species, the Poisson gives the better fit, indicating that pairing is quasi-continuous across a large number of synaptic sites (Table 1). By contrast, neither distribution can be taken to represent the data from Einkorn

Table 1. Variation in the incidence of pairing partner exchange, identified by pachytene spreading of three autopolyploids, compared first with binomial and then with Poisson expectations (italics). For the binomials, i is taken to be the minimum number of synaptic sites commensurate with the maximum number of pairing partner exchanges.

Organism	2n	No. of pairing partner exchanges (r)										Sets	Goodness of fit
		0	1	2	3	4	5	6	7	8	>8		
Allium sphaerocephalon (Loidl & Jones, 1986)	3x 24	1	9	28	28	19	13	8	2	3	0	111	
i=9, ρ=0.423		(9)	21	30	28	16	(	7		)			$x^2(3) = 10.7$ $P < 0.05$
r=3.387, V=2.785, Sr=0.16 α=0.0090, Sα=0.0089		(17)	22	24	21	14	8	(	6	)			$x^2(5) = 5.5$ $P > 0.30$
Allium vineale (Loidl, 1986)	4x 32	17	33	21	10	3	1	0	0	0	0	85	
i=6, ρ=0.287		16	32	25	(			12			)		$x^2(1) = 1.2$ $P > 0.20$
r=1.435, V=1.249, Sr=0.12 α=0.200, Sα=0.043		20	29	21	10	(		5			)		$x^2(3) = 1.3$ $P > 0.70$
Triticum monococcum (Gillies et al., 1987	4x 28	65	75	7	4	3	0	0	0	0	0	154	
i=5, ρ=0.184		68	62	21	(			3			)		$x^2(1) = 16.0$ $P < 0.001$
r=0.734, V=0.680, Sr=0.07 α=0.422, Sα=0.040		74	54	19	(			6			)		$x^2(2) = 17.5$ $P < 0.001$

wheat. As the variance in r is less than its mean (0.68 vs 0.73), this may be a case of positive interference, which would be in keeping with observations of extensive asynapsis around at least some exchange regions (Gillies, Kuspira & Bhambhani 1987).

Autonomous synaptic sites

While a small number of pairing sites may remain independent if widely spaced, they are ever less likely to do so as their proximity increases with number. Loss of independence is not, however, the same as loss of autonomy. An autonomous synaptic site may be defined as: the smallest region of a chromosome with the capacity to initiate pairing partner exchange in an autopolyploid regardless of the likelihood of such exchange.

Mechanics of multivalent formation

The formation of pachytene multiples may be considered from a mechanical point of view where pairing is governed by i autonomous synaptic sites separated by i–1 intervals each with a constant likelihood, ρ, of including a pairing partner exchange. Because pachytene bivalents arise in the absence of exchange, their frequency (α) in these circumstances would be $(1-\rho)^{i-1}$. Taking logarithms onboth sides we have: $\ln\alpha = (i-1)\ln(1-\rho)$, (1). In odd-numbered polyploids showing pairwise synapsis, a set of homologues forming bivalents would also include a univalent (Fig. 1). This approach has the advantage that ρ can account for both differential affinity and inter-dependence. It can be extended to any size of association, any ploidy and any number of pairing sites, including the limit as i tends to infinity and ρ tends to zero (Callow & Gladwell 1984, 1988).

Tests of this model require more PMCs than can be studied realisitically by the pachytene spreading method at present. The material must also be autopolyploid and of sufficient ploidy that sets of homologues can form at least three pairing patterns: one more than the number of unknown parameters i and ρ. Sixty high chiasma frequency sets of six homologous large acrocentrics from autohexaploid *Crepis capillaris* form three bivalents, a bivalent plus a quadrivalent or a hexavalent with frequencies that correspond with those predicted by the model, both with i=2 and in the continuous case (Table 2). A much larger sample would be required to distinguish between these extremes.

Pairing mechanics in relation to chromosome size

In autopolyploids with a wide range of chromosome sizes the larger chromosomes tend to form more multivalents than the smaller, both because of their higher incidence of pairing partner exchange and their greater chiasma frequencies (John & Henderson 1962). Influences of chiasma variation can be reduced by restricting samples to PMCs with sufficient chiasmata for each set of homologues to be represented by the maximum size of multivalent.

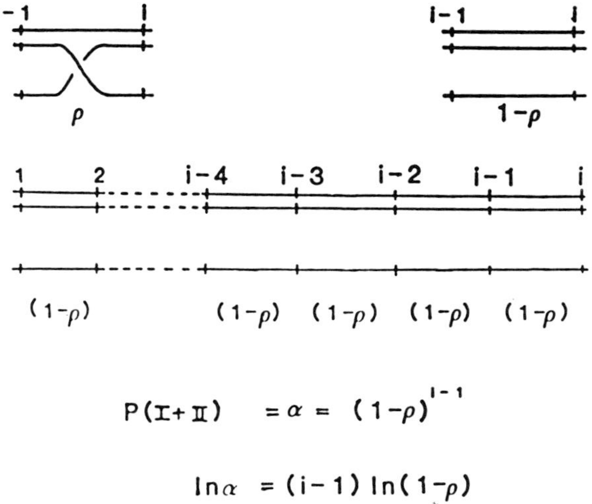

$$P(I+II) = \alpha = (1-\rho)^{i-1}$$

$$\ln\alpha = (i-1)\ln(1-\rho)$$

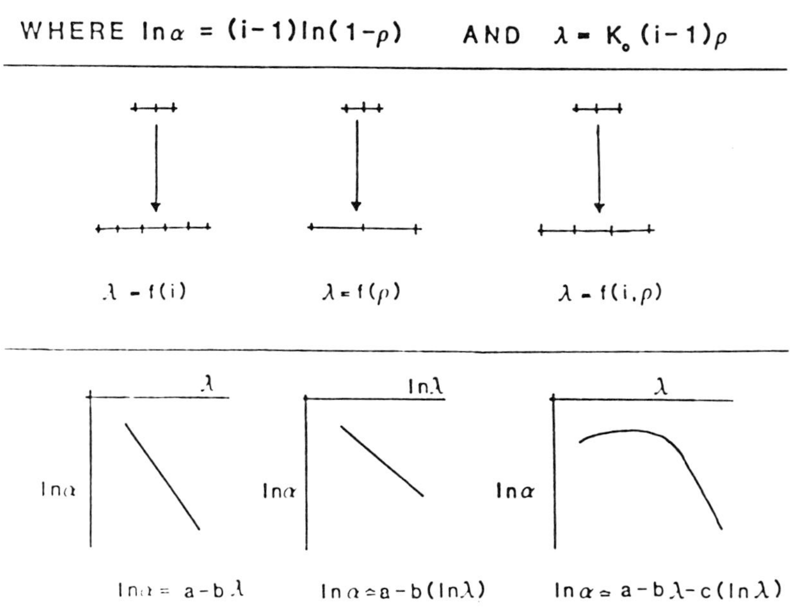

Figure 1. Diagrams of the model relating bivalent frequency (α) to chromosome length (λ), both features being expressed in terms of pairing at i synaptic sites separated by i–1 synaptic intervals each with the same probability of pairing partner exchange (ρ).

Detailed evidence is provided by chromosome-specific studies of autotriploids and autotetraploids of *Crepis capillaris* and *C. rupra*. All three sets of homologues can be distinguished at meiosis in *C. capillaris* on the basis of size and centromere position. Chromosomes 1 and 3 are acrocentrics whereas chromosome 2 is subtelocentric. They display a distinct series of mitotic lengths (4.8, 4.0, 2.6 μm) (Fig. 3). *C. rubra* has two large acrocentrics (5.3 and 4.7 μm), a subtelocentric (3.2 μm) and two small metacentrics which are indistinguishable in length (2.9 μm). In all, there are three size groups whose pairing may be analysed in *C. capillaris* and four such groups in *C. rubra*. Before we examine this evidence let us consider what we might expect from the pairing model.

We have already seen that bivalent frequency (α) is related to the number of synaptic intervals (i-1) and probability of pairing exchange (ρ) by equation (1): $\ln\alpha = $ (i-1) ln (1-ρ). Taking ρ to be a measure of each synaptic interval, we may assume that chromosome length, λ, is proportional to the product of the number and size of synaptic intervals: λ = K0 (i-1) ρ, (2), where K0 is constant. We now consider three possibilities relating to variation in length (Fig. 1) (Callow & Gladwell 1980). Firstly, increases in length may result from increases in the number but not the size of synaptic intervals such that λ is a function of i (λ = f(i)). Under these circumstances the logarithm of pachytene bivalent frequency is expected to decline in linear fashion with increases in chromosome length: ln α =-K1 λ, (3). The simple alternative is that length reflects the size of synaptic

Table 2.   Behaviour of Chromosome 1 in high chiasma frequency sets of hexaploid
Crepis capillaris, compared with expectations based on discrete and continuous
synapsis (Callow & Gladwell 1984). $\alpha = (1-\rho)^{i-1} = 0.2556$, $\theta = (\,2(1-\rho)^2 \,/\, (2-\rho)\,)^{i-1}$.

		Discrete		Continuous	
Pairing	Sets				
pattern	5–6 Xta	i = 2 $\rho = 1-\alpha^{1/2}$		lim i to inf. ρ to 0	
		Probability	Sets	Probability	Sets
3II	7	θ	(6.3)	$\alpha 3/2$	(7.8)
1II 1IV	25	$3(\alpha-\theta)$	(27.3)	$3\alpha(1-\alpha^{1/2})$	(22.7)
1VI*	28	$1-3\alpha+2\theta$	(26.4)	$(1-\alpha^{1/2})^2(1+2\alpha^{1/2})$	(29.5)
		$\chi^2(1) = 0.368$ P > 0.50		$\chi^2(1) = 0.391$ P > 0.50	

* includes 3(1I 1V) and 2(2III), each assumed to result from partial desynapsis
of a pachytene hexavalent.

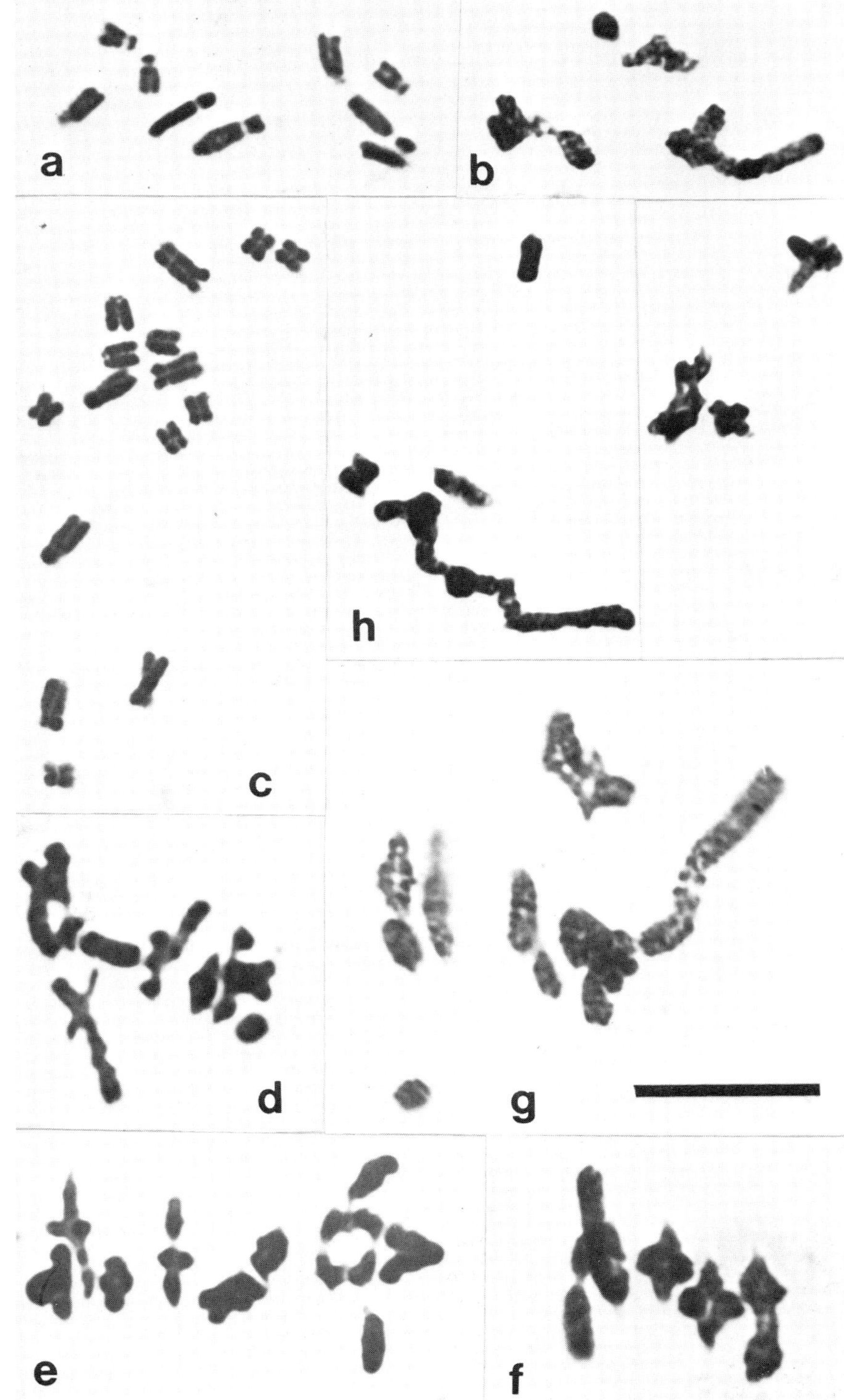

intervals rather than their number ($\lambda = f(\rho)$): $\ln \alpha = K2 + K3 \ln (K4 - \lambda)$, (4), K3 being a positive constant and K4 being always greater than λ. Finally, we may put these expressions together and consider the situation where both the number and size of synaptic intervals vary with length ($\lambda = f(i, \rho)$): $\ln \alpha = K5 + K6 \lambda + K7 \ln (K8 - \lambda)$, (5). As the constant K8 has to be greater than λ, we may take it to represent either total mitotic length or total nuclear DNA, depending on how we measure λ. The other constants (K5, K6 and K7) can be estimated by the techniques of multiple regression but, as K6 and K7 are disproportionately influenced by the choice of K8, we cannot gauge the relative importance of number and size of synaptic interval as influences on pairing behaviour. We therefore use the simpler relation: $\ln \alpha = a - b \lambda - c \ln \lambda$, (6), which provides a good approximation to equation (5) within the range of λ being measured. Similarly, equation (4) may be replaced by: $\ln \alpha = a - b \ln \lambda$, (7) (Fig. 1).

Returning to the *Crepis* data, we see that reliable estimates of α have been obtained from large samples of high chiasma frequency sets of homologues (Table 3). Chromosome size has been estimated by apportioning estimates of total nuclear DNA (Bennett & Smith 1976) according to mitotic length. While not as satisfactory as using photometric observations, this approach does at least permit direct comparisons between chromosomes of the two species. At each ploidy, the logarithm of bivalent frequency declines significantly with increasing chromosome length, indicating an increase in the number of synaptic intervals (Table 3). The improvement in fit achieved by the multiple regression is also highly significant in each case. Clearly, length must be a function of both the number and sizes of synaptic intervals. The relative sizes of the linear and logarithmic partial regression coefficients (b, c) suggest that both influences are of similar magnitude. The coefficients are of opposite sign, possibly indicating that the number and sizes of synaptic interval do not increase consistently.

In *C. capillaris*, multivalent frequencies are similar at each ploidy for the large acrocentric, lower in triploids for the subtelocentric but lower in tetraploids for the small acrocentric (Fig. 3). In *C. rubra*, by contrast, each chromosome group forms significantly more trivalents in autotriploids than it does quadrivalents in autotetraploids. The signs of the partial regression coefficients in these triploids are the reverse of those in the related tetraploids so that the curve of best fit is inverted. These findings in *C. rubra* correspond with the observations of higher frequencies of pairing partner exchange in triploid *Allium sphaerocephalon* by comparison with tetraploid *A. vineale* (3.4 vs 1.4) (Table 1). One possible

Figure 2. Chromosomes of autopolyploid *Crepis*. **a** Complement of triploid *C. capillaris* (2n=9). **b** Metaphase I in triploid *C. capillaris*: III_1, III_2, I_3, II_3. **c** Complement of triploid *C. rubra* (2n=15). **d** Metaphase I in triploid *C. rubra*: III_1, I_2, II_2, III_3, $(I, II, III)_{4, 5}$. **e** Metaphase I in tetraploid *C. rubra*: $2II_1$, $2II_2$, $2II_3$, $(2II, IV)_{4, 5}$. **f** Metaphase I in tetraploid *C. capillaris*: IV_1, $2II_2$, IV_3. **g,h** Metaphase I in hexaploid *C. capillaris*. **g** II_1, IV_1, I_2, II_2, III_2, $2I_3$, $2II_3$. **h** VI_1, I_2, II_2, III_2, $3II_3$. Homologous groups identified as subscripts. Bar = 10 μm.

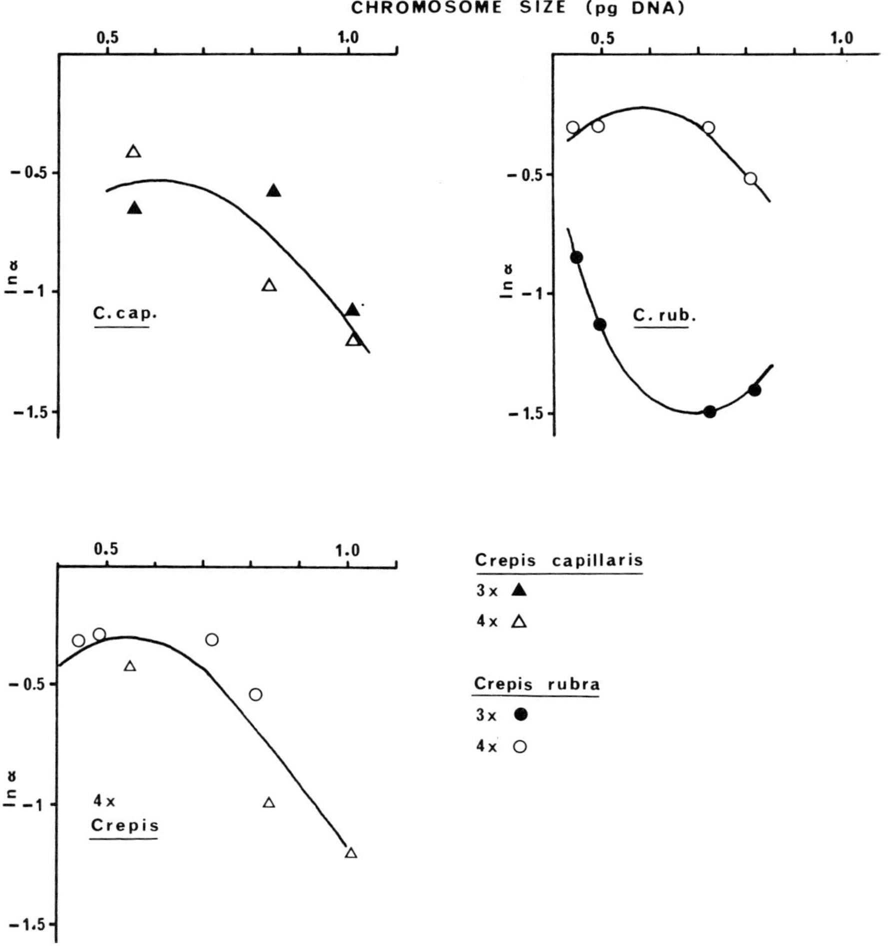

Figure 3. Multiple regressions relating the logarithm of bivalent frequency (1 nα) to chromosome length (λ) in triploid and tetraploid *Crepis*.

explanation is that pairing is slower in triploids because one synaptic site interferes with the association of its two homologues. Such circumstances may favour pairing partner exchange. The differential behaviour of the chromosomes of *C. capillaris* may reflect the relatively advanced position of this species in the genus (Babcock 1947).

Table 3. Estimates of bivalent frequency (α) from high chiasma frequency sets in autopolyploid Crepis (Roman) together with multiple regression analyses of their relations to chromosome length (λ) (Italics).

Chr.	λ DNA (pg)	Triploids α	$S\alpha$	Sets	Tetraploids α	$S\alpha$	Sets
Crepis capillaris							
1	1.01	0.335	0.021	486	0.300	0.021	482
2	0.84	0.556	0.027	340	0.370	0.029	278
3	0.55	0.510	0.031	259	0.660	0.029	272

3x+4x	$\ln\alpha = 0.235 - 1.326\,\lambda,$	$F_1/_{2115} = 5688.6,\ P < 0.001$
	$\ln\alpha = 5.246 - 6.369\,\lambda + 3.821\,\ln\lambda,$	$F_1/_{2114} = 590.8,\ P < 0.001$

Chr.	λ DNA (pg)	Triploids α	$S\alpha$	Sets	Tetraploids α	$S\alpha$	Sets
Crepis rubra							
1	0.81	0.245	0.061	49	0.589	0.025	392
2	0.72	0.222	0.062	45	0.732	0.027	272
3	0.49	0.323	0.080	34	0.739	0.034	165
4/5	0.44	0.429	0.062	63	0.732	0.023	369

3x:	$\ln\alpha = -0.232 - 1.580\,\lambda,$	$F_1/_{189} = 986.1,\ P < 0.001$
	$\ln\alpha = -12.361 + 11.480\,\lambda - 7.869\,\ln\lambda,$	$F_1/_{188} = 117091.6,\ P < 0.001$
4x:	$\ln\alpha = -7.203 - 0.490\,\lambda,$	$F_1/_{1196} = 1764.8,\ P < 0.001$
	$\ln\alpha = 6.533 - 7.589\,\lambda + 4.291\,\ln\lambda,$	$F_1/_{1195} = 5603.1,\ P < 0.001$

Because of the small number of points being fitted, the validity of these analyses rests on the low standard errors (Table 3) and on the form of the fitted curve which has been determined *a priori* on theoretical grounds. The next steps must be to extend this analysis not only to other species of *Crepis* but also to other genera. An obvious candidate is *Lathyrus*: a large, predominantly diploid genus displaying extensive variation in chromosome size (Fig. 4), largely in relation to moderately repetitive DNA sequences (Narayan 1983). Such sequences may augment synaptic intervals or even constitute new synaptic sites. Both possibilities are testable by the multiple regression technique.

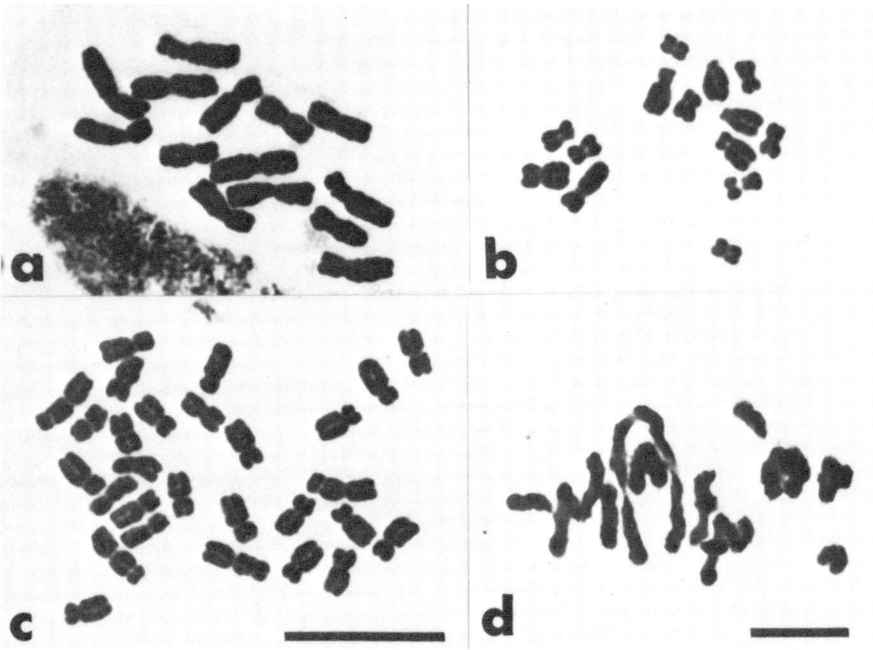

Figure 4. a Complement of diploid *Lathyrus sylvestris* (2n=14). **b** Complement of diploid *Lathyrus articulatus* (2n=14). **c** Complement of tetraploid *Lathyrus pratensis* (2n=28). **d** Metaphase I in tetraploid *Lathyrus pratensis*: 2I, 5II, 4IV. Bar = 10 µm, Scale a = b, c.

Acknowledgements

We thank SERC for financial support and Mrs S. Crowe for technical help.

References

Babcock, E.B. 1947. Cytogenetics and speciation in *Crepis. Adv. Genet.* 1, 69–93.

Bennett, M.D. and J.B. Smith 1976. Nuclear DNA amounts in Angiosperms. *Phil. Trans. R. Soc. B* 274, 227–274.

Callow, R.S. and I. Gladwell 1980. Chromosome synapsis in hexaploids. *J. theor. Biol.* 87, 703–722.

Callow, R.S. and I. Gladwell 1984. A general treatment of chromosome synapsis in even-numbered polyploids. *J. theor. Biol.* 106, 455–494.

Callow, R.S. and I. Gladwell 1988. Chromosome synapsis in odd-numbered polyploids. *J. theor. Biol.* (in press).

Callow, R.S., Y. Hamey and S.M. Pattrick 1984. Pairing of identical chromosomes in an isogenic tetraploid. *Heredity* 53, 107–111.

Comings, D.E. and T.A. Okada 1971. Triple chromosome pairing in triploid chickens. *Nature* 231, 119–121.

Darlington, C.D. 1929a. The significance of chromosome behaviour in polyploids for the theory of meiosis. In *Conference on polyploidy* (John Innes Horticultural Institute), pp. 42–44.

Darlington, C.D. 1929b. Meiosis in polyploids. II. Aneuploid hyacinths. *J. Genetics* 21, 17–56.

Dollin, A.E. and J.D. Murray 1984. Triple chromosome pairing in an aneuploid bull spermatocyte. *Can. J. Genet. Cytol.* 26, 782–783.

Flemming, W. 1887. Neue Beiträge zur Kenntniss der Zelle. *Arch. mikrosk. Anat.* 29, 389–463.

Gillies, C.B., S. Kuspira and R.N. Bhambhani 1987. Genetic and cytogenetic analyses of the A genome of *Triticum monococcum*. IV. Synaptonemal complex formation in autotetraploids. *Genome* 29, 309–318.

Giraldez, R. and J.L. Santos 1981. Cytological evidence for preferences of identical over homologous but non-identical meiotic pairing. *Chromosoma* 82, 447–451.

Grégoire, V. 1910. Les cinèses des maturation dans les deux règnes. L'unité essentielle du processus mèiotique. (Second memoire). *Cellule* 26, 221–422.

Hotta, Y., S. Tabata and H. Stern 1984. Replication and nicking of zygotene DNA sequences. Control by a meiosis-specific protein. *Chromosoma* 90, 243–253.

John, B. and S.A. Henderson 1962. Asynapsis and polyploidy in *Schistocerca paranensis*. *Chromosoma* 13, 111–147.

Klingstedt, H. 1937. On some tetraploid spermatocytes in *Chrysochraon dispar*. *Memor. Soc. Fauna et Flora Fenn.* 12, 194–209.

Loidl, J. 1986. Synaptonemal complex spreading in *Allium*. II. Tetraploid *A. vineale*. *Can. J. Genet. Cytol.* 28, 754–761.

Loidl, J. and G.H. Jones 1986. Synaptonemal complex spreading in *Allium*. I. Triploid *A. sphaerocephalon*. *Chromosoma* 93, 420–428.

Moens, P.B. 1969. The fine structure of meiotic chromosome pairing in the triploid *Lilium tigrinum*. *J. Cell Biol.* 40, 273–279.

Narayan, R.K.J. 1983. Chromosome changes in the evolution of *Lathyrus* species. In *Kew Chromosome Conference II*, P.E. Brandham and M.D. Bennett, eds, 243–250. London UK. George Allen and Unwin.

Newton, W.C.F. and C.D. Darlington 1929. Meiosis in polyploids. I. Triploid and pentaploid Tulips. *J. Genetics* 21, 1–15.

Sved, J.A. 1966. Telomere attachment of chromosomes, genetical and cytological consequences. *Genetics* 53, 747–756.

Wallace, B.M.N. and M.A. Hultén 1983. Triple chromosome synapsis in oocytes from a human foetus with trisomy 21. *Ann. Hum. Genet.* 47, 271–276.

Weismann, A. 1887. *Ueber die Zahl der Richtungskörper und über ihre Bedeutung für die Verebung.* Jena.

Wilson, E.B. 1912. Studies on chromosomes VIII. Observations on the maturation phenomena in certain hemiptera and other forms, with considerations on synapsis and reduction. *J. Exp. Zool.* 13, 345–448.

Winiwarter, H. von 1900. Recherches sue l'ovogenèse et l'organogènse de l'ovaire des Mammifères (Lapin et Homme). *Archs Biol., Paris* 17, 33–199.

Genetic control of chromosome pairing in polyploids

G.M. Evans

Department of Agricultural Botany, University College of Wales, Penglais, Aberystwyth, Dyfed SY23 3DD, UK

The term "polyploid" is used to denote a range of variants which contain more than two sets of chromosomes. The relationship between these sets ranges from that where all are identical to that where they are all different, although the latter condition would be unlikely except under controlled crossing. In natural tetraploids the two extremes would normally be an autotetraploid with four identical sets of chromosomes and an allotetraploid with two distinct sets (genomes) each represented twice. Intermediate types where the two basic genomes are partially differentiated are not uncommon, especially in synthetic polyploids. In fact this is the expectation in most synthetic interspecific polyploids.

Although an increase in chromosome number often has a dramatic effect on the morphological and phsyiological characteristics of plants within a species or genus, the behaviour of chromosomes at meiosis is often equally important in determining the chances of survival of a new polyploid. This is on account of the profound modification of chromosome association at meiosis, which will inevitably occur when more than two homologous or homoeologous sets of chromosomes are available for pairing at meiosis, and the consequences of this on fertility and stability of subsequent generations. In artificially-created polyploids, ranging from those based on a single species or genotype to those based on two or more, there is almost invariably a degree of "irregularity" in the nature of chromosome association and separation at meiosis. This is not normally seen in natural polyploids. They have evolved a genetic system which ensures that meiosis proceeds in a regular and predictable manner, usually by the formation of homologous bivalents which will separate equationally at anaphase I of meiosis.

There is no doubt that such systems, although not perfect, do arise from time to time in artificially induced polyploids (Evans & Macefield 1973; Taylor & Evans 1976; Lewis 1980; Evans & Davies 1983; Taing Aung & Evans 1985). This short review sets out to examine the extent of this regulatory mechanism in artificial polyploids with particular reference to those within the *Lolium/Festuca* complex. An attempt is also made to compare these systems with that present in a natural polyploid.

Autopolyploids

The classical concept of meiosis in such polyploids is that of random association of chromosome arms within a homologous group. The frequency and distribution of chiasmata will then determine the frequency of the different configurations at metaphase I of meiosis. In autotetraploid rye (4x=28), in which the chiasmata are mainly distal, the theoretical expectation at metaphase I of meiosis in cells with 28 chiasmata, one allocated to each paired arm, would be on average 4.67 bivalents and 4.67 quadrivalents (Timmis & Rees 1971). This is seldom achieved. A disproportionate number of bivalents are normally formed. Indeed there is evidence from other work that the ratio of bivalents to quadrivalents can be under genetic control even in autotetraploids. Hazerika & Rees (1967) showed that there was a significant difference in the number of quadrivalents relative to bivalents formed in PMCs in different tetraploid inbred lines of *Secale cereale* originally isolated from the same population. Although this was in part due to variation in the number of chiasmata, itself under genetic control, there was a further component of variation in the frequency of quadrivalents which was independent of PMC chiasma frequency. Similar differences in quadrivalent frequency per PMC were shown to exist by Crowley & Rees (1968) between C2 and C7 generations of autotetraploid *Lolium perenne* (4x=28) and by Charpentier *et al.* (1986) between two genotypes of *Agropyron elongatum.*

The effect of B chromosomes is of interest in this context although they are normally associated with modification of pairing in interspecific polyploids. It was clearly established by Evans & Macefield (1976) that B chromosomes increase the number of bivalents and decrease the number of multivalents per cell at metaphase I of meiosis in autotetraploid *L. perenne*. Again, this was independent of any change in PMC chiasma frequency.

Several models aimed at assessing the relative affinity at meiosis of the different genomes in polyploid material are available for use on data of these types. That due to Jackson & Casey (1982) for assessing pairing in autopolyploid material is typical of these. It predicts the frequency of the various configurations that would be expected at metaphase I of meiosis assuming random association of chromosome arms and bearing in mind the number of paired arms actually observed at this stage. A statistical analysis of the deviation of the observed from the expected gives an indication of the nature of chromosome association at first metaphase of meiosis.

When the Jackson/Casey model is used to analyse the data of Evans & Macefield (1976) the effect of B chromosomes is clearly seen to increase the deviation from random association (Table 1). In fact the pattern of meiosis in the 0B plants themselves does not show a very good fit to that expected from the model. One possible explanation for this is that *Ph* type genes have been identified in derivatives of this population.

Table 1.   Analysis of observed and expected frequencies of first metaphase configurations in autotetraploid L. perenne using the Jackson-Casey 4:0 model. (OB values are the mean of 9 plants while +B values are the mean of 15 plants).

Chromosome Number		Univ.	Biv.	Triv.	Quad.	c	S.S.
4x=28 (OB)	Obs.	1.37	8.87	0.46	1.89	0.754	
	Exp.	1.71	5.24	0.65	3.47		8.24
4x=28+ Bs	Obs.	1.38	10.62	0.07	1.29	0.774	
	Exp.	1.42	5.16	0.57	3.64		19.13

C = Number of paired arms/28

Interspecific polyploids — Allopolyploids

Many of the natural polyploids, and in particular polyploid crop plants, are allopolyploids based on two or more diploid species. Chromosome association in such polyploids is invariably restricted to bivalents at metaphase I of meiosis, with the result that meiosis is mechanically efficient. Indeed, as would be expected, the evidence from genetic studies using marker genes indicates that the bivalents must be of homologous chromosomes. Meiosis is therefore genetically efficient as well in that there is no dissipation of the interspecific hybridity over sexual generations. The restriction of "effective pairing" to homologous bivalents in such cases is the result of a combination of structural differences between the corresponding chromosomes of the component genomes and the presence of pairing suppressor genes, the so-called *Ph* genes. Divergence between homoeologous chromosomes, whether involving structural rearrangements or changes in DNA amount, is usually not sufficient on its own to achieve complete diploidisation. Such differences would certainly lead to a preponderance of bivalents at metaphase I of meiosis, the amount of preferential pairing being dependent on the degree of divergence between the genomes. Complete diploidisation is normally the result of the action of one or more pairing control genes superimposed on differentiation of genomes. It is not at all clear how much genomic divergence is needed for the *Ph* genes to be effective.

The *Lolium* genus affords a good opportunity to examine the effect of genetic suppressors in relation to the divergence between genomes. This small genus comprises two groups of species differing in chromosome size and DNA content (Hutchinson *et al.* 1979). The inbreeders, *L. temulentum*, *L. remotum* and *L. loliaceum* have approximately 50% larger chromosomes that the outbreeders, *L. perenne*, *L. multiflorum* and *L. rigidum*. Within each group there is very little variation in chromosome size and indeed in overall karyotype (Malik & Thomas

1966). Interspecific hybrids are readily obtained both within and between groups although embryo rescue is essential to obtain hybrids between the inbreeders and outbreeders (Evans & Macefield 1973).

Hybrids between similar species

Hybrids between species which show very little difference in karyotype and in the amount of nuclear DNA are included under this heading. In the genus *Lolium* hybrids between the three outbreeders are easily obtained.

Chromosome pairing in diploid hybrids between the three species, *L. perenne*, *L. rigidum* and *L. multiflorum* is normally complete with 7 bivalents at metaphase I of meiosis. Meiosis in the tetraploid is normally not substantially different from that in autotetraploids although there is evidence of some preferential pairing of homologous chromosomes in tetraploid *L. perenne* × *L. multiflorum* (Breese & Thomas 1976). Nevertheless there is still extensive homoeologous association in the form of quadrivalents and trivalents.

Again, as in the autotetraploids, B chromosomes can be shown to reduce the number of multivalent associations at metaphase I of meiosis. Genotypes of both *L. perenne* and *L. rigidum*, both containing B chromosomes, can be crossed to give progeny with 0, 1 *rigidum*, 2 *perenne* and 1 *rigidum* plus 2 *perenne* Bs. On chromosome doubling these give tetraploids with 0, 2R, 4P and 2R+4P B chromosomes respectively. This series of crosses has been used not only to assess the effect of Bs in changing the pattern of association at meiosis but also to determine the effect of different B chromosomes and of different number of Bs. As with the autotetraploid data, models are available to assist in the interpretation of results. The models of Kimber & Alonso (1981) have been developed to calculate the expected pattern of chromosome association according to different levels of genomic affinities using the observed number of paired arms as a basis. Although the models of this type are never completely satisfactory they nevertheless can be used legitimately to assess the effect of pairing regulatory factors in changing the apparent affinity between genomes. In "doubled diploid" hybrids of the type used here, goodness of fit to the 2:2 model of Kimber and Alonso would be the obvious test.

Table 2 gives the observed and expected values together with the optimised x and y values for the series of *L. rigidum* × *L. perenne* hybrids referred to earlier. The variable x denotes the relative affinity of the most closely related genomes and can range from 0.5, which denotes random association of both homoeologues and homologues, to 1.0, indicating bivalent association only. The variable y is fixed as $1-x$ and denotes the relative affinity of the less closely related genomes. With the 2:2 model the nearer y is to zero, the greater is the separation of the two genomes at meiosis. The overall pattern in this series of hybrids is for the B chromosomes to effect a decrease in the relative affinity of the two less closely related genomes, which it is reasonable to assume are the *perenne* and *rigidum* genomes. Although the results indicate no difference between the effect of 2 *rigidum* Bs and 4 *perenne* Bs it appears that their effects are additive with the estimated *y* value being depressed by the sum of the effects of the 2 *rigidum* and 4 *perenne* Bs when present separately. It is worth noting as well that the relative affinity of the two genomes, even in the absence of Bs, is not high. There is clearly some preferential association as bivalents.

Table 2. Analysis of observed and expected frequencies of the various configurations at first metaphase of meiosis in tetraploid hybrids of L. rigidum x L. perenne using the Kimber-Alonso 2:2 model.

Hybrid	n		Uni v.	Biv.	Triv.	Quad.	x	y	S.S.
				Frequency of				2:2 Model	
4x=28(OB)	13	Obs.	1.36	9.82	0.15	1.65			
		Exp.	1.71	9.25	0.53	1.55	0.808	0.192	2.607
4x=28+2B(R)	10	Obs.	1.28	10.92	0.11	1.14			
		Exp.	1.28	10.74	0.81	1.09	0.859	0.141	0.981
4x=28+4B(P)	17	Obs.	1.03	11.10	0.07	1.14			
		Exp.	1.08	10.81	0.30	1.10	0.858	0.142	0.852
4x=28+6B(2R+4P)	6	Obs.	1.84	11.42	0.08	0.78			
		Exp.	1.85	11.46	0.30	0.57	0.916	0.084	0.750

n=Number of plants

That *Ph* genes on the A chromosomes can also suppress multivalent formation in tetraploid hybrids between species of similar karyotype was clearly shown by Taing Aung & Evans (1985) in hybrids between *L. perenne* and *L. multiflorum* where "pairing control" genes were present in both parents. Again the Kimber-Alonso 2:2 model can be used to compare genomic homologies in different crosses. Table 3 gives the observed frequencies of the different configurations in two contrasting hybrids of tetraploid *L. multiflorum* × *L. perenne*. The hybrid *L. multiflorum* S.22 and *L. perenne* Ba 8973 from Evans & Macefield (1976) is as far as is known without *Ph* genes, while that between *L. multiflorum* Bb 1232/4 and *L. perenne* Lp 19 has *Ph* genes on both genomes (Taing Aung & Evans 1985). The effect is clear; the *x/y* values are changed from 0.779/0.221 to 0.948/0.052 in the hybrid containing *Ph* genes.

Hybrids between dissimilar species

Extensive information on the suppression of homoeologous association at metaphase I of meiosis in both diploid and tetraploid hybrids between the inbreeder *L. temulentum* and the outbreeder *L. perenne* was given by Evans & Davies (1983, 1985). It is sufficient here to summarise the main conclusions briefly. B chromosomes from *L. perenne* significantly reduced the degree of homologous chiasmate association in both diploid and tetraploid hybrids. *Ph* genes on the A chromosomes of *L. perenne* also had a similar effect, which moreover was additive to that of the B chromosomes, with the result that first metaphase of meiosis in some of the diploid hybrids with B chromosomes was characterised by 14 univalents and no bivalents while corresponding tetraploids had no multivalents. Moreover, it was confirmed through the use of marker genes that pairing in these tetraploid hybrids was indeed between homologous chromosomes.

Table 3. Effect of Ph genes in changing the pairing pattern at first metaphase of meiosis in L. multiflorum x L. perenne tetraploid hybrids (4x=28).

Hybrid	Frequency of				2:2 Model		
	Univ.	Biv.	Triv.	Quad.	x	y	S.S.
S.22 x Ba 8973	0.92	9.49	0.13	1.94	0.779	0.221	4.353
Bb1232/4 x Lp19	1.00	12.16	0.20	0.52	0.948	0.052	1.489

2:2 Model of Kimber and Alonso (1981).

Diploid/natural allopolyploid hybrids

The evolution of natural allopolyploids is almost invariably associated with the presence of pairing control genes. This has been clearly shown for wheat (Riley & Chapman 1958; Sears & Okamoto 1958), for oats (Rajhathy & Thomas 1972) and is almost certainly true for the hexaploid *Festuca arundinacea*, which is a bivalent former with disomic inheritance (Lewis *et al.* 1980). Hybrids are readily obtained between it and the diploid *Lolium* with the result that extensive homoeologous association occurs not only between the *Lolium* genome and the *Festuca* chromosomes but also between homoeologous chromosomes from within the *Festuca arundinacea* complement. Evans & Taing Aung (1986) showed that the genomic affinities in the tetraploid hybrid (ABCP genomes) between *F. arundinacea* and the diploid *L. perenne* could be changed drastically by the presence of genetic factors from the diploid *L. perenne*. In the straight hybrid intergenomic association was extensive with bivalents and multivalents predominant and with the frequency of paired arms per complement being, on average, 17.67. When either B chromosomes or the *Ph* genes from *L. perenne* were present the frequency of paired arms was reduced to 13.47 and 12.10 respectively. Finally, when both B chromosomes and *Ph* genes were present at the same time homoeologous association was reduced to an average of 5.4 arms per complement. According to the models of Kimber and Alonso the pattern of association changes from a 2:2 to a 2:1:1 type of genomic affinity.

Discussion

Three separate genetic systems of diploidisation are described here as operating within the *Lolium/Festuca* complex. Two originated from diploids while the third was identified in a natural polyploid. The question that remains to be answered is whether they are variants of the same genetic system. At least the effects of the first two, i.e. the B chromosomes and the diploid *Ph* genes, seem to be additive. Moreover, these interact with the *Ph* genes from the natural hexaploid *F. arundinacea*. The *Ph* genes and B chromosome "genes" of *L. perenne* could be related. Indeed, there is the possibility that one might have given rise to the other. It is most likely that they both came from the same introduction, a natural population of *L. perenne* from Algeria.

The broad basis of the mechanics of this control system has been described by Jenkins (1985; 1986) and by Jenkins & Scanlon (1987). However their work did not attempt to compare the fine detail of B chromosome and *Ph* mediated control although there did not seem to be any obvious differences between a low pairing diploid hybrid with B chromosomes and *Ph* genes and one with only B chromosomes. Basically homoeologous association does occur in the early stages of meiosis but this is corrected by pachytene with the result that chiasmate association in tetraploid hybrids is overwhelmingly between homologous bivalents. In corresponding diploids there is a high level of desynapsis. In fact the desynaptic nature of meiosis in diploids *L. temulentum* × *L. perenne* with Bs was first noted by Evans & Macefield (1973) using the light microscope.

What is somewhat surprising is that B chromosomes are shown to reduce multivalent formation in the autotetraploid *L. perenne*. In no way could it be argued that this is diploidisation. In the interspecific tetraploids the increasing effect of B chromosomes can be clearly related to the increased differentiation of homoeologous chromosomes, with their effect being less in *L. rigidum* × *L. perenne* and *L. multiflorum* × *L. perenne* than in *L. temulentum* × *L. perenne* and *L. remotum* × *L. perenne*.

Finally the nature of the interaction of the *L. perenne Ph* genes and of B chromosomes with the *Ph* genes of *F. arundinacea* is indicative of a close relationship between all three. However there is no hard evidence that the polyploid and diploid *Ph* genes are allelic.

References

Breese, E.L. and A.C. Thomas 1976. Monitoring chromosome segregation in *Lolium multiflorum* × *L. perenne* amphiploids. *Rep. Welsh Pl. Breed. St.* 1976, 29–32.

Charpentier, A., M. Feldman and Y. Cauderon 1986. Genetic control of meiotic chromosome pairing in tetraploid *Agropyron elongatum*. I: Pattern of pairing in natural and induced tetraploid and F_1 triploid hybrids. *Canad. J. Genet. Cytol.* 28, 783–788.

Crowley, J.G. and H. Rees 1968. Fertility and selection in tetraploid *Lolium*. *Chromosoma* 24, 300–308.

Evans, G.M. and E.W. Davies 1983. Fertility and stability of induced polyploids. In *Kew Chromosome Conference II*. P.E. Brandham and M.D. Bennett, eds, 139–146.

Evans, G.M. and E.W. Davies 1985. The genetics of chromosome pairing in *Lolium temulentum* × *L. perenne* tetraploids. *Theor. Appl. Genet.* 71, 185–192.

Evans, G.M. and A.J. Macefield 1973. The effect of B chromosomes on homoeologous pairing in species hybrids. I: *Lolium temulentum* × *L. perenne*. *Chromosoma* 41, 63–73.

Evans, G.M. and A.J. Macefield 1976. The effect of B chromosomes on meiosis in autotetraploid *Lolium perenne*. *Heredity* 36, 393–397.

Evans, G.M. and Taing Aung 1986. The influence of the genotype of *Lolium perenne* on homoeologous chromosome association in hexaploid *Festuca arundinacea*. *Heredity* 56, 97–103.

Hazerika, M.H. and H. Rees 1967. Genotypic control of chromosome behaviour in rye. X: Chromosome pairing and fertility in autotetraploids. *Heredity* 22, 317–332.

Hutchinson, J., H. Rees and A.G. Seal 1979. An assay of the activity of supplementary DNA in *Lolium*. *Heredity* 43, 411–421.

Jenkins, G. 1985. Synaptonemal complex formation in hybrids of *Lolium temulentum* × *Lolium perenne*. I: High chiasma frequency diploid. *Chromosoma* 92, 81–88.

Jenkins, G. 1986. Synaptonemal complex formation in hybrids of *Lolium temulentum* × *Lolium perenne*. III: Tetraploids. *Chromosoma* 93, 413–419.

Jenkins, G. and M. Scanlon 1987. Chromosome pairing in a *Lolium temulentum* × *L. perenne* diploid with low chiasma frequency. *Theor. Appl. Genet.* 73, 516–522.

Kimber, G. and L.C. Alonso 1981. The analysis of meiosis in hybrids. III: Tetraploid hybrids. *Canad. J. Genet. Cytol.* 23, 235–254.

Lewis, E.J. 1980. Chromosome pairing in tetraploid hybrids between *Lolium perenne* and *L. multiflorum*. *Theor. Appl. Genet.* 58, 137–143.

Lewis, E.J., M.W. Humphreys and M.P. Caton 1980. Disomic inheritance in *Festuca arundinacea* Schreb. *Pflanzenzuchtg.* 84, 335–341.

Malik, C.P. and P.T. Thomas 1986. Karyotypic studies in some *Lolium* and *Festuca* species. *Caryologia* 19, 167–196.

Rajhathy, T. and H. Thomas 1972. Genetic control of chromosome pairing in hexaploid oats. *Nature N.B.* 239, 217–219.

Riley, R. and V. Chapman 1958. Genetic control of cytologically diploid behaviour in hexaploid wheat. *Nature* 182. 713–715.

Sears, E.R. and M. Okamoto 1958. Intergenomic chromosome relationships in hexaploid wheat. *Proc. Int. Cong. Genet.* 2, 258–259.

Taing Aung and G.M. Evans 1985. The potential for diploidizing *Lolium multiflorum* × *L. perenne* tetraploids. *Canad. J. Genet. Cytol.* 27, 506–509.

Taylor, I.B. and G.M. Evans 1977. The genetic control of homoelogous chromosome association in *Lolium temulentum* × *L. perenne*. *Chromosoma* 62, 57–67.

Timmis, J.N. and H. Rees 1971. A pairing restriction at pachytene upon multivalent formation in autotetraploids. *Heredity* 26, 269–275.

Chromosome pairing in *Lolium* hybrids

G. Jenkins

Department of Agricultural Botany, School of Agricultural Sciences, U.C.W. Aberystwyth, Penglais, Aberystwyth, Dyfed. SY23 3DD, UK

The chromosomes of the annual ryegrass *Lolium temulentum* (2n=2x=14) are longer and contain on average about 50% more DNA than those of the closely related perennial ryegrass *L. perenne* (2n=2x=14+B) (Hutchinson *et al.* 1979). Despite this difference, homoeologous chromosomes are paired effectively at first metaphase of meiosis in both the diploid and tetraploid interspecific hybrids, the latter forming multivalents and behaving essentially as an autotetraploid. However, in the presence of supernumerary B chromosomes and "diploidising" genes located on the normal A chromosome complement the association of homoeologues at metaphase I is greatly reduced. This results in a high frequency of univalents in the diploid hybrid and largely homologous bivalent formation in the tetraploid hybrid (Evans & Macefield 1973, Taylor & Evans 1977). The latter hybrid has been effectively "diploidised" by these factors and behaves more like an allotetraploid. The factors operate in much the same way as the Ph locus of wheat, which suppresses homoeologous association at first metaphase and ensures exclusive homologous bivalent formation (Riley & Chapman 1958). This effect is achieved, not by preventing synapsis between homoeologous chromosomes during zygotene and pachytene of meiosis, but by ensuring that illegitimate pairing configurations are resolved before metaphase I (Hobolth 1981, Jenkins 1983, Holm 1986).

The corollary was that the "diploidising" factors of the *Lolium* hybrids operate in the same way. This possibility was investigated by employing the method of chromosome reconstruction from serial electron micrographs (Jenkins 1985a, Jenkins & Scanlon 1987). Also, it was hoped that a study of the details of chromosome synapsis during zygotene and pachytene would reveal the way in which the large disparity in DNA amount between homoeologous chromosomes is accommodated within the synaptonemal complex (SC).

Diploid hybrids

In the absence of B chromosomes and "diploidising" genes the homoeologues form seven bivalents at metaphase I in the majority of pollen mother cells (Table 1), with high chiasma frequencies per bivalent comparable with those of the *L. perenne* parent. This regularity of chromosome association is reflected in the production of seven SCs in each of five reconstructed pachytene nuclei (Jenkins 1985a). The four largest SCs of the complements could be identified positively

Table 1. The chromosome and genetic constitutions, frequency of pairing configurations and non-homologous pairing (NH) in the hybrid <u>Lolium</u> cells at zygotene (Z), pachytene (P), diplotene (D) and metaphase I (M) from Jenkins <u>1985a</u> (1), Jenkins 1985b (2), Jenkins 1986 (3) and Jenkins & Scanlon 1987 (4).

Ploidy	No. Bs	"Diploid-ising" genes	Meiotic stage	No. cells	Mean frequency pairing configurations per cell					NH	Source
					I	II	III	IV	V+		
2x	0	–	Z	2	2.5	5.0	0.5	0.0	0.0	+	1
			P	5	0.0	7.0	0.0	0.0	0.0	–	
			M	35	0.3	6.9	0.0	0.0	0.0	–	
2x	2	+	P	2	4.0	3.5	1.0	0.0	0.0	+	4
			D	2	14.0	0.0	0.0	0.0	0.0	–	
			M	240	11.0	1.5	0.0	0.0	0.0	–	
3x	2	–	Z	1	6.0	6.0	1.0	0.0	0.0	+	2
			P	1	7.0	7.0	0.0	0.0	0.0	+	
			M	50	7.0	7.0	0.0	0.0	0.0	–	
4x	0	–	Z	1	0.0	7.0	0.0	1.0	1.0	+	3
			M	20	1.3	9.6	0.2	1.7	0.0	–	
4x	4	+	Z	1	1.0	7.0	1.0	0.0	2.0	+	3
			P	2	1.5	10.0	0.0	1.0	0.5	+	
			M	30	7.3	10.3	0.0	0.0	0.0	–	

since they are morphologically distinct and bear close resemblance to those of the *L. perenne* parent (Jenkins 1985a) and its mitotic karyotype (Seal & Rees 1982). In spite of a considerable difference in size between homoeologous chromosomes, 40% of the SCs are uninterrupted from telomere to telomere, showing no loops, buckles or overlaps of the lateral components. In other words, the difference has been completely accommodated at least at the level of the SC. In the remaining 60% of bivalents the lateral component of the longer chromosome forms a single loop, which is contiguous with that within the SC and which folds back upon itself to form an SC indistinguishable from that of the main bivalent axis (Fig. 1).

Mapping the positions of the loops within the bivalents failed to reveal the sites of "extra" DNA belonging to *L. temulentum* since the loops appeared in only 60% of the bivalents, accounted for only 14% of the expected length difference and occupied variable positions even within the same bivalent observed in different cells. Clearly, much of the "extra" DNA appears not to contribute to the lateral component length and is, therefore, not involved in the assembly of the SC. A similar observation was made in a partially asynaptic *Festuca* hybrid in which much of the difference in DNA amount was manifested as asymmetry in the distribution of chromatin about the SC (Jenkins & Rees 1983). Alternatively, loops may be formed which are subsequently accommodated by synaptic adjustment, as has been shown by a study of the disappearance of duplication loops during pachytene in mouse (Moses & Poorman 1981). The variation in loop position within the same bivalent may be envisaged as a consequence of the

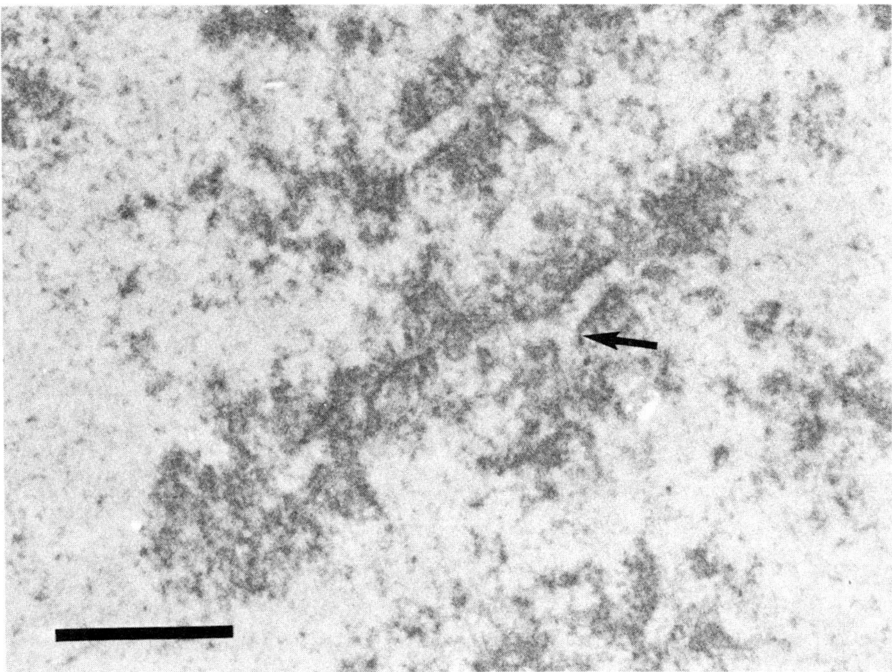

Figure 1. Electron micrograph showing a synaptonemal complex and the exact site where a lateral component loop leaves the main axis (arrow) in a diploid hybrid. Bar represents 1 μm.

convergence of synapsis from opposite ends of the chromosomes, such that long segments remain unaccommodated i.e. as loops, at the point of convergence (Fig. 2). Changes in the relative rates of synapsis from the telomeres would alter the point of convergence and the position of the loop and would inevitably incorporate non-homologous segments of lateral component into the SC. Indeed, the high incidence (67%) of mispaired centromeres indicates that non-homologous SC formation occurs at least within the vicinity of the centromeres and probably in other regions too. It is worth emphasising, though, that neither loop formation nor complete accommodation appear to affect the ability of the SC to support crossing-over.

In contrast to pachytene, pairing at early zygotene is by no means as regular. In fact, reconstruction of two nuclei at this stage revealed that in addition to some two-by-two pairing of homoeologues, chromosomes may pair either within themselves, forming foldback loops, or may pair non-homologously between themselves forming multiple associations (Jenkins 1985a). In view of the strict bivalent formation at pachytene and metaphase I, it must be concluded that even in the absence of "diploidising" factors non-homologous association is illegitimate, ineffective in terms of chaisma formation and will be transformed some time before or during pachytene.

At first metaphase and diplotene of meiosis in the diploid hybrid containing B chromosomes and "diploidising" genes chiasmate association is virtually absent (Table 1). However, at pachytene, as in the other diploid, pairing is permitted between not only homoeologous but also non-homologous chromosome segments, resulting in the formation of bivalents, multivalents and foldback loops (Jenkins & Scanlon 1987). Since only non-homologous SC is inherently ineffective, diploidisation by these factors must be achieved in addition by correction of homoeologously paired regions, as in wheat.

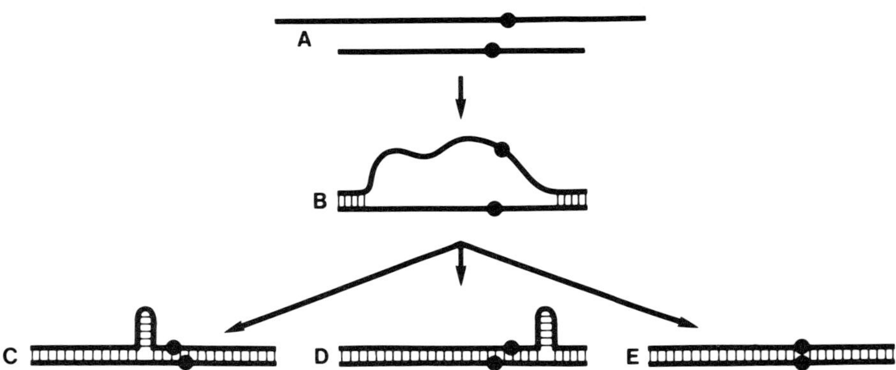

Figure 2. Diagram to explain the variation in lateral component loop position within the same homoeologous bivalent of different cells. Lateral components are represented by thick lines, SCs by crosshatching and centromeres by filled circles. Two homoeologues of unequal length (**a**) begin pairing preferentially at the telomeres during zygotene (**b**). Lateral component loop position at pachytene will differ and centromeres will mispair depending on the relative rates of synapsis from the telomeres (**c** and **d**). The disparity in length may be completely accommodated (**e**).

Tetraploid hybrids

In the colchicine-doubled amphidiploid without "diploidising" factors homoeologues are permitted chiasmate association at metaphase I, forming trivalents and quadrivalents (Table 1), but at late zygotene SCs form between non-homologous chromosomes also, with the result that multiple configurations involving more than four chromosomes are observed (Jenkins 1986). As in its diploid hybrid counterpart, the conclusion is that in the tetraploid there is an element of correction of non-homologous SCs, even in the absence of "diploidising" factors. If B chromosomes and "diploidising" genes are introduced there is a dramatic decrease in the frequency of multivalents at metaphase I and a commensurate increase in the number of bivalents (Table 1). This effect is not achieved by a suppression of synapsis, since at early zygotene and pachytene both homoeologous and non-homologous chromosomes pair in multiple associations (Jenkins 1986). The conclusion is the same as in the diploid hybrid; "diploidising" factors are responsible for the transformation of homoeologous SCs. How this may be achieved is considered below.

Mechanics of diploidisation

In wheat the observation of multivalents at zygotene and only bivalents at pachytene led to the conclusion that the correction of multivalents was the means of exclusive bivalent formation (Hobolth 1981). Furthermore, it was assumed that this process must be completed before crossing-over, otherwise the crossovers themselves would consolidate any illegitimate pairing configuration, which as a consequence would persist to metaphase I (Rasmussen & Holm 1979). Similar pairing behaviour of chromosomes in the diploid *Lolium* hybrid without "diploidising" factors (Jenkins 1985a) and a triploid hybrid (Jenkins 1985b) appeared to confirm that the correction of multivalents depended entirely upon its timing relative to crossing-over. However, further studies in *Lolium* (Jenkins 1986; Jenkins & Scanlon 1987) and wheat (Holm 1986) have shown that some multivalents may indeed persist beyond the period of crossing-over, yet are not consolidated by this process and are still resolved by metaphase I. The most likely explanation for this observation is that some SCs involved in the multiple configurations do not support crossing-over, which leads to the inevitable breakdown of the multivalents upon entry into diplotene. It follows that B chromosomes and "diploidising" genes probably exert their effect in the *Lolium* hybrids by preventing crossing-over within all SCs that are not strictly homologously paired.

A model of chromosome pairing in the *Lolium* hybrids based upon observations during meiotic prophase and metaphase I is given in Fig. 3. The figure shows an association of five chromosomes, comprising two homologues, their two homoeologous partners and a non-homologous chromosome. This multivalent represents those seen during meiotic prophase in the *Lolium* hybrids studied. Under the influence of "diploidising" factors, only strictly homologously paired segments are predisposed to crossing-over during zygotene. The nature of the predisposition is not known, but may involve, for example, the modification of the molecular structure of the SC or the attachment of a recombination apparatus (such as recombination nodules) to the SC. As the multivalent enters pachytene, it may undergo early transformation, whereby those SC segments not destined to cross over break down and are reassembled between only homologous regions. This would be typical of the pairing behaviour observed in the triploid hybrid (Jenkins 1985b). Alternatively, the multivalent may persist to late pachytene and break down upon entry into diplotene, except in the homologous regions stabilised by crossovers. The timing of the transformation process relative to crossing-over is not important since the correction is directed simply by the position of crossovers determined during zygotene. In any event, the outcome at metaphase I will be the same, two rod bivalents and one univalent.

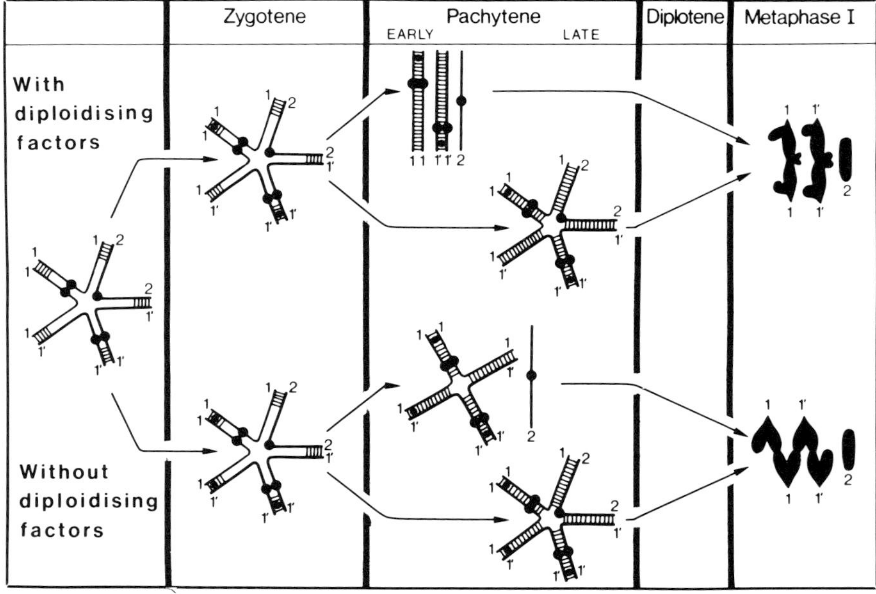

Figure 3. Diagram to show the correction of a typical "pentavalent" during meiotic prophase in both the presence and absence of B chromosomes and "diploidising" genes. The sites of predisposition to crossing-over and crossing-over itself are represented by small filled circles. Other symbols are as in Figure 2. A full explanation of the correction process is given in the text.

In the absence of "diploidising" factors, homoeologous as well as homologous segments are predisposed to crossing-over. Non-homologous regions appear to be incapable of genetic exchange presumably due to lack of sequence homology between participating DNA molecules. Therefore, at early or late pachytene the "pentavalent" is transformed into a quadrivalent and univalent, characteristic of those seen at metaphase I in the tetraploid hybrid.

The model attempts to explain the mode of action of the B chromosomes and "diploidising" genes in terms of the resolution of illegitimate configurations according to a strictly localised chiasma distribution. It fails to explain, however, how the "diploidising" factors actually influence the siting of crossovers, and how some chromosomes of the complement can form quite legitimate associations at the beginning of synapsis, apparently without the need of any correction mechanism. The full understanding of their effect would undoubtably require a detailed analysis of the biochemistry of chromosome pairing and recombination.

References

Evans, G.M. and A.J. Macefield 1973. The effect of B chromosomes on homoeologous pairing in species hybrids 1. *Lolium temulentum × Lolium perenne. Chromosoma* 41, 63–73.

Hobolth, P. 1981. Chromosome pairing in allohexaploid wheat var. Chinese Spring. Transformation of multivalents in bivalents, a mechanism for exclusive bivalent formation. *Carlsberg. Res. Commun.* 46, 129–173.

Holm, P.B. 1986. Chromosome pairing and chiasma formation in allohexaploid wheat, *Triticum aestivum*, analysed by spreading of meiotic nuclei. *Carlsberg Res. Commun.* 51, 239–294

Hutchinson, J., H. Rees and A.G. Seal. 1979. An assay of the activity of supplementary DNA in *Lolium. Heredity* 43, 411–421.

Jenkins, G. 1983. Chromosome pairing in *Triticum aestivum* cv. Chinese Spring. *Carlsberg Res. Commun.* 48, 255–283

Jenkins, G. 1985a. Synaptonemal complex formation in hybrids of *Lolium temulentum × Lolium perenne* I. High chiasma frequency diploid. *Chromosoma* 92, 81–88.

Jenkins, G. 1985b. Synaptonemal complex formation in hybrids of *Lolium temulentum × Lolium perenne* II. Triploid. *Chromosoma* 92, 387–390.

Jenkins, G. 1986. Synaptonemal complex formation in hybrids of *Lolium temulentum × Lolium perenne* III. Tetraploid. *Chromosoma* 93, 413–419.

Jenkins, G. and H. Rees. 1983. Synaptonemal complex formation in a *Festuca* hybrid. In *Kew Chromosome Conference II*, P.E. Brandham and M.D. Bennett, eds. 233–242. London: Allen and Unwin.

Jenkins, G. and M.J. Scanlon. 1987. Chromosome pairing in a *Lolium temulentum × Lolium perenne* diploid hybrid with a low chiasma frequency. *Theor. Appl. Genet.* 73, 516–522.

Moses, M.J. and P.A. Poorman. 1981. Synaptonemal complex analysis of mouse chromosome rearrangements II. Synaptic adjustment in a tandem duplication. *Chromosoma* 81, 519–535.

Rasmussen, S.W. and P.B. Holm. 1979. Chromosome pairing in autotetraploid *Bombyx* females. Mechanism for exclusive bivalent formation. *Carlsberg Res. Commun.* 44, 101–125.

Riley, R. and V. Chapman. 1958. Genetic control of the cytologically diploid behaviour of hexaploid wheat. *Nature* 182, 713–715

Seal, A.G. and H. Rees. 1982. The distribution of quantitative DNA changes associated with the evolution of diploid Festuceae. *Heredity* 49, 179–190.

Taylor, I.B. and G.M. Evans. 1977. The genotypic control of homoeologous chromosome association in *Lolium temulentum × Lolium perenne* interspecific hybrids. *Chromosoma* 62, 57–67.

Genetic control of bivalent pairing in common wheat: the mode of *Ph1* action

M. Feldman* and L. Avivi**

*Plant Genetics Dept., The Weizmann Insitute of Science, Rehovot, Israel
**Human Genetics Dept., Sackler School of Medicine, Tel-Aviv University, Israel

The *Ph1* gene of common wheat, which was discovered about 30 years ago, has had a great impact on basic and applied aspects of wheat cytogenetics. It is located on the long arm of chromosome 5B (5BL), about 1.0 cM from the centromere (Sears 1984), and is principally responsible for the cytologically diploid-like behaviour of hexaploid common wheat (Sears 1976). Several theories have been suggested to explain the mode of action of *Ph1*, all of which fall into two main categories: those assuming that *Ph1* exerts its effect during pre-meiotic stages, before the commencement of synapsis, thereby controlling chromosomal spatial relationships (Feldman 1966; Feldman & Avivi 1973, 1984; Yacobi & Feldman 1983) — and those assuming that this gene operates exclusively after synapsis has already commenced, affecting processes involved in crossing-over (Driscoll *et al.* 1979; Hobolth 1981; Holm 1986). This paper reviews the different effects of extra dosage of *Ph1* on chromosomal pairing and presents evidence that all these effects can be phenocopied by pre-meiotic treatment with the anti-microtubule agent colchicine.

Effect of *Ph1* on chromosomal pairing

Effect of Ph1 on pairing of homologous chromosomes.

While homologous pairing is hardly affected by the absence of *Ph1*, plants exhibit partial asynapsis with four doses of this gene, and more asynapsis with six doses (Feldman 1966; Yacobi 1984; Table 1). Studies of C-banded first meiotic metaphases of di-isosomic 5BL (four doses of *Ph1*) have shown that most of the unpaired chromosomes belong to the B genome (Table 2). Similar evidence that *Ph1* affects the pairing of the B-genome chromosomes more strongly than that of the A- and D-genome chromosomes was recently obtained by Naranjo *et al.* (1987) in hybrids with rye. This stronger effect of *Ph1* on the B-genome chromosomes indicates that the ancestral genomic organisation within the wheat nucleus is still maintained and that *Ph1* can recognise this organisation, presumably through some genome-specific DNA sequence(s).

To distinguish between pairing disturbances occurring prior to the

Table 1. Mean arm-pairing frequency and probabilities of pre-synaptic homologous association (<u>a</u>) and chiasma formation (<u>c</u>) as calculated from frequencies of chromosome-pairing configurations at first meiotic metaphase of plants with different doses of <u>Ph1</u>. Data from Feldman (1966) and Yacobi (1984); cells with multivalents not included.

Genotype	Dose of Ph1	Mean arm-pairing*	Calculated pairing probabilities**	
			a	c
Nullisomic 5B	0	0.90	1.00	0.88
ph mutant	0	0.89	1.00	0.84
Disomic 5B	2	0.97	1.00	0.97
Tetrasomic 5B	4	0.91	0.99	0.90
Di-isosomic 5BL	4	0.90	1.00	0.90
Tri-isosomic 5BL	6	0.43	0.68	0.56

* The ratio between the number of paired arms to the total number of arms.
**Calculated by the formula of Driscoll <u>et al</u>. (1979).

Table 2. The allocation to genomes of unpaired chromosomes in 50 C-banded first meiotic metaphases of plants having four doses of <u>Ph1</u> (di-isosomic 5BL plants). Data from Lukaszewki & Feldman (unpublished).

		Number of univalents	
	Total	From A+D genomes*	From B genome**
Observed	139	58	81
Expected	139.0	99.3	39.7

$$\chi^2 = 60.14; \ P < 0.001.$$

* Includes chromosomes 1A to 3A, 4B, 5A to 7A, 1D to 7D, and iso-5BL (included because of identification difficulties). The data exclude the iso-5BL ring univalents.
**Includes chromosomes 1B, 2B, 3B, 4A, 6B, and 7B (in accord with recent evidence that 4A belongs to the B genome).

commencement of meiotic pairing and those occurring later, Driscoll *et al.* (1979) constructed a mathematical model bearing upon chromosomal configuration frequency at first meiotic metaphase. Two parameters were defined: the probability of pre-synaptic homologous chromosome association (*a*); and the efficiency of chiasma formation (*c*). When metaphase pairing is complete, each parameter equals 1.0. While pre-synaptic disturbances reduce the value of *a*, post-synaptic disturbances reduce exclusively the value of *c*, which may also be reduced due to reduction in *a*. Accordingly the value of *a* enables the distinction between events occurring prior to the commencement of meiotic pairing and those occurring after this stage. Yacobi & Feldman (1983) obtained maximum values for *a* (1.0) and slight reduction in the *c* values for disomic- and nullisomic-5B plants (Table 1; similar values were obtained by Driscoll *et al.* 1979). On the other hand, plants with six doses of *Ph1* exhibited a drastic reduction in the *a* value (0.68; Table 1), indicating that the partial asynapsis caused by this dosage of *Ph1* had been brought about by affecting pre-synaptic events.

Effect of Ph1 on inter- and intrachromosomal pairing of isochromosomes.

While the interchromosomal pairing of the three isochromosomes 5 BL (in tri-isosomic 5BL) was reduced by six doses of *Ph1* to a degree the same as that of conventional homologous chromosomes (to about 50% of the normal pairing), the degree of intrachromosomal pairing of these isochromosomes was not reduced at all and was even increased (Table 3). The increase presumably

Table 3. Expected and observed frequencies of inter- and intra-chromosomal pairing of the three isochromosomes at first meiotic metaphase of tri-isosomic 5BL plants having six doses of Ph1. Data from Feldman (1966) and Yacobi (1984).

Pairing frequency of the possible configurations**

	VVV	VVO	VOO	OOO	⊐V	⊐O	℃V	℃O	ΛΛV	⊖ (ring tri)
Expected when mean arm pairing equals 1.00	0.000	0.000	0.000	0.070	0.000	0.000	0.000	0.400	0.000	0.530
Expected when mean arm pairing equals 0.50	0.125	0.075	0.030	0.008	0.300	0.122	0.061	0.050	0.162	0.067
Observed when the mean arm pairing of the conventional homologues was 0.43	–	–	0.110	0.130*	–	–	–	–	0.120	0.100

* Significant deviation at .01 level from the corresponding expected data.
**Unpaired isochromosome: V open univalent.

Intrachromosomal pairing: O ring univalent.

Interchromosomal pairing: ⊐ open bivalent;   ℃ ring bivalent;

ΛΛV open trivalent;   ⊖ ring trivalent.

resulted from decreased competition with interchromosomal pairing. Hence, when the homologous arms are connected by a common centromere and cannot be spatially separated, their pairing is not suppressed by six doses of *Ph1*.

Effect of Ph1 on bivalent and quadrivalent pairing of the four homologous chromosomes in tetrasomic lines.

The frequency of cells showing quadrivalents at first meiotic metaphase of tetrasomic lines of common wheat (carrying four homologous chromosomes) was affected by the *Ph1* gene. Yacobi & Feldman (1983) and Yacobi (1984) produced aneuploid lines that were tetrasomic for chromosomes of homoeologous group 7 and contained different (0–4) doses of *Ph1*. They found that the frequency of cells with multivalents at first meiotic metaphase was negatively correlated with *Ph1* dose (Table 4). Elevation of the dose of this gene

Table 4. Frequency of first meiotic metaphases with a multivalent (a quadrivalent or, rarely, a trivalent) formed from the four homologous chromosomes in tetrasomics of homoeologous group 7 having different doses of Ph1. Data from Yacobi (1984). Only cells with complete pairing of the 20 disomes (mean arm-pairing frequency above 0.92) were included.

Genotype	Dose of Ph1	Frequency of cells with a multivalent		
		Tetra 7A	Tetra 7B	Tetra 7D
ph1 mutant	0	0.65	0.72	–
Monosomic 5B	1	0.44	0.38	0.44
Disomic 5B	2	0.42	0.34	0.38
Monosomic – mono-isosomic 5BL	3	0.24	0.19	0.29
Di-isosomic 5BL	4	–	0.15	0.08

from zero to four scarcely reduced the level of pairing of these homologues, yet their pattern of pairing was strongly affected, changing from quadrivalent to bivalent pairing.

A similar bivalent-producing effect by *Ph*-like genes was found in wild relatives of wheat; namely in an induced autotetraploid line of *Aegilops longissima* (Avivi 1976a) and in natural autotetraploid *Agropyron elongatum* (Charpentier *et al.* 1986).

Effect of Ph1 on pairing of homoeologous chromosomes.

The average frequency of homoeologous pairing per cell, as deduced from the frequency of multivalents and heteromorphic bivalents at first meiotic metaphase, is 0.63 in plants deficient for *Ph1* and 0.77 in plants with six doses of *Ph1* (Table 5). The corresponding values for homoeologous pairing per paired chromosomal arm are 0.017 and 0.042 respectively (Table 5). Interestingly, in spite of the reduction in pairing of homologues, six doses of *Ph1* induce more than twice as much homoeologous pairing as occurs with zero dosage of this gene.

Table 5. Mean homoeologous pairing and mean interlocked bivalents at first meiotic metaphase of plants with different doses of Ph1. Data from Feldman (1966) and Yacobi (1984).

Genotype	Dose of Ph1	Homoeologous pairing*		Interlocked bivalents	
		per cell	per paired arm	per cell	per ring bivalent
ph1 mutant	0	0.63	0.017	0.54	0.040
Monosomic 5B	1	0.00	0.000	0.20	0.010
Disomic 5B	2	0.00	0.000	0.04	0.002
Monotelosomic – monoisosomic 5BL	3	0.00	0.000	0.78	0.050
Di-isosomic 5BL	4	0.00	0.000	1.40	0.080
Tri-isosomic 5BL	6	0.77	0.042	0.94	0.135

* Combined frequencies of multivalents and heteromorphic bivalents.

Effect of Ph1 on interlocking of bivalents.

Interlocking bivalents are maintained at first meiotic metaphase of common wheat only when at least one of the interlocking partners is a ring bivalent. Thus, the frequency of interlocking at first metaphase is affected by the number of ring bivalents and should be expressed relative to the frequency of these. Calculations based on this inference show a proportional increase in the frequency of interlocking events per ring bivalent when the dose of *Ph1* is elevated from two to six (Table 5). Reducing the dose of *Ph1* to one and zero also increases the frequency of interlocking per ring bivalent, but to a lesser extent. However, in zero dose of *Ph1*, interlocking configurations are composed of only two or three bivalents (as predicted if interlocking in zero dose of *Ph1* involves only homoeologous bivalents). In plants with four and six doses of *Ph1*, more than three bivalents (up to seven) are capable of forming interlocking configurations, indicating that in extra dosage of *Ph1* non-homoeologous bivalents may also interlock with each other (Yacobi et al. 1982; Yacobi & Feldman 1983).

Effect of pre-meiotic treatment with colchicine on chromosomal pairing

Effect of colchicine on pairing of homologous chromosomes.

Pre-meiotic treatment with colchicine induces partial asynapsis of homologous chromosomes (Driscoll *et al.* 1967; Dover & Riley 1973; Table 6). If the drug is applied in a concentration above 5×10^{-4}M the degree of pairing reduction is more than 50%, similar to the effect of six doses of *Ph1*. The colchicine treatment reduces the value of the pairing parameter *a*, indicating that it interferes with pre-meiotic events (Driscoll *et al.* 1979; Table 6).

Table 6. Mean arm-pairing frequency of conventional homologues and of an isochromosome, and probabilities of pre-synaptic homologous association (a) and chiasma formation (c) as calculated from frequencies of chromosomal pairing configurations at first meiotic metaphase of plants treated with 7.5 x 10⁻⁴ M colchicine 7 days before meiosis. (Adapted from Driscoll & Darvey 1970; Driscoll et al. 1979.)

| Pre-meiotic treatment | Conventional homologues | | | Isochromosome 5DL |
| | Mean arm pairing | Probabilities | | Mean arm pairing |
		a	c	
Contol	0.96	1.00	0.95	0.97
Colchicine	0.44	0.49	0.82	0.96

Effect of colchicine on intra-chromosomal pairing of an isochromosome.

As with six doses of *Ph1*, pre-meiotic treatment with colchicine, which reduces conventional pairing to about 45%, does not affect the intra-chromosomal pairing of an isochromosome (Driscoll & Darvey 1970; Table 6).

Effect of colchicine on bivalent and quadrivalent pairing.

If colchicine is applied before the last pre-meiotic mitosis, chromosomes are doubled in number, but only bivalent pairing takes place (Driscoll *et al.* 1967;

Avivi 1976b). In spite of the fact that each chromosome is present in four doses and pairing is complete, quadrivalents were never observed. It is assumed that the pattern of pairing is largely determined by the pattern of pre-meiotic spatial relationships and that colchicine causes a random distribution of the original homologues at the end of the last pre-meiotic mitosis, while the homologues that derived from sister chromatids remain closely associated (Avivi 1976b).

Effect of colchicine on pairing of homoeologous chromosomes.

Pre-meiotic treatment of common wheat with colchicine induced homoeologous pairing (multivalents and heteromorphic bivalents) at first meiotic metaphase (Driscoll *et al.* 1967; Feldman & Avivi unpublished). The frequency of such pairing varied with the colchicine concentration. Usually, low concentrations that did not cause high frequencies of asynapsis (5×10^{-4}M and below) induced the most homoeologous pairing (an average of 0.16 per cell and 0.005 per paired chromosomal arm). Homoeologous pairing was also observed in other polyploids of *Triticum* and *Aegilops* that were pre-meiotically treated with low colchicine concentrations (Feldman & Avivi unpublished).

Effect of colchicine on interlocking of bivalents.

Pre-meiotic treatment of common wheat with colchicine induced a relatively high frequency of interlocked bivalents at first meiotic metaphase (Feldman & Avivi, unpublished). Also here, the most effective treatments were those with low colchicine concentrations causing only a slight asynapsis. The frequency of interlocking bivalents induced by these treatments was 1.03 per cell and 0.055 per ring bivalent. The number of interlocked bivalents per configuration was as high as four, indicating that non-homoeologous bivalents may interlock.

Discussion

Extra doses of *Ph1* suppress pairing of homologous chromosomes but not of homologous arms that are attached to each other, as in isochromosomes. This indicates that *Ph1* neither interferes with processes of pairing itself nor with those of crossing-over but rather, as was suggested by Feldman (1966), that it disrupts a pre-synaptic event which is a prerequisite condition for regular meiotic pairing. That *Ph1* exerts its effect at a pre-synaptic stage is also apparent from the reduction in the calculated probability of pre-meiotic chromosomal association (*a*) in the presence of six doses of *Ph1*. It is also in accord with recent evidence (Sears, unpublished) that two doses of *Ph1* fail to suppress completely homoeologous pairing and recombination between a segment of chromosome 3Ag of *Agropyron elongatum* and its wheat 3D homoeologous segment, when the 3Ag segment has been translocated into an interstitial position in wheat chromosome 3D. The conclusion that *Ph1* controls a pre-synaptic event can also easily explain the paradoxical effects of extra dosage of this gene (see below), namely the reduction in pairing of homologues and, at the same time, the induction of homoeologous pairing and interlocking of bivalents. Also the suppression of quadrivalent pairing in tetrasomic lines by four doses of *Ph1*, without decreasing the total number of chiasmata, can be easily accounted for by the conclusion that this gene controls a pre-synaptic event.

As with extra dosage of *Ph1*, pre-meiotic colchicine treatment reduces the pairing of conventional homologues but does not affect the intra-chromosomal pairing of an isochromosome. This is in accord with the calculated reduced probability of pre-meiotic chromosomal association (*a*) resulting from such treatment. Concomitantly with partial asynapsis, colchicine induces homoeologous pairing and interlocking of bivalents. When present before the last pre-meiotic mitosis, it induces almost exclusive bivalent pairing in meiocytes with doubled chromosome number. Thus, all the different effects of extra dosage of *Ph1* can be phenocopied by colchicine if applied during pre-meiotic stages. Consequently, colchicine, like *Ph1*, does not suppress the processes of pairing or crossing-over, but rather interferes with a pre-synaptic event which is necessary for the regularity of meiotic pairing.

The assumption that *Ph1* postpones recombination processes to late pachytene until correction of pairing is completed or that it suppresses recombination between homoeologous chromosomes (Hobolth 1981; Jenkins 1983; Holm 1986) is neither compatible with all the above effects of *Ph1* nor with the fact that *Ph1* action can be phenocopied by pre-meiotic treatment with colchicine.

The only possibility to explain the effects of *Ph1* and colchicine on chromosomal pairing at meiosis is to assume that: (1) premeiotic association of pairing partners is a necessary prerequisite for regular meiotic pairing, and (2) the pre-meiotic alignment of homologous and homoeologous chromosomes is the pre-synaptic event which is affected by both *Ph1* and pre-meiotic treatment with colchicine (Feldman 1966; Feldman & Avivi 1973, 1984).

The differential effect of *Ph1* on chromosomes of the B genome is in accord with the hypothesis (Avivi & Feldman 1980; Feldman & Avivi 1984) that this gene controls the association of chromosomes by manipulating chromosomal sets rather than individual chromosomes. Hence, in plants with zero dose of *Ph1* all the six chromosomal sets of hexaploid wheat, homologous as well as homoeologous, are assumed to be closely associated at pre-synaptic stages. This results in some homoeologous pairing at meiosis, superimposed on the homologous pairing, and in interlocking of homoeologous bivalents that fail to form a multivalent. In the normal two doses of *Ph1* the pre-synaptic association of homoeologous sets is suppressed (genomic separation), while homologous sets stay close to one another, resulting in regular and exclusive pairing of homologues at meiosis. With six doses of *Ph1* the homologous sets are also separated, resulting in a partial asynapsis of conventional homologues, in pairing of homoeologues that happen to lie close to each other and in interlocking of non-homoeologous bivalents as a result of pairing between slightly separated partners. Because of the physical connection between the two arms of an isochromosome, the intrachromosomal pre-synaptic association of these arms is not affected by extra dosage of *Ph1*, and consequently their meiotic pairing with each other is regular.

Indeed, the above assumed spatial relationships between homologous and homoeologous sets in plants with zero or two doses of *Ph1* were observed in somatic cells (review in Feldman & Avivi 1984). The specific non-random distribution of chromosomes in plants with two doses of *Ph1* was confirmed recently in root-tip interphases by *in situ* hybridisation with genome-specific DNA probes (Rayburn & Gill 1987). Colchicine, similarly to extra dosage of *Ph1*, affects chromosomal distribution and suppresses the association of homologous chromosomes in somatic cells (Avivi *et al.* 1969), indicating that this drug interferes with the non-random chromosomal arrangement.

The great similarity between the effects of *Ph1* and colchicine points to the molecular target of *Ph1* action. Since colchicine interacts specifically with microtubular proteins, it is inferred that microtubules are the most likely molecular target of *Ph1*. Thus microtubules, which are the main components of the spindle fibres and as such connect centromeres to specific sites at the spindle poles during metaphase and anaphase, are assumed also to be responsible for chromosomal attachment to the nuclear membrane during telophase, interphase and prophase. *Ph1* controls spatial relationships between chromosomal sets by affecting some microtubule characteristics. In fact, studies have shown that *Ph1* affects microtubule response to colchicine as well as to other anti-microtubule agents (Avivi *et al.* 1970; Avivi & Feldman 1973; Ceoloni *et al.* 1984). Other genes, with *Ph*-like effect, were also found to affect spindle sensitivity to colchicine in common wheat and related species, namely *Ph2* in wheat (Ceoloni & Feldman 1987) and *Ph*-like genes in *Aegilops longissima* (Avivi unpublished) and *Agropyron elongatum* (Charpentier *et al.* 1986). Genes with effects opposite to *Ph1*, such as that of chromosome arm 5BS of common wheat or that of 5RL of *Secale cereale*, also affect spindle characteristics (Avivi *et al.* 1970; Ceoloni *et al.* 1984; Lukaszewski *et al.* 1987). *Ph1* may stabilise microtubules and reduce their sensitivity to anti-microtubule drugs by coding for an isoform of alpha- or beta-tubulin or for a microtubule-associated protein (MAP) which stabilises a specific microtubule structure. How *Ph1*, by affecting microtubule stability, determines the spatial relationships among chromosomal sets, thus affecting the regularity and pattern of meiotic pairing, is currently being studied.

Acknowledgements

The authors are grateful to Dr. E. R. Sears for critical reading of the manuscript and for stimulating discussions.

References

Avivi, L. 1976a. The effect of genes controlling different degrees of homoeologous pairing on quadrivalent frequency in induced autotetraploid lines of *Triticum longissimum*. *Can. J. Genet. Cytol.* 18, 357–364.

Avivi, L. 1976b. Colchicine induced bivalent pairing of tetraploid microsporocytes in *Triticum longissimum* and *T. speltoides*. *Can. J. Genet. Cytol.* 18, 731–738.

Avivi, L. and M. Feldman 1973. The mechanism of somatic association in common wheat *Triticum aestivum* L. IV. Further evidence for modification of spindle tubulin through the somatic association genes as measured by vinblastine binding. *Genetics* 73, 379–385.

Avivi, L. and M. Feldman. 1980. Arrangement of chromosomes in the interphase nucleus of plants. *Hum. Genet.* 55, 281–295.

Avivi, L., M. Feldman and W. Bushuk 1969. The mechanism of somatic association in common wheat *Triticum aestivum* L. I. Suppression of somatic association by colchicine. *Genetics* 62, 745–752.

Avivi, L., M. Feldman and W. Bushuk 1970. The mechanism of somatic association in common wheat *Triticum aestivum* L. II. Differential affinity for colchicine of spindle microtubules of plants having different doses of the somatic-association suppressor. *Genetics* 65, 585–592.

Ceoloni, C. and M. Feldman. 1987. Effect of *Ph2* mutants promoting homoeologous pairing on spindle sensitivity to colchicine in common wheat. *Genome* 29 (in press).

Ceoloni, C., L. Avivi and M. Feldman. 1984. Spindle sensitivity to colchicine of the *Ph1* mutant in common wheat. *Can. J. Genet. Cytol.* 26, 111–118.

Charpentier, A., M. Feldman and Y. Cauderon. 1986. Genetic control of meiotic chromosome pairing in tetraploid *Agropyron elongatum*. I. Pattern of pairing in natural and induced tetraploids and in F_1 triploid hybrids. *Can. J. Genet. Cytol.* 28, 783–788.

Dover, G. A. and R. Riley. 1973. The effect of spindle inhibitors applied before meiosis on meiotic chromosome pairing. *J. Cell Sci.* 12, 143–161.

Driscoll, C. J. and N. J. Darvey. 1970. Chromosome pairing: effect of colchicine on an isochromosome. *Science* 169, 290–291.

Driscoll. C. J., L. M. Bielig and N. L. Darvey. 1979. An analysis of frequencies of chromosome configurations in wheat and wheat hybrids. Genetics 91, 755–767.

Driscoll, C. J., N. L. Darvey and H. N. Barber. 1967. Effect of colchicine on meiosis of hexaploid wheat. *Nature* 216, 687–688.

Feldman, M. 1966. The effect of chromosome 5B, 5D and 5A on chromosomal pairing in *Triticum aestivum*. *Proc. Natl. Acad. Sci., USA,* 55, 1447–1453.

Feldman, M. and L. Avivi. 1973. The pattern of chromosomal arrangement in nuclei of common wheat and its genetic control. *Proc. 4th Inter. Wheat Genet. Symp., Columbia, Mo.,* 675–684.

Feldman, M. and L. Avivi. 1984. Ordered arrangement of chromosomes in wheat. In *Chromosomes Today, vol. 8,* M. D. Bennett, A. Gropp and U. Wolf, eds, 181–189. London UK: George Allen and Unwin.

Hobolth, P. 1981. Chromosome pairing in allohexaploid wheat var. Chinese Spring. Transformation of multivalents into bivalents, a mechanism for exclusive bivalent formation. *Carlsberg Res. Commun.* 46, 129–173.

Holm, P. B. 1986. Chromosome pairing and chiasma formation in allohexaploid wheat, *Triticum asestivum* analyzed by spreading of meiotic nuclei. *Carlsberg Res. Commun.* 51, 239–294.

Jenkins, G. 1983. Chromosome pairing in *Triticum aestivum* cv. Chinese Spring. *Carlsberg Res. Commun.* 48, 255–283.

Lukaszewski, A. J., B. Apolinarska, J. P. Gustafson and K.-D. Krolow. 1987. Chromosome pairing and aneuploidy in tetraploid triticale. II. Unstablized karyotypes. *Genome* 29, (in press).

Naranjo, T., A. Roca, P. G. Goicoechea and R. Giraldez. 1987. Arm homoeology of wheat and rye chromosomes. *Genome* (in press).

Rayburn, A. L. and B. S. Gill. 1987. Use of repeated DNA sequences as cytological markers. *Amer. J. Bot.* 74, 574–580.

Sears, E. R. 1976. Genetic control of chromosome pairing in wheat. *Ann. Rev. Genet.* 10, 31–51.

Sears, E. R. 1984. Mutations in wheat that raise the level of meiotic chromosome pairing. In *Gene Manipulation in Plant Improvement*, J. P. Gustafson, ed., 295–300. New York, Plenum Press.

Yacobi, Y. Z. 1984. *Chromosomal arrangement and genetic control of meiotic pairing in common wheat.* Ph.D. Thesis, The Weizmann Institute of Science, Rehovot, Israel (in English).

Yacobi, Y. Z. and M. Feldman. 1983. The control of the regularity and pattern of chromosome pairing in common wheat by the *Ph1* gene. *Proc. 6th Inter. Wheat Genet. Symp., Kyoto, Japan,* 1113–1118.

Yacobi, Y. Z., T. Mello-Sampayo and M. Feldman. 1982. Genetic induction of bivalent interlocking in common wheat. *Chromosoma* 87, 165–175.

An ultrastructural analysis of the effect of chromosome 5B on chromosome pairing in allohexaploid wheat

P.B. Holm, Xingzhi Wang* and B. Wischmann

*Department of Physiology, Carlsberg Laboratory, 10 Gamle Carlsberg Vej, DK–2500 Copenhagen Valby, Denmark, and *Institute of Genetics, Academia Sinica, Beijing, China*

The disomic inheritance of allohexaploid wheat, *Triticum aestivum*, is known to be controlled primarily by the Ph locus on the long arm of chromosome 5B (5BL; Sears 1976, 1977). In the absence of this chromosome arm (Okamoto 1957, Riley & Chapman 1958, Riley & Kempanna 1963) and in the ph1b mutant line (Sears 1977), the disomic behaviour breaks down, and crossing-over occurs between the homoeologues of the A, B and D genomes in allohexaploid wheat, trihaploid wheat and hybrids between wheat and closely related species.

The present paper reports an investigation of chromosome pairing in wheat at the ultrastructural level to elucidate the effect of chromosome 5B. It summarises evidence obtained from spreading meiotic nuclei of allohexaploid wheat (Holm 1986), trihaploids with and without chromosome 5B (Xingzhi Wang, in preparation) and allohexaploid wheat monosomic and nullisomic for 5B or carrying two or three copies of an isochromosome for 5BL (Holm, in preparation; Wischmann 1986). Previous work on allohexaploid wheat using serial sectioning and three-dimensional reconstruction are presented in Hobolth (1981, 1983) and Jenkins (1983, see von Wettstein *et al.* 1984 for review).

Chromosome pairing in allohexaploid wheat

The analysis of euploid wheat involved tracings of lateral components (LCs) and synaptonemal complexes (SCs) of 93 spread nuclei ranging in stage from the beginning of zygotene to late diplotene. The major findings can be summarised as follows:

1) Interlockings are extremely frequent at zygotene. In one mid-zygotene nucleus, where the interlocking frequency was analysed in detail, about 20 interlockings were identified. At late zygotene (mean of 93% pairing) there was a mean of five bivalent interlockings and a few chromosome interlockings in the seven nuclei analysed. Some interlockings persisted up to pachytene and a few were apparent at early diplotene.

2) Analyses of nuclei at the leptotene/zygotene transition revealed the formation of a typical chromosome bouquet and initiation of synapsis at a few of the telomeres. Several interstitial regions were in the process of alignment and synapsis. Tracings of five early zygotene nuclei (mean of 39% pairing, range from 27% to 45%) revealed that the pairing was primarily into bivalents. It is thus apparent that pairing and SC formation preferentially involve homologues. Multiple assocations of lateral components due to pairing partner exchange were, however, regularly seen at these and later stages of zygotene, although at a low frequency (Table 1). In these nuclei as well as in the zygotene nuclei of the other genotypes the associations were often quite complex and a classification into different types was not attempted with the exception of associations in which a lateral component exchanges pairing partner over a short distance (0.5–1.0 μm) and then switches back to its original partner (type II exchanges in Table 1). As pairing between partly homologous segments probably occurs more readily than between segments of low or no homology, several if not most pairing partner exchanges are probably between homoeologous chromosomes. Supporting evidence for this interpretation was found at late zygotene, where the multiple associations appear in most cases to involve six chromosomes, i.e., probably the six members of a homoeologous group.

3) Nearly all multiple associations are corrected into bivalents during the course of the zygotene stage (Hobolth 1981, Holm 1986, see von Wettstein *et al.* 1984 for a detailed description and discussion of the phenomenon of pairing correction). As shown in Table 1 the number of LCs in multiple associations decreases from a mean of 8 at early zygotene (mean of 39% pairing) to a mean of 5 at mid- and late zygotene, while at pachytene only a mean of 0.6 LCs or LC fragments are involved in multiple associations. This gradual reduction in the number of pairing partner exchanges during zygotene thus reveals that pairing partner exchanges are detected and resolved as pairing and SC formation proceeds. Central regions combining lateral components in these regions are removed and re-form between the alternative combination of LCs, the outcome being bivalent formation at the expense of the multiple associations.

4) Synapsis is usually incomplete. The mean degreee of 97% synapsis at pachytene is obtained from nuclei selected for having almost complete pairing. If measured from unselected nuclei at early diplotene, the mean degree of synapsis achieved amounts to 85%.

5) The associations retained at pachytene are probably not corrected any further. The pachytene stage appears to be very short as judged by the rare occurrence of nuclei at this stage and possible pairing partner exchanges have in fact been found at early diplotene. As multivalents are virtually absent at metaphase I in euploid wheat cv. 'Chinese Spring', crossing-over does not occur between homoeologous chromosomes in multivalents present at pachytene, which subsequently can resolve into bivalents during diplotene when the SC is degraded.

On the basis of this evidence the following three processes were suggested to be essential for ensuring the disomic inheritance of allohexaploid wheat (Holm 1986):

Table 1. Degree of pairing, LC length of one complement ($\frac{1}{2}$ total LC length) and the number of LCs or LC segements in multiple associations in wheat microsporocytes nullisomic or monosomic for chromosome 5B, euploid wheat and wheat di-isosomic for the long arm of chromosome 5B. The number of multiple associations in parentheses are associations of type II (see text for explanation). ±, standard deviation. ND, not determined.

Stage	Number of nuclei	% pairing	$\frac{1}{2}$ LC length µm	Number of LCs in multiple associations	
Euploid wheat					
Early zygotene	5	39 ± 7	1984 ± 342	8 ± 4	(6 ± 3)
Mid zygotene	10	62 ± 7	2040 ± 164	5 ± 6	(3 ± 3)
Late zygotene	7	93 ± 7	1806 ± 160	5 ± 5	(0)
Pachytene	20	97 ± 7	1474 ± 214	0.6	(0.3)
Wheat monosomic for 5B					
Early zygotene	3	42	2342	15	(4)
Mid-late zygotene	6	79 ± 8	1887 ± 254	10 ± 4	(0)
Pachytene	17	90 ± 5	1322 ± 220	9 ± 3	(1)
Wheat nullisomic for 5B					
Zygotene-pachytene	12	37 ± 13	2185 ± 493	19 ± 7	(6 ± 5)
Pachytene-diplotene	6	37 ± 15	1722 ± 177	ND	
Early diplotene	7	33 ± 17	1722 ± 360	ND	
Wheat di-isosomic for 5BL					
Zygotene-pachytene	7	31 ± 6	1940 ± 222	6 ± 5	(2 ± 4)
Early diplotene	6	40 ± 12	1702 ± 185	ND	

i) A high stringency of chromosome pairing at zygotene, whereby pairing and SC formation preferentially occur between homologues.

ii) A transformation of multiple associations into bivalents during zygotene, i.e., before crossing-over occurs.

iii) A suppression of crossing-over between paired segments of homoeologous chromosomes.

Chromosome pairing in trihaploid wheat

Euhaploid plants were obtained by anther culture of 'Chinese Spring' wheat monosomic for 5B. The presence of chromosome 5B in these plants was confirmed by N-banding. A total of 68 nuclei ranging in stage from early zygotene to late diplotene were analysed in detail. At the beginning of zygotene a typical bouquet is organized and as in allohexaploid wheat synapsis is initiated distally and interstitially. Up to 40% of the complement pairs in 'Chinese Spring' trihaploids (Fig. 1a), while nearly 90% synapses in trihaploids of the chinese winter wheat variety 'Kedong'. Associations of more than three chromosomes were observed in all 31 zygotene nuclei traced. The mean number of LCs or LC fragments engaged in multiple associations decreased as zygotene progressed, while the frequency of unpaired LCs or associations of two or three LCs or LC fragments increased (Table 2). Correction of pairing does accordingly also take place in trihaploid wheat. Although none of the chromosomes could be identified, it is considered likely that a substantial proportion of the pairing and SC formation into bivalents and trivalents, in particular at mid-late zygotene, reflects a synapsis between homeologous chromosomes. As haploid wheat is virtually achiasmate (one bivalent in about every six metaphase I complements), crossing-over must accordingly be suppressed between homoeologous chromosomes in this genotype.

Chromosome pairing in nullisomic and monosomic 5B wheat

The study of plants nullisomic and monosomic for chromosome 5B comprised an analysis of 25 nuclei of the nullisomic in the range from early zygotene to early diplotene and 30 nuclei of the monosomic spanning the same stages. The nullisomic material was obtained from two plants and the monosomic from several and the absence/presence of chromosome 5B was confirmed by N–banding of root tip metaphase complements. In the nullisomic material the mean frequency of multivalents (generally tri- or quadrivalents) at metaphase I amounted to 0.5 per cell. Several cells appeared, however, to contain more multivalents, but these were generally not analysable as the chromosomes were too intermingled to permit a quantitative study. In contrast, multivalents were

Figure 1. Idiograms of spread, silver stained microsporocyte nuclei of trihaploid wheat (**a**, 38% synapsis) and wheat monosomic for 5B (**b**, 42% synapsis). The synapsed regions are shown as filled areas and the actual length of the LCs are represented by the vertical lines. Breaks of LCs are denoted by crosses. Multiple associations are shown first and bivalents or bivalent fragments are arranged after decreasing length.

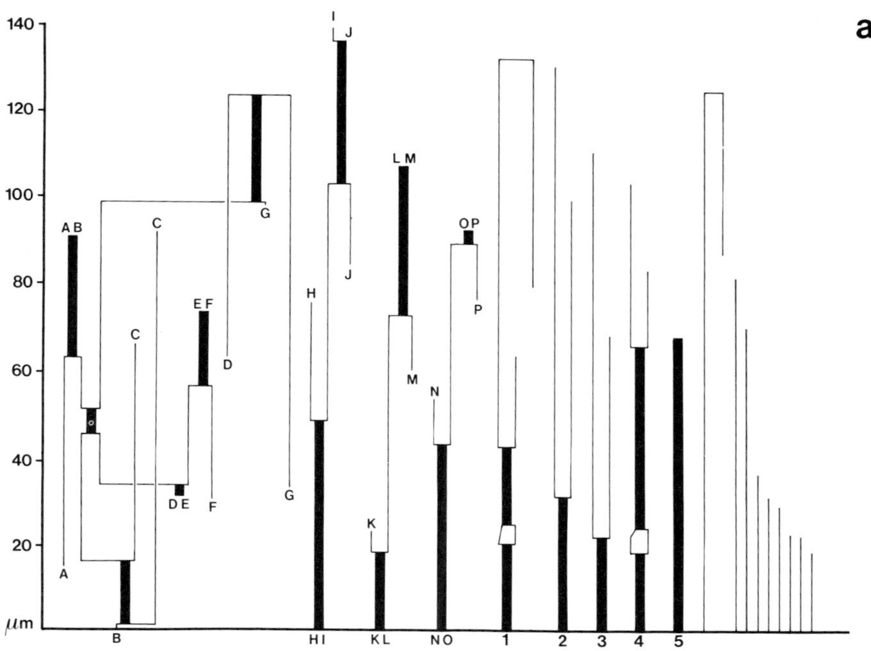

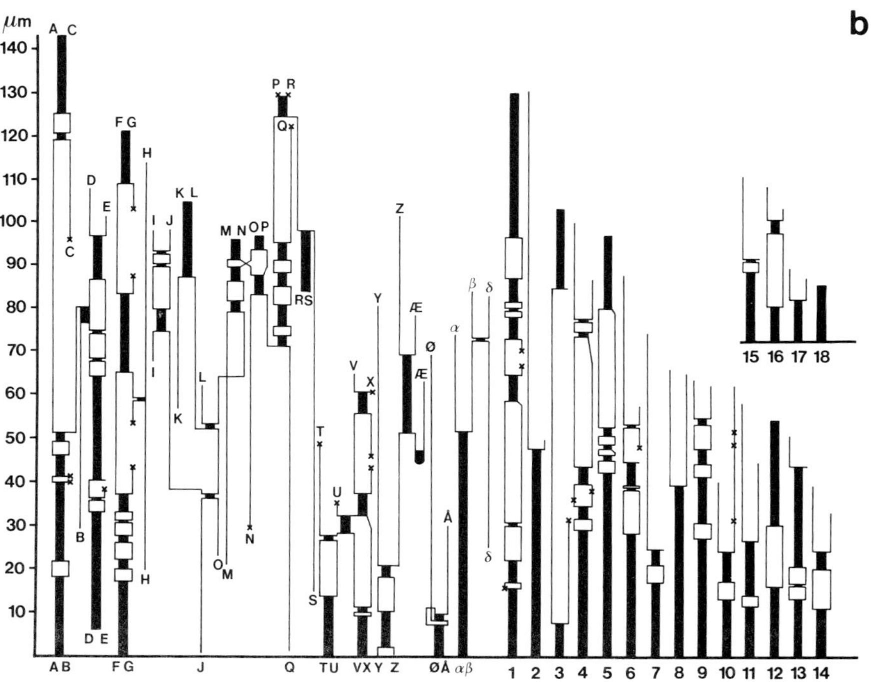

virtually absent at metaphase I in the monosomic material, where the complements were present almost exclusively as 20 bivalents and a single univalent chromosome 5B. The major conclusions from these studies are as follows:

Table 2. Degree of pairing, LC complement length and the mean number per nucleus of LCs or LC fragments in multiple associations in ten zygotene nuclei of microsporocytes of trihaploid wheat. The ten zygotene nuclei had less than 35 LC pieces. III, association of three LCs or LC fragments, II, association of two LCs or LC fragments, I, unpaired LC or LC fragment. The pairing range is given in parenthesis. ±, standard deviation.

Stage	Number of nuclei	% pairing	LC length μm	Mean number of LCs in multiple associations			
				>III	III	II	I
Early zygotene	5	7 ± 3 (3%–10%)	2283 ± 257	16.0	1.8	2.8	4.4
Mid-late zygotene	5	23 ± 12 (10%–36%)	2314 ± 400	13.2	3.0	6.4	8.2

1) In nullisomic 5B wheat pairing is arrested when an average of 35% of the complement is paired (37% synapsis at the pachytene/diplotene transition and 33% at early diplotene, Table 1). There were large differences in the degree of synapsis achieved among the nuclei, the range observed at early diplotene being from 18% to 66% pairing. In contrast, synapsis in monosomic 5B was nearly as complete as in euploid wheat (90% at pachytene and 84% at early diplotene, Table 1).

2) The number of LCs or LC fragments engaged in multiple associations in nullisomic and monosomic 5B is 2.5 times and 2 times higher than were observed in euploid wheat at a similar stage of pairing (Table 1). In agreement with these data for the nullisomic material, limited data for trihaploid wheat nullisomic for chromosome 5B reveal an approximately threefold increase in the number of pairing partner exchanges compared to the euhaploid at a similar stage of pairing. The frequency of the multiple associations appeared in the nullisomic material to be the same from the onset of synapsis to its arrest (mean of 20 LCs or LC fragments in multiple associations in the pairing range 13% to 37% and 18 LCs or LC fragments in the pairing range from 42% to 61%). Thus there appears to be a limited correction of the pairing configurations in this genotype. This absence of correction may be attributed to the arrest of pairing at an early stage. Exchanges of pairing partner often occurred close to the telomeres (LCs number M-P in Fig. 1b). Such exchanges were not seen in

euploid wheat and only in one case in monosomic 5B wheat. In monosomic 5B material the number of LCs involved in multiple associations decreased by a factor of nearly two from early zygotene to pachytene indicating that correction occurs in this genotype (Table 1). Several of the retained associations appeared to involve 5 LCs, and although the chromosomes cannot be identified it is conceivable that these associations are pentavalents consisting of the 5 chromosomes of group 5 (2 × 5A + 2 × 5D + 5B). As in the euhexaploid wheat, these multivalents did not appear to undergo further correction. Hence, a large number of multivalents combining homoeologous chromosomes persist in plants monosomic for chromosome 5B through the period where crossing-over occurs, but in spite of this multivalents are virtually absent at metaphase I.

The present data suggest that in the absence of chromosome 5B, or if one copy is present, pairing and SC formation occur more readily between chromosomes that are not homologous than if two copies of 5B are present. In the absence of 5B, pairing and SC formation are arrested at an early stage of zygotene. Crossing-over is also affected as recombination between homoeologues can occur if 5B is absent, but is suppressed by one copy of 5B.

Dosage effect of the long arm of chromosome 5B

In order to ascertain the effect of extra copies of the long arm of chromosome 5B on pairing and chiasma formation, spread microsporocytes from plants di-isosomic for 5BL (Table 1) and tri-isosomic for 5BL (Wischmann 1986) were analysed. As illustrated in Fig. 2a an increase in the number of copies of 5BL from two to four to six leads to an almost linear reduction in the extent of pairing, the mean degree of pairing in di-isosomic and tri-isosomic 5BL plants being 40% and 25%. Hence, an excess in the number of copies of 5BL or the complete absence of 5B leads to pairing arrest at an early stage of zygotene. These results also explain the correlation observed between the number of copies of 5BL and the frequency of interlocking at metaphase I reported by Yacobi *et al.* (1982). As discussed in detail elsewhere (Holm 1986), resolution of interlockings depends on completion of synapsis, the resolving mechanism being triggered by an intimate contact between the interlocked LCs. Interlockings may therefore escape detection and resolution in unsynapsed regions. The presence and frequency of interlocking at metaphase I in wheat can thus be considered to be a qualitative and to a large extent quantitative measure of the extent of synapsis achieved during zygotene.

A dosage effect of 5BL on the frequency of pairing partner exchange at early zygotene is also apparent from the decrease in the frequency of shifts of pairing partner as the number of copies of 5BL increases from zero to four (Fig. 2b). The relationship does not appear to be linear, as the frequency of LCs in multiple associations in plants di-isosomic for 5BL is almost as high as in euploid wheat. These nuclei contain, however, four homologous chromosome arms of 5BL and a significant proportion of the associations may result from pairing partner exchange between these four arms. The frequency of LCs in multiple associations between the chromosomes that are not homologous may therefore actually be lower than in euploid wheat, and a linear relationship may be valid for up to four copies of 5BL. It should be noted, though, that the data compared are from early zygotene, where complete tracings of the nuclei are seldom possible, and further that the number of nuclei examined in detail is relatively small. Furthermore, the high frequency of pairing partner exchange in monosomic 5B plants may be attributed in part to a disturbing effect of the univalent chromosome 5B on the synapsis of chromosomes 5A and 5D by competition for a pairing partner.

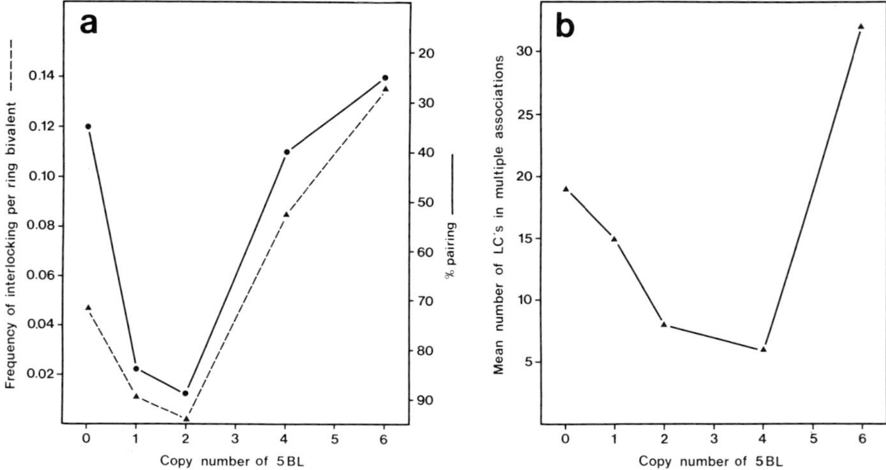

Figure 2. a. Diagrammatic presentation of the relationship between the number of copies of 5BL (or 5B in the nullisomic, monosomic and euploid genotypes) and the mean degree of synapsis achieved as estimated from early diplotene nuclei (the data for tri-isosomic 5BL are the mean pairing observed at zygotene). The data of Yacobi *et al.* (1982) on the interlocking frequency as a function of different copy number of 5BL are included for comparison (broken line). **b.** Relationship between the number of copies of 5BL (or the entire 5B in the nullisomic, monosomic and euploid genotypes) and the mean number of LCs per nucleus in multiple associations at early zygotene in the different genotypes.

A dosage effect is not apparent for plants with six copies of 5BL (Fig. 2B). The data in Fig. 2b are derived from only one serially-sectioned nucleus, but analyses of several spread nuclei revealed the presence of extensive pairing partner exchange in each one. In contrast to the situation in plants with one to four copies of 5BL, where crossing-over occurs only between homologues, homoeologues can recombine in these plants as can be judged from the presence of heteromorphic bivalents and multivalents at metaphase I.

Discussion

The present investigation has shown that chromosome 5B affects the frequency of multiple associations formed during zygotene. In its absence, or if six copies of the long arm are present, several multiple configurations are formed, while increasing the number of copies of the arm from zero to one to two to four leads to a proportional reduction in the number of pairing partner exchanges. These differences are already apparent from the very start of pairing and SC formation, suggesting that synapsis occurs under different stringency in the different genotypes. Chromosome 5B also has an effect on the extent of synapsis. Whereas SC formation is almost complete in monosomic and euploid material, it is arrested at an early stage of zygotene in the three other genotypes. Finally, crossing-over can occur between homoeologues if zero or six copies of 5BL are present while recombination is restricted to homologues if one to four copies are present.

It is thus apparent that genes on 5BL somehow affect the stringency of pairing as well as the control of whether crossing-over between homoeologues is permitted or not. Recent data (Gillies in press) indicate that this control indeed resides in the Ph locus (loci) on 5BL. Spreading of microsporocytes of hybrids between *T. aestivum* and *T. kotschyii* with and without the ph1b mutation revealed differences similar to those reported above. A detailed analysis of three nuclei (one with the wild type and two with the mutant allele) at the same stage of pairing thus disclosed a two- to three-fold increase in the frequency of LCs involved in multiple associations in hybrids carrying the ph allele when compared to hybrids carrying the wild type allele Ph. As initially reported by Sears (1977), the ph1b mutation leads to extensive chiasma formation between the homoeologues in hybrids of the two species. Analogous results have also been reported for diploid and allotetraploid hybrids of *Lolium temulentum* and *Lolium perenne*, where the presence of accessory B chromosomes has an effect similar to that of the ph gene in wheat. In the presence of four B chromosomes the allotetraploid hybrid behaves as a diploid showing only bivalents at metaphase I, and only univalents are seen at metaphase I in the diploid hybrid carrying two B chromosomes. In the absence of B chromosomes the allotetraploid shows multivalents at metaphase I and in the diploid the complement is at this stage almost exclusively present as bivalents. Serial sectioning and three-dimensional reconstructions of zygotene and pachytene nuclei of the four genotypes (Jenkins 1985, 1986, Jenkins & Scanlon 1987) thus indicated that, at least the diploid hybrids, SC formation between the homoeologues appeared to be more extensive in the absence of accessory B chromosomes (almost complete bivalent formation) than in their presence (incomplete pairing in the two late pachytene nuclei analysed).

The mechanisms underlying the Ph dosage-dependent changes in pairing stringency and crossing-over remain unknown at present. An attractive hypothesis could be that these processes were somehow interrelated. It might thus be assumed that the absence of chromosomes 5B, besides affecting the pairing specificity, somehow provides for a type of synapsis between homoeologues (effective pairing), which can induce the recombination machinery operating in these segments. Causal relationships between pairing and crossing-over at the biochemical level have been obtained by Hotta, Stern and co-workers (Hotta *et al.* 1979. Hotta & Stern 1981). In the achiasmatic *Lilium* hybrid 'Black Beauty' nicking of the pachytene small nuclear DNA, assumed to relate to crossing-over, did not take place even though a normal amount of the meiotic endonuclease was synthesised and a rather elaborate pairing and SC formation took place between the homoeologues. The absence of nicking was presumably due to a lack of the pachytene small nuclear RNA, which has to bind to the DNA together with a protein before nicking can occur. In the allotetraploid, however, the endonuclease activity was restored. Hence, in this case the biochemical machinery required for crossing-over is triggered only if there is abundant or perhaps a nearly complete pairing of homologous chromosomes. In analogy with these findings, it may be suggested that, in the absence of chromosome 5B or in the presence of six copies of 5BL, synapsis between homoeologues can trigger the synthesis of regulating elements such as the pachytene small nuclear RNA and thereby elicit crossing-over in homoeologous regions.

Acknowledgements

We are indebted to Professor Diter von Wettstein and Dr. Søren W. Rasmussen for discussion and review of the manuscript. The Royal Danish Academy of Sciences and Letters are thanked for having provided one of us (P.B.H.) with a Niels Bohr Scholarship. The work was supported by grant B16–E–168–DK from the Commission of the European Communities.

References

Gillies, C.B. 1987. The effect of Ph gene alleles on synaptonemal complex formation in *Triticum aestivum* × *T. kotschyii* hybrids. *Theor. Appl. Genet.* in press.

Holbolth, P. 1981. Chromosome pairing in allohexaploid wheat var. 'Chinese Spring'. Transformation of multivalents into bivalents, a mechanism for exclusive bivalent formation. *Carlsberg. Res. Commun.* 46, 129–173.

Holm, P.B. 1986. Chromosome pairing and chiasma formation in allohexaploid wheat, *Triticum aestivum* analyzed by spreading of meiotic nuclei. *Carlsberg Res. Commun.* 51, 239–294.

Hotta, Y. and H. Stern 1981. Small nuclear RNA molecules that regulate nuclease accessibility in specific chromatin regions of meiotic cells. *Cell* 27, 309–319.

Hotta, Y., M.D. Bennett, L.A. Toledo and H. Stern 1979. Regulation of R-protein and endonuclease activities in meiocytes by homologous chromosome pairing. *Chromosoma* 72, 191–201.

Jenkins G. 1983. Chromosome pairing in *T. aestivum* cv. 'Chinese Spring'. *Carlsberg Res. Commun.* 48, 255–283.

Jenkins, G. 1985. Synaptonemal complex formation in hybrids of *Lolium temulentum* × *Lolium perenne* I. High chiasma frequency diploid. *Chromosoma* 92, 81–88.

Jenkins, G. 1986. Synaptonemal complex formation in hybrids of *Lolium temulentum* × *Lolium perenne* III. Tetraploid. *Chromosoma* 93, 314–419.

Jenkins, G. and M.J. Scanlon 1987. Chromosome pairing in a *Lolium temulentum* × *Lolium perenne* diploid hybrid with a low chiasma frequency. *Theor. Appl. Genet.* 73, 516–522.

Riley, R. and V. Chapman 1958. Genetic control of the cytologically diploid behaviour of hexaploid wheat. *Nature* 182, 713–715.

Riley, R. and C. Kempanna 1963. The homoeologous nature of the non-homologous meiotic pairing in *Triticum aestivum* deficient for chromosome V (5B), *Heredity* 18, 287–306.

Okamoto, M. 1957. Asynaptic effect on chromosome V. *Wheat Inf. Serv.* 5, 6–7.

Sears, E.R. 1976. Genetic control of chromosome pairing in wheat. *Ann. Rev. Genet.* 10, 31–51.

Sears, E.R. 1977. An induced mutant with homoeologous pairing in common wheat. *Can. J. Genet. Cytol.* 19, 585–593.

Wettstein, D. von, S.W. Rasmussen and P.B. Holm 1984. The synaptonemal complex in genetic segregation. *Ann. Rev. Genet.* 18, 331–413.

Wischmann, B. 1986. Chromosome pairing and chiasma formation in wheat plants tri-isosomic for the long arm of chromosome 5B. *Carlsberg Res. Commun.* 51, 1–25.

Yacobi, Y.Z., T. Mello-Sampayo and M. Feldman 1982. Genetic induction of bivalent interlocking in common wheat. *Chromosoma* 87, 165–175.

Chromosome pairing in foetal oocytes of mouse inversion heterozygotes

C. Tease and G. Fisher

M.R.C. Radiobiology Unit, Chilton, Didcot, Oxon OX11 0RD, UK

Homologous chromosome pairing in an inversion heterozygote should result in the formation of an inversion loop at pachytene, but it is often found that not all heterozygous pachytene cells contain loops. For example, in a recent study of synaptonemal complexes in mice heterozygous for the large, X-linked inversion In(X)1H, Tease & Fisher (1986) observed an inversion loop in only about 37% of early pachytene oocytes. Although synaptic adjustment may have been one cause of the deficit, a second factor was also found to be involved. Examination of C-banded pachytene chromosomes revealed the occasional occurrence of a single, large bivalent with reverse pairing. Tease & Fisher (1986) suggested that this was the X chromosome pair which was displaying homologous pairing of the long inverted segment and non-homologous association of centromeric and telomeric ends. Reverse chromosome pairing has also been observed in another large inversion in the mouse (Moses *et al.* 1984). Little information is available so far on the circumstances in which an inversion may undergo reverse pairing or on the frequency of this pattern of homologue association.

The work described here was undertaken to provide a more full account of reverse chromosome pairing in oocytes of inversion heterozygotes. Three particular aspects of the phenomenon have been examined here, namely: (i) the importance of the size of the inversion; (ii) the relationship, if any, to synaptic adjustment; and (iii) the effect of a second chromosome rearrangement on the non-inverted homologue.

Material and methods

Four paracentric inversions, In(X)1H, In(5)9Rk, In(2)2H and In(2)5Rk, and an insertional translocation Is(In7;X)1Ct were used here (see Green (1981) for details of these rearrangements).

Foetal female mice heterozygous for one or other of the inversions were generated by mating female mice of the F_1 hybrid strain C3H/HeH × 101/H to males of the appropriate genotype. Double heterozygotes for In(X)1H and Is(In7;X)1Ct were produced by mating females homozygous for the inversion to translocation-carrying males. Pregnant females were killed on day 16 or 18 after mating (the morning of the vaginal plug = day 0). Both ovaries were removed from each female foetus and processed together for pachytene chromosome

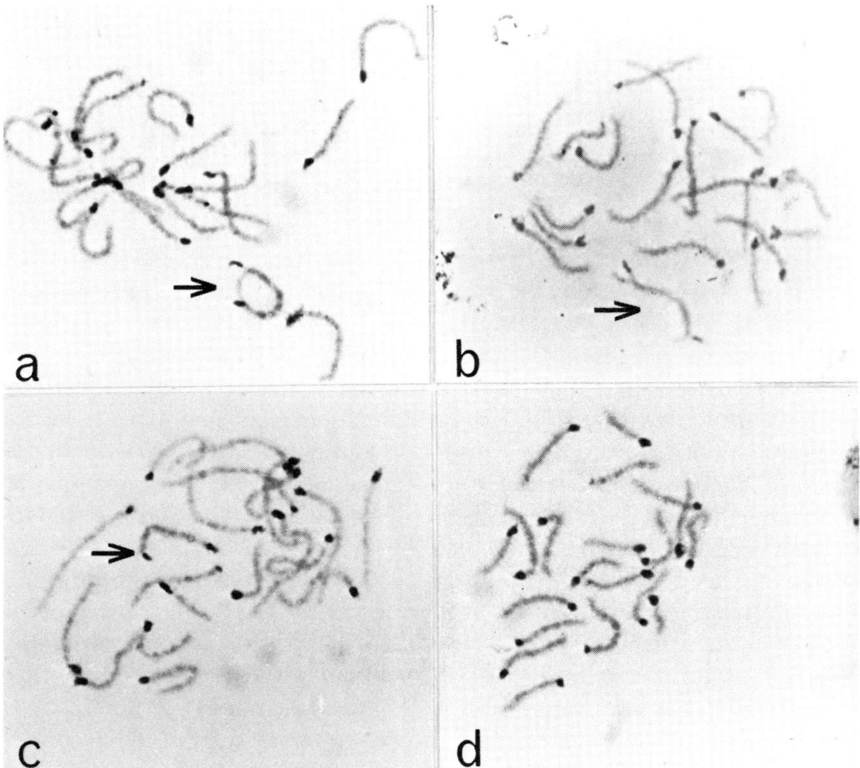

Figure 1. C-banded pachytene chromosomes from an In(X)1H heterozygote. **a** inversion loop with unpaired centromeric regions, **b** reverse chromosome pairing with unpaired centromeric and telomeric ends, **c** reverse chromosome pairing with non-homologous association of centromeres and telomeres, **d** non-homologous pairing across the inversion, the X bivalent cannot be identified.

preparations. Ovaries were given a 4 hr hypotonic treatment with 0.56% KC1 and fixed overnight in 3:1 ethanol-acetic acid. Chromosome preparations were made by dissociating the ovaries in 6:3:1 acetic acid-water-ethanol. C-banding was performed using the procedure of Sumner (1971). Intact pachytene oocytes were scored for (i) the presence of an inversion loop (and, in the case of In(X)1H/Is(In7;X)1Ct double heterozygotes, the additional presence of an insertion loop), (ii) the occurrence of reverse pairing or (iii) non-homologous pairing of the inverted segment (Figs 1, 2).

Results and discussion

Reverse chromosome pairing was observed in pachytene oocytes of mice heterozygous for In(X)1H or In(5)9Rk but was not found in In(2)2H or In(2)5Rk heterozygotes (Table 1). The principal difference between these two groups is

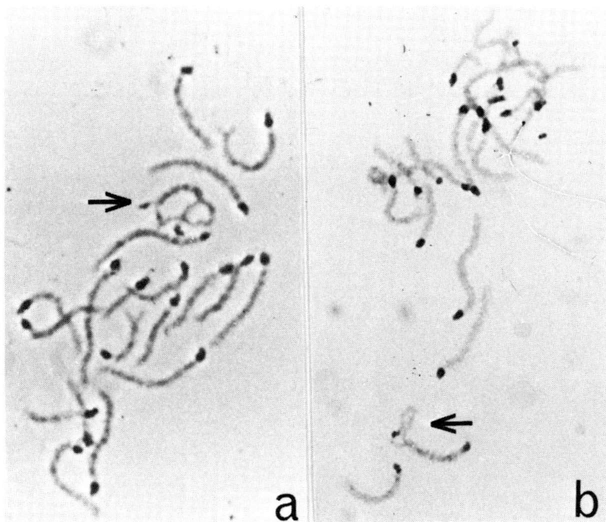

Figure 2. C-banded pachytene chromosomes from an In(X)1H/Is(In7;X)1Ct double heterozygote. **a** inversion loop with a smaller insertion loop, the centromeric segments are unpaired, **b** reverse paired X bivalent with an insertion loop.

the relative size of the inversion. In(X)1H and In(5)9Rk are large inversions occupying approximately 85 and 90% of the X and chromosome 5 respectively, whereas In(2)2H and In(2)5Rk each occupy about 40% of chromosome 2 (see Green (1981) for details). It would therefore seem likely that relative length of the inversion is the crucial determinant of whether or not reverse pairing occurs. The slightly higher incidence of reverse pairing in In(5)9Rk heterozygotes compared to In(X)1H heterozygotes provides tentative support for this conclusion as the former is the relatively larger inversion.

The pattern of pairing in In(X)1H heterozygotes differed considerably between 16 and 18 days. There was a significant decrease with age in the proportion of cells with a loop ($X_{(1)}^2$=9.8, p=0.002) and a correlated rise in the proportions with reverse pairing or non-homologous pairing across the inversion (Table 1). The decrease in frequency of cells with a loop almost certainly reflects synaptic adjustment. At 16 days, early pachytene oocytes predominate whereas at 18 days an increasingly large proportion of later pachytene cells are present (Speed 1982; Dietrich 1986). Overall, therefore, the population of pachytene cells analysed at 18 days would have had a greater opportunity for synaptic adjustment than at 16 days. Although the incidence of reverse pairing rose from 16 to 18 days, the increase was not significant ($X_{(1)}^2$=0.81, p= 0.37); therefore the available data do not suggest that bivalents with an inversion loop can transform to reverse paired configurations as an alternative to synaptic adjustment.

Table 1. The proportions of C-banded pachytene oocytes adjudged to show an inversion loop, reverse chromosome pairing or non-homologous pairing of the inverted segment in inversion heterozygotes.

Genotype	Foetal age (days)	Prop. (%) oocytes with			Total no. of cells
		inversion loop	reverse pairing	non-homologous pairing	
In(X)1H/+	16	22.4	35.1	42.4	245
"	18	11.5	39.9	48.6	296
In(5)9Rk/+	16	23.1	53.8	23.1	130
In(2)2H/+	16	20.7	0	79.3	87
In(2)5Rk/+	16	32.6	0	67.4	43

The proportions of cells adjudged to have an inversion loop, reverse pairing or non-homologous association across the inversion in the doubly heterozygous combination of In(X)1H and Is(In7;X)1Ct are given in Table 2. This combination was produced to examine the effect of a second chromosome

Table 2. The proportions of oocytes from two 16 day In(X)1H/Is(In7;X)1Ct foetuses adjudged to show an inversion loop, reverse chromosome pairing or non-homologous pairing of the inverted segment.

Prop. (%) oocytes with			Total no. of cells
inversion loop	reverse pairing	non-homologous pairing	
(11.3)[a] (21.8)[b]	(9.8)[a] (18.0)[b]		
33.1	27.8	39.1	133

a. With insertion loop.   b. No insertion loop.

rearrangement on the non-inverted X on the pattern of X chromosome pairing. Comparison with the proportion of cells in each of the three categories in In(X)1H heterozygotes suggests that the additional rearrangement did not have a significant effect ($X_{(5)}^2$=5.3, p=0.38). There does, however, appear to be a slight tendency towards a higher incidence of inversion loops and a lower incidence of reverse pairing in the double heterozygote. Further data are required to determine the reality of this trend.

The X bivalent in double heterozygotes could not be identified unambiguously in about 39% of pachytene cells which were therefore classified as showing non-homologous association of the inversion (Table 2). In slightly more than half of these cells, one of the longer bivalents showed a varying degree of asynapsis/desynapsis. It therefore appears that the structural heterozygosities of the X chromosome may occasionally cause partial failure of pairing. Even where the X bivalent could be identified, an insertion loop was not always present (Table 2). The lack of an insertion loop in these instances indicates that the chromosomes are in some fashion able to accommodate the additional material. A somewhat similar observation has been made by Mahadevaiah *et al.* (1984) who examined synaptonemal complexes in spermatocytes and oocytes of mice carrying the insertional translocation Is(7;1)4OH. They found that loops were occasionally not visible and suggested that their absence was a consequence of synaptic adjustment.

References

Dietrich, A.J.J. 1986. The influence of hypotonic treatment on the morphology of meiotic stages. II. Prophase of the first meiotic division of female mice up to dictyotene. *Genetica* 70, 161–165.

Green, M.C. (ed.) 1981. *Genetic Variants and Strains of the Laboratory Mouse.* Gustav Fisher Verlag. Stuttgart, New York.

Mahadevaiah, S., U. Mittwoch and M.J. Moses 1984. Pachytene chromosomes in male and female mice heterozygous for the Is(7;1)4OH insertion. *Chromosoma* 90, 163–169.

Moses, M.J., M.E. Dresser and P.A. Poorman 1984. Composition and role of the synaptonemal complex. *Symp. Soc. Exp. Biol.* 38, 245–270.

Speed, R.M. 1982. Meiosis in the foetal mouse ovary. I. An analysis at the light microscope level using surface-spreading. *Chromosoma* 85, 427–437.

Sumner, A.T. 1972. A simple technique for demonstrating centromeric heterochromatin. *Exptl. Cell Res.* 75, 304–306.

Tease, C. and G. Fisher 1986. Further examination of the production-line hypothesis in mouse foetal oocytes. I. Inversion heterozygotes. *Chromosoma* 93, 447–452.

An analysis of synaptic adjustment, length, and twisting of synaptonemal complexes in maize

S. Stack and L. Anderson

Department of Biology and the Cell and Molecular Biology Program, Colorado State University, Fort Collins, CO 80523, USA

The development of techniques for spreading synaptonemal complexes (SCs) in two dimensions has eased the effort required to examine synapsis at the ultrastructural level compared to the older technique of three-dimensional reconstruction from serial thin sections. As a result it has become possible to examine a variety of synaptic phenomena in large numbers of cells.

Synaptic adjustment

McClintock (1931, 1933) first described synapsis in inversion heterozygotes based on squash preparations of primary microsporocytes of maize. She determined that the inverted segments could synapse homologously to form a loop, not synapse at all (asynapse), or synapse non-homologously to form a straight bivalent. This remained the standard description until recently, when Moses and co-workers (Moses 1977; Moses *et al.* 1982) examined surface spreads of SCs from primary spermatocytes of inversion heterozygotes of laboratory mice. They found that an inversion loop occurred in every set of SCs from early pachytene. As pachytene progressed, desynapsis occurred at both ends of the loop followed by resynapsis of the segments non-homologously so that the loops became progressively smaller until no loops remained. This process was called synaptic adjustment (Moses 1977), and subsequently Moses *et al.* (1985) have suggested that synaptic adjustment may be a general phenomenon among many different species.

Only two species of plants, maize and *Allium ursinum*, that were heterozygous for an inversion have been examined using SC spreads (Gillies 1981; Loidl 1987). Although comprehensive studies of inversion loops at different substages of pachytene were not performed, both authors decided that synaptic adjustment did not occur. Maguire (1981), using squashes of maize chromosomes primarily at late pachytene, also reached this conclusion. However, her conclusions were questioned by Moses *et al.* (1982) because, "the critical information, a quantitative comparison at early, mid- and late pachytene, is difficult to obtain in maize and the extent of loop adjustment has yet to be determined." This reservation plus the question of the universality of synaptic adjustment led us to investigate synaptic adjustment in inversion heterozygotes of maize using a hypotonic bursting technique to spread SCs from primary microsporocytes that ranged from early to late pachytene (Stack & Anderson 1987).

Identification of early, middle, and late pachytene

Critical to Moses and co-workers' discovery of synaptic adjustment in the laboratory mouse was their ability to identify substages of pachytene in SC spreads based on cytological criteria. Many plants, including maize, provide an excellent opportunity for accurate determination of meiotic substages of SC spreads because from leptotene to pachytene the microsporocytes from all the anthers within a single bud are usually in the same substage of meiosis. Thus, one anther may be used to make a conventional squash for determination of the meiotic substage of its primary microsporocytes, and then the remaining anthers can be used for hypotonic bursting to spread SCs. Using this method, the following cytological markers for early, middle, and late pachytene were identified in hypotonically burst, AgI-stained primary microsporocytes of maize (Stack & Anderson 1987): zygotene and early pachytene were characterised by closely associated SCs (presumably a manifestation of the synizetic knot seen in squash preparations) and a large, darkly stained, round nucleolus (Fig. 1); middle pachytene was characterised by less tightly clumped SCs and a smaller, less regularly-shaped nucleolus that often seemed to be disintegrating into dark granules (Fig. 2); in late pachytene and early diplotene SCs tended to be well separated, and the nucleolus was no longer visible, except for scattered dark granules (Fig. 3).

Frequency of inversion loops

Based on cytological criteria for identifying early, middle, and late pachytene, we have determined the frequencies of different patterns of synapsis in maize plants grown from caryopses kindly supplied by Dr Greg Doyle of the University of Missouri. The caryopses were obtained by selfing plants that were heterozygous for one of four different inversions. Two of these inversions were paracentric (In3f and In7a), and two were pericentric (In7b, In7e). For this study we used four populations of 24 plants each. At least half of the plants in each population were sampled to make SC spreads.

Over 1200 complete sets of SCs were analysed in which the pattern of synapsis was completely interpretable. Each set was placed in only one of three categories. (1) Sets of SCs with an inversion loop took precedence in that these sets may also have had varying amounts of asynapsis, but because of the presence of a loop, they were placed in the loop category only. (2) Sets with neither inversion loops nor asynapsis were placed in the straight synapsis category. (3) Sets with any amount of asynapsis but without an inversion loop

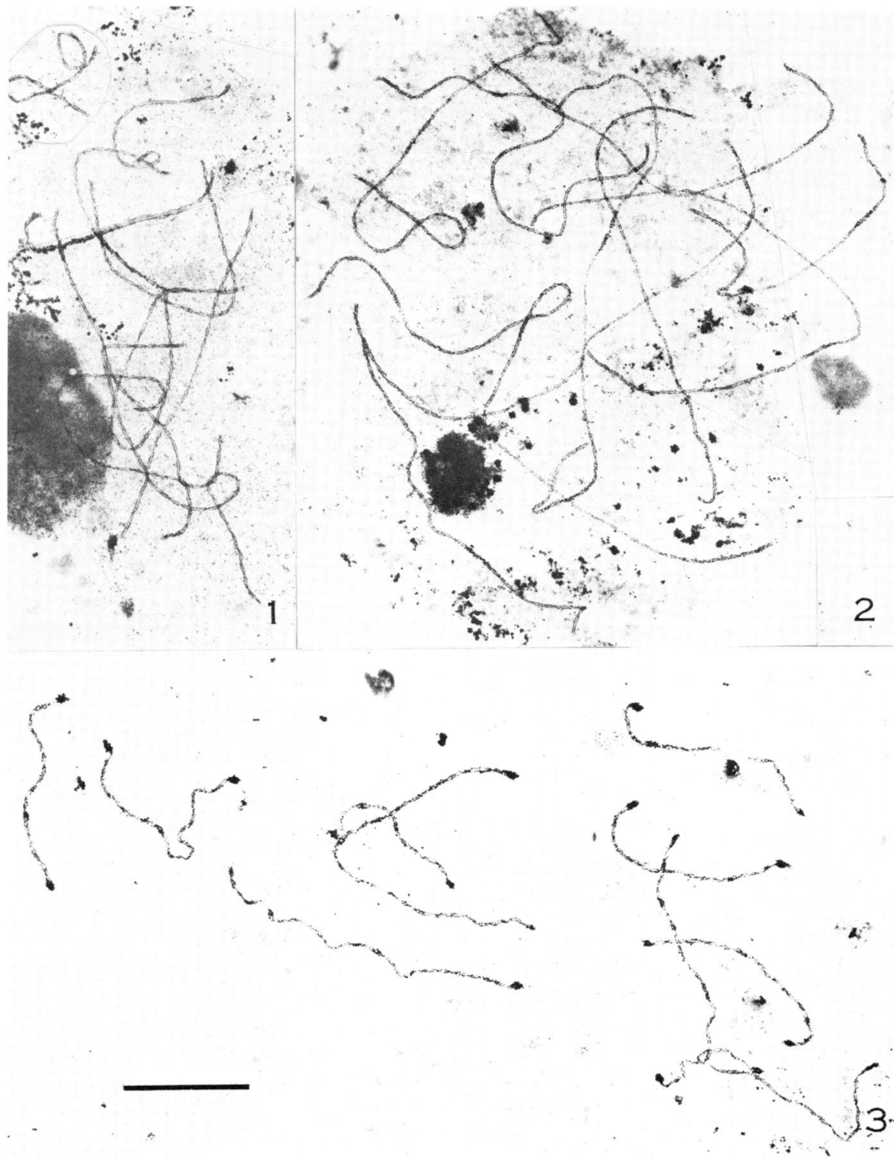

Figures 1–3. Representative spreads of SCs from primary microsporocytes of maize in early, middle and late pachytene. Although the plants were inversion heterozygotes, each of these spreads shows only straight synapsis, i.e., no inversion loop or asynapsis. **Figure 1.** In7e early pachytene. Note the large, dark-staining nucleolus to the lower left. The SCs tend to be clumped together. The SC at the upper left was cut out and moved to that position from just out of the figure at centre right. **Figure 2.** In7e middle pachytene. The nucleolus to the lower left is smaller and the SCs more separated than at early pachytene. **Figure 3.** In7a late pachytene. The nucleous has disintegrated to the point that it is represented only by a few dark granules, but desynapsis has not yet begun. The SCs tend to be well separated. Bar = 10 μm.

were placed in the asynapsis category. Since more than one SC in a set can show asynapsis, it is clear that not all asynapsis was related to inversion heterozygosity. Unfortunately, kinetochores were not consistently stained in our preparations, so it was impossible to identify individual chromosomes and to determine whether inverted segments were preferentially involved in asynapsis. However, this seems likely, based on squash preparations (McClintock 1931, 1933; Maguire, 1966).

Since most plants used in this study were not confirmed to be inversion heterozygotes based on the independent criteria of observing inversion loops or frequent bridges and fragments in squash preparations, the observed frequencies of inversion loops per set of SCs (Table 1) are almost certainly lower

Table 1. Frequencies of synaptic configurations found during early, middle and late pachytene in complete sets of SCs from four populations of maize plants carrying different inversions.

		Synaptic configuration						
		Loop		Straight		Asynapsis		Total
Inv.	Pachytene substage	No.	%	No.	%	No.	%	
3f	early	2	4.3	15	32.6	29	63.1	46
	middle	4	1.8	51	23.2	165	75.0	220
	late	5	5.0	19	19.0	76	76.0	100
	Total	11	3.0	85	23.2	270	73.8	366
7a	early	6	13.1	14	30.4	26	56.3	46
	middle	16	7.7	23	11.0	170	81.3	209
	late	19	9.8	38	19.7	136	70.5	193
	Total	41	9.2	75	16.7	332	74.1	448
7b	early	2	6.5	8	25.8	21	67.7	31
	middle	8	11.4	20	28.6	42	60.0	70
	late	9	13.4	23	34.3	35	52.3	67
	Total	19	11.3	51	30.4	98	58.3	168
7e	early	9	24.3	12	32.4	16	43.3	37
	middle	20	18.5	31	28.7	57	52.8	108
	late	26	28.6	22	24.2	43	47.2	91
	Total	55	23.3	65	27.5	116	49.2	236

than the true frequencies of the inversion loops in pure populations of inversion heterozygotes. However, this does not interfere with an analysis of synaptic adjustment because the frequencies of inversion loops should fall to zero in late pachytene if maize adjusts inversion loops as the laboratory mouse does (Moses et al. 1982). For In3f, In7b, and In7e, the frequency of inversion loops increased slightly from early to late pachytene. For In7a, a slight decrease in loop frequency from early to late pachytene was observed. In3f, In7a, and In7e have a lower percentage of loops at middle pachytene than at either early or late pachytene. The rather erratic variation in the frequency of inversion loops at different substages of pachytene is probably related to sample size and low loop frequency. In spite of this variation, frequencies of all the inversion loops remained essentially constant throughout pachytene. Thus, synaptic adjustment apparently did not occur.

For every population except that containing In7b, there was a decrease in the percentage of sets with straight synapsis that was roughly paralleled by an increase in sets with asynapsis from early to late pachytene (Table 1). Again this is at variance with the expectations of synaptic adjustment where a decrease in asynapsis, not an increase, is expected as pachytene progresses.

Inversion loop characteristics

Each inversion loop was characterised regarding position and size (Table 2). However, because kinetochores did not usually stain with silver, it was impossible to identify individual SCs or to locate the position of inversion loops in relation to kinetochores. As an alternative, the positions of loops were characterized by measuring from each end of an SC to the nearest margin of the inversion loop (taken to be the beginning of asynapsis at the synaptic partner exchange). The shortest SC length thus measured was used as the reference to describe the position of the inversion loop. This method of describing loop postion is reasonable because all four of the inverted segments are asymmetrically located with respect to the ends of the chromosomes. Of these, In7e is least asymmetrical because of its large size. The total size of a loop was determined by measuring the length of the homologously synapsed portion of the loop together with the asynapsed portions on either side (Table 2 — total loop size). Because some of the asynapsis could represent areas adjacent to but not part of the inversion, loop size was also determined by measuring only the homologously synapsed portion of the loop (Table 2 — synapsed loop size).

The length of each bivalent with an inversion loop was also measured. The absolute length of these SCs varied both within and between substages of pachytene for each inversion heterozygote. Variations in length of individual SCs may be related to the variations in overall length of complete sets of SCs that have been observed in maize (Gillies 1981; Anderson *et al.* 1985; and see the section entitled 'Lengths of sets of synaptonemal complexes'). Because of this, loop position and loop size are most usefully presented as percentages of the length of the SC carrying the loop (Table 2).

Table 2. Characteristics of loops observed in SCs from inversion heterozygotes for In3f, In7a, In7b and In7e expressed as per cent of the total length of the SC.

	Pachytene substage	Number[a]	Loop position[b]	Total loop size[c]	Synapsed loop size[d]
In3f	early	2	11.0 ± 5.7	14.5 ± 7.8	5.0 ± 0.0
	middle	5	9.6 ± 8.3	14.6 ± 8.3	4.8 ± 4.3
	late	6	10.8 ± 6.5	17.5 ± 19.2	5.2 ± 1.7
	Total	13			
In7a	early	7	3.6 ± 1.5	66.0 ± 5.3	46.1 ± 6.1
	middle	15	5.6 ± 6.4	58.7 ± 13.4	34.7 ± 12.2
	late	23	3.8 ± 2.3	62.5 ± 11.3	41.1 ± 10.3
	Total	45			
In7b	early	2	13.5 ± 3.5	51.0 ± 2.8	39.5 ± 7.8
	middle	9	11.6 ± 4.2	60.0 ± 10.3	31.2 ± 11.2
	late	10	11.8 ± 4.8	56.8 ± 15.8	30.8 ± 9.9
	Total	21			
In7e	early	9	1.4 ± 1.2	95.4 ± 2.5	84.0 ± 13.2
	middle	21	3.9 ± 2.9	93.8 ± 6.3	68.7 ± 21.8
	late	29	5.8 ± 8.5	88.8 ± 16.0	68.3 ± 24.3
	Total	59			

[a]Includes all loops that were interpretable regardless of whether or not the complete set of SCs was interpretable.
[b]Shortest length from one end of the SC to the beginning of asynapsis of the inversion loop. Inversion loops with one end of the SC unsynapsed were omitted from calculations of the average loop position.
[c]The length of the homologously synapsed portion of the loop plus the asynapsed portions on either side.
[d]The length of the synapsed portion of the loop only.

In3f. This is a previously uncharacterised paracentric inversion that has arisen recently from a maize line that concomitantly lost In3c (Greg Doyle, personal communication). This inversion involves only 5% (synapsed loop size) or 9% (total loop size) of the length of the SC (Table 2, Fig. 4). Marjorie Maguire has also observed these small inversion loops in this maize line using squash preparations (personal communication).

Thirteen inversion loops from In3f heterozygotes were examined (Table 2). Two of the loops were from sets of SCs that otherwise were not completely interpretable, and were thus not included in the frequency determinations (Table 1). For In3f, the position of the loop, the size of the synapsed loop, and the size of the total loop all remained essentially unchanged from early to late pachytene.

In7a. According to examination of heterozygotes for In7a in squashes, this inversion is paracentric and involves most of the long arm (Greg Doyle, personal communication) or about 74% of the total length of the chromosome. According to data from our SC spreads, the inversion involves about 40% (synapsed loop size) or 60% (total loop size) of the length of the SC (Table 2, Fig. 5).

Forty-five inversion loops from In7a heterozygotes were examined (Table 2). Four of the loops were from sets of SCs that were not otherwise completely interpretable and so were not included in frequency determinations (Table 1). The position of the loop did not change appreciably through the substages of pachytene. Loops with the short end asynapsed were common, perhaps due to the location of the inversion so close to one end of the SC. In these cases, the positions of the loops were not determined, and so these SCs were excluded from the calculation of average loop position. Synapsed loop size and total loop size remained essentially the same throughout pachytene.

In7b. Based on squash preparations, In7b is a pericentric inversion that involves 32% of the short arm and 30% of the long arm (Greg Doyle, personal communication) or 31% of the total length of the bivalent. Data from our SC spreads indicate that 32% (synapsed loop size) or 58% (total loop size) of the SC is involved in the inversion (Table 2, Fig. 6).

Twenty-one inversion loops were analysed (Table 2). Two loops were from sets of SCs that otherwise were not completely interpretable and so were not included in the frequency determinations (Table 1). Loop position was stable and loop size varied little during pachytene.

In7e. According to determinations made from squashed chromosomes (Greg Doyle, personal communcation), the pericentric loop in In7e includes 88% of the short arm and 93% of the long arm. This is equivalent to 92% of the total length of the bivalent. According to our SC spreads, 71% (synapsed loop size) or 92% (total loop size) of the SC is involved in the inversion (Table 2, Fig. 7).

Fifty-nine inversion loops from In7e were analysed (Table 2). Of these, four were from sets of SCs that were not completely interpretable and were not included in frequency determinations (Table 1). Loop positions changed slightly with substage of pachytene (Table 2). Like In7a, one end of the SC was often asynapsed, presumably due to the location of the loop so close to the ends of the SC. In these cases, the positions of the loops were not determined and these SCs were excluded from the calculation of the average loop position. The data indicate that while the total size of the loop changed little from early to late pachytene, there is an increase in the amount of asynapsis during middle and

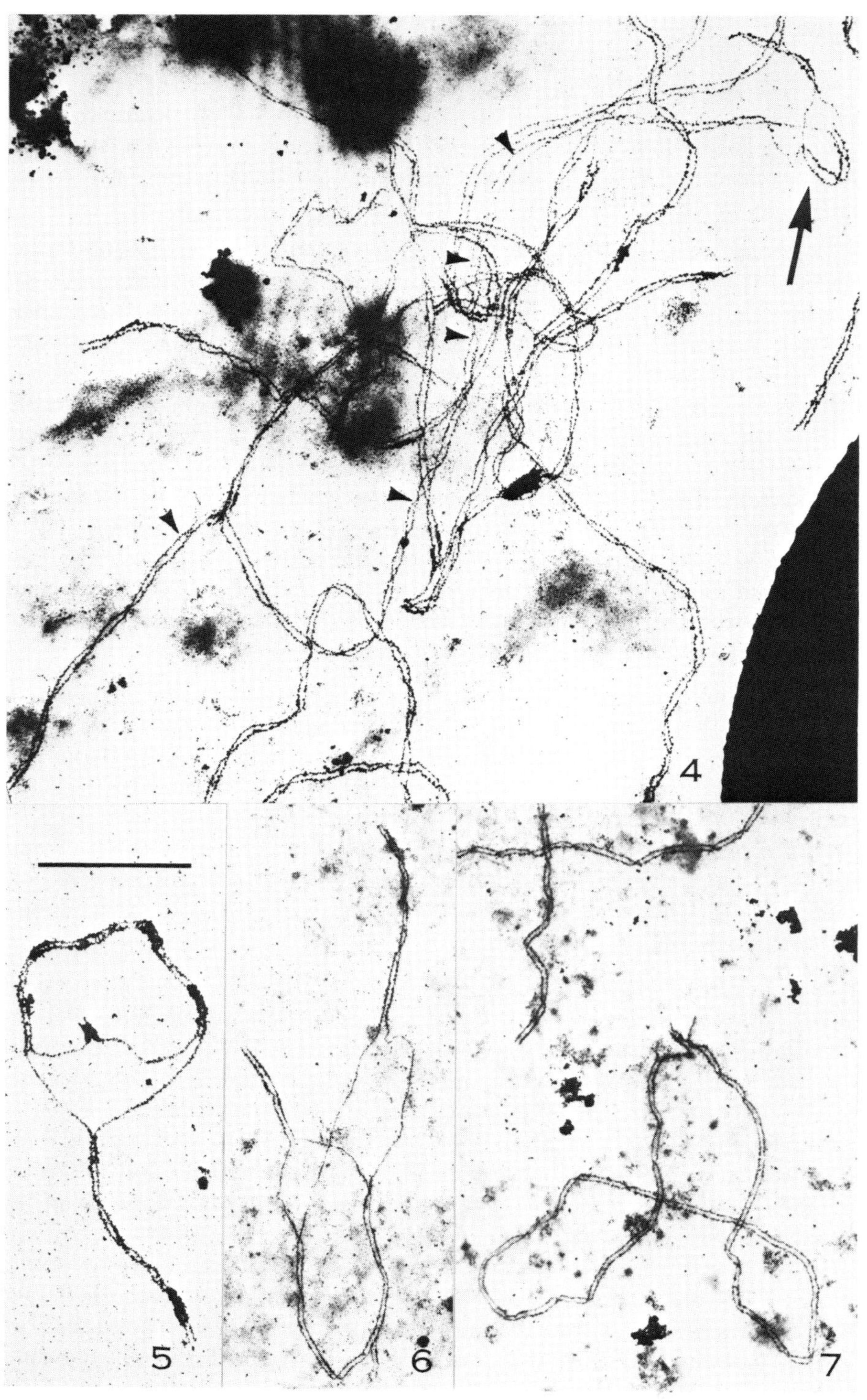

late pachytene at the expense of homologously synapsed portions of the loop, i.e., the size of the synapsed loop decreased.

Conclusion

A narrow definition of synaptic adjustment is the progressive conversion during pachytene of certain homologous synaptic configurations, e.g. inversion loops, to straight non-homologous synapsis (Moses *et al.* 1982). Adjustment is due to two-phase synapsis. The first phase occurs in zygotene and is confined to strictly homologous synapsis that continues until early pachytene. The second phase occurs during mid- and late pachytene and favours straight synapsis whether homologous or not. Second phase synapsis results in progressive loss of inversion loops in favour of straight, non-homologous synapsis. Synaptic adjustment has been shown to occur in certain inversion heterozygotes of the laboratory mouse. It may be widespread, but this remains to be demonstrated in other species by identifying different pachytene substages and analyzing synapsis. Enthusiasm for synaptic adjustment has led some authors to broaden its definition to include a change from asynapsis to synapsis during pachytene, to assume that any non-homologous synapsis is the result of synaptic adjustment when prior homologous synapsis was not demonstrated, and that synaptic adjustment occurred in zygotene when non-homologous synapsis extends throughout pachytene. This is confusing because it allows synaptic adjustment to embrace everything including no change in synapsis during pachytene, which is ostensibly what synaptic adjustment is all about. If the term synaptic adjustment is to be used, a clear definition must be generally agreed upon. One possibility is to broaden and simplify the definition to include any demonstrated change in synapsis during pachytene.

In our investigation of four different inversion heterozygotes of maize, we find no consistent change in the pattern of synapsis throughout pachytene. There is roughly the same amount of asynapsis, non-homologous straight synapsis, and inversion loop synapsis in early, middle, and late pachytene. Also, inversion loops appear neither to change size nor position during pachytene. Thus, regardless of whether the narrow or the broad definition is used, we find no evidence of synaptic adjustment of inversion loops in maize.

Lengths of sets of synaptonemal complexes

There are only a few generalisations that can be made about the length of SCs

Figures 4–7. SCs with inversion loops from maize primary microsporocytes that are inversion heterozygotes. **Figure 4.** In3f late pachytene. The small inversion loop is at the upper right (large arrow). The remainder of the SC from bivalent 3 is marked with arrowheads. **Figure 5.** In7a late pachytene. Although synapsed here, the short distal end often is not synapsed. **Figure 6.** In7b late pachytene. **Figure 7.** In7e late pachytene. Only one of the two ends is synapsed, but often both ends are synapsed. This is the only spread shown in which kinetochores are visible. Bar = 5 μm.

with any confidence. For instance, there seems to be a strong correlation between genome size and average total length of a set of SCs among angiospermous plants (Anderson *et al.* 1985), and this relationship may also occur among grasshoppers and locusts (Croft & Jones 1986). A strong correlation between SC length and the amount of crossing-over has been reported in several species (Holm & Rasmussen 1977, but see Wallace & Hultén, 1985, and Bojko, 1985, for differing interpretations of the relation of SC lengths to crossing-over in human primary oocytes). On the other hand, a trend for shortening SCs from zygotene to diplotene might be expected, based on reports of chromosome shortening during pachytene (Darlington 1935) and the obvious shortening of chromosomes that occurs during diplotene. Indeed, SC shortening has been observed in several species where substages of pachytene were identified (Rasmussen & Holm 1979, for tetraploid silkworm oocytes; Carmi *et al.* 1978, for *Schizophyllum commune*; Holm *et al.* 1981, for *Coprinus*; Holboth 1981 and Holm 1986, for wheat; Abirached-Darmency *et al.* 1983, for rye; Gillies 1983, for maize; Moses & Poorman 1984, for mouse spermatocytes and oocytes). However, in other cases the SCs may lengthen during pachytene (Rasmussen 1976, for diploid silkworm oocytes; Rasmussen 1977, for triploid silkworm oocytes; Zickler 1977, for *Sordaria macrospora*; Rasmussen & Holm 1978, for human spermatocytes; Gillies 1979, for *Neurospora*; Bogdanov *et al.* 1986, for sex bivalents in spermatocytes from *Ellobius talpius*) or shorten and then lengthen during pachytene (Carpenter 1975, for *Drosophila melanogaster*; Moses *et al.* 1977, for Chinese hamster spermatocytes; Greenbaum *et al.* 1986, for *Peromyscus maniculatus*), or remain the same length (Goldstein & Triantaphyllou 1982, for *Meloidogyne hapla*; Goldstein & Slaton 1982, for *Caenorhabditis elegans*; Bojko 1985 for human oocytes), or lengthen from zygotene to pachytene (Wallace & Hultén 1985, for human oocytes).

For an analysis of SC length in maize, we selected sets of SCs from our study of synaptic adjustment that showed no evidence of stretching. In plants heterozygous for In3f, In7a, and In7b, SCs tended to shorten from early to late pachytene (Table 3). However, the shortening is statistically significant only for In7a and In7b. In plants heterozygous for In7a, SCs lengthen and then shorten significantly. In plants heterozygous for In7e, SCs tended to lengthen during pachytene, but the differences were not statistically significant (Table 3). An outstanding feature of these observations is the substantial variability in total SC length that was encountered in every substage. Such variability has been observed regularly in other species also. Variation in SC length could be introduced during preparation (Moens & Ashton 1985). In addition, small sample sizes may give misleading impressions of differences in SC length between substages. If the observed patterns of changes during pachytene are real and not due solely to preparative techniques or small sample sizes, the extremely divergent patterns between species and even within species (e.g., maize and silkworms) obscure any obvious functional role for progressive changes in SC length.

Table 3. Comparison of lengths and number of twists in complete sets of SCs during substages of zygotene for four inversions in maize.

Inversion		Early pachytene	Middle pachytene	Late pachytene
3f	length	460 ± 46µm (n = 5)	426 ± 62µm (n = 16)	399 ± 62µm (n = 20)
	No. twists	146 ± 34 (n = 18)	148 ± 30 (n = 43)	147 ± 28 (n = 46)
7a	length	403 ± 73µm (n = 9)[a]	435 ± 90µm (n = 14)[b]	337 ± 56µm (n = 23)[a,b]
	No. twists	147 ± 30 (n = 24)[d]	164 ± 22 (n = 44)[d,e]	144 ± 21 (n = 75)[e]
7b	length	494 ± 140µm (n = 7)[c]	424 ± 73µm (n = 8)	366 ± 63µm (n = 9)[c]
	No. twists	173 ± 20 (n = 10) [f,g]	139 ± 29 (n = 28)[f]	134 ± 28 (n = 33)[g]
7e	length	434 ± 130µm (n = 9)	461 ± 79µm (n = 9)	481 ± 98µm (n = 12)
	No. twists	204 ± 40 (n = 14)	222 ± 29 (n = 18)	202 ± 47 (n = 29)

a, b, c, d, e, f, g   Comparisons were made only in the same rows. Those lengths or number of twists that are statistically different at P>0.95 bear the same superscript.

Twisting of the synaptonemal complex

Twisting of pachytene chromosomes was suspected by light microscopists (Wilson 1928: Manton 1950), but only electron microscopy has the resolution to observe this phenomenon with certainty. With the development of serial thin sectioning (Wettstein & Sotelo 1967) and surface spreading of SCs (Counce & Meyer 1973), observations of twisted SCs became common. The nature of twists can vary from tight spirals (Moens 1972) to the gentle twists that are much more commonly reported. In spite of the common occurrence of twisting, we know of only two reports in which the level of twisting was related to pachytene substages, and in both cases surface spread SCs from mammalian primary spermatocytes were used. Moses *et al.* (1977) reported that in Chinese hamsters the number of twists per SC complement increases throughout pachytene, but the SCs first shorten and then lengthen. Moses *et al.* (1977) concluded that twists cannot account for the length changes observed in pachytene and that intertwining of SCs is random and proportional to SC length. On the other hand, the SCs of *Peromyscus maniculatus* show similar changes in SC length during pachytene, but twisting increases during early pachytene and decreases in late pachytene (Greenbaum *et al.* 1986).

We have made a similar analysis of twisting in spread preparations of maize SCs from heterozygotes for In3f, In7a, In7b, and In7e. We find that the pattern of twists varies among the different lines (Table 3). For instance, the level of twisting does not change significantly during pachytene in In3f plants, goes up and then down significantly in In7a plants, goes down significantly in In7b plants, and goes up and then down but not significantly in In7e plants. Only about 30% of the variation in number of twists per set can be related to length of SCs (r^2 ranges from 0.22 to 0.45). The distribution of total twists per set of SCs for each inversion heterozygote is indistinguishable from a normal curve, implying a random distribution of twists per set of SCs.

Compared to three-dimensional reconstructions, surface spreads have the advantage that large numbers of SCs can be examined but the disadvantage that true twists usually cannot be distinguished from lateral elements that cross during spreading and drying. Still, true twists will always be counted and preparation-induced crossing of lateral elements will form a background of apparent twisting that should be proportional to SC length. It is possible that some or all of the 30% of the variation in total twists that can be related to SC length is due to preparation-induced crossing of lateral elements, while the remainder is related to the mechanism of synapsis that causes a certain level of homologue intertwining as synapsis occurs. In support of the latter possibility, Moens (1972, 1978) observed that all of the SCs twisted clockwise in three-dimensional reconstructions of nuclei from primary microsporocytes of *Rhoeo spathacea* and counter-clockwise in three-dimensional reconstructions of nuclei from primary spermatocytes of *Rattus norvegicus*, an unlikely coincidence if twisting were not imposed on the SCs.

Acknowledgements

We thank Dr. Greg Doyle at the University of Missouri, Columbia, Missouri for supplying caryopses for the maize inversion heterozygotes and giving us the position and sizes of In7a, In7b, and In7e. This research was supported in part by NSF grant PCM–8302924, USDA grant 5901–0410 — 9–0350–0, and a Biomedical Research Support Grant from CSU.

References

Abirached-Darmency, M.D., D. Zickler and Y. Cauderon 1983. Synaptonemal complex and recombination nodules in rye (*Secale cereale*). *Chromosoma* 88, 299–306.

Anderson, L., S. Stack, M. Fox and C. Zhang 1985. The relationship between genome size and synaptomenal complex length in higher plants. *Exp. Cell Res.* 156, 367–378.

Bogdanov, Y., O. Kolomiets, E. Lyapunova, I. Yanina and T. Mazurova 1986. Synaptonemal complexes and chromosome chains in the rodent *Ellobius talpinus* heterozygous for ten Robertsonian translocations. *Chromosoma* 94, 94–102.

Bojko, M. 1985. Human meiosis IX. Crossing over and chiasma formation in oocytes. *Carlsberg Res. Commun.* 50, 43–72.

Carmi, P., P. Holm, Y. Koltin, S. Rasmussen, J. Sage and D. Zickler 1978. The pachytene karyotype of *Schizophyllum commune* analysed by three-dimensional reconstruction of synaptonemal complexes. *Carlsberg Res. Commun.* 43, 117–132.

Carpenter, A. 1975. Electron microscopy of meiosis in *Drosophila melanogaster* females. I. Structure, arrangement and temporal change of the synaptonemal complex in wild-type. *Chromosoma* 51, 157–182.

Counce, S. and G. Meyer 1973. Differentiation of the synaptonemal complex and the kinetochore in *Locusta* spermatocytes studied by whole mount electron microscopy. *Chromosoma* 44, 231–253.

Croft, J. & G. Jones 1986. Surface spreading of synaptonemal complexes in locusts I. Pachytene observations. *Chromosoma* 93, 483–488.

Darlington, C. 1935. The internal mechanics of chromosomes. III. Relational coiling and crossing-over in *Fritillaria. Proc. Roy. Soc. London B* 118, 74–96.

Gillies, C. 1979. The relationship between synaptonemal complexes, recombination nodules and crossing-over in *Neurospora crassa* bivalents and translocation quadrivalents. *Genetics* 91, 1–17.

Gillies, C. 1981. Electron microscopy of spread maize pachytene synaptonemal complexes. *Chromosoma* 83, 575–591.

Gillies, C. 1983. Spreading plant synaptonemal complexes for electron microscopy. In *Kew Chromosome Conf. II*, P. Brandham and M. Bennett, eds, 115–122. London UK: George Allen and Unwin.

Goldstein, P. and D. Slaton 1982. The synaptonemal complexes of *Caenorhabditis elegans.* Comparison of wild-type and mutant strains and pachytene karyotype analysis of wild-type. *Chromosoma* 84, 585–597.

Goldstein, P. and A. Triantaphyllou 1982. The synaptonemal complexes of *Meloidogyne*: Relationship of structure and evolution of parthenogenesis. *Chromosoma* 87, 117–124.

Greenbaum, I., D. Hale and K. Fuxa 1986. The mechanism of autosomal synapsis and the substaging of zygonema and pachynema from deer mouse spermatocytes. *Chromosoma* 93, 203–212.

Hobolth, P. 1981. Chromosome pairing in allohexploid wheat var. Chinese Spring. *Carlsberg Res. Commun.* 46, 129–173.

Holm, P. 1986. Chromosome pairing and chiasma formation in allohexaploid wheat, *Triticum aestivum* analyzed by spreading of meiotic nuclei. *Carlsberg Res. Commun.* 51, 239–294.

Holm, P. and S. Rasmussen 1977. Human meiosis. I. the human pachytene karyotype analyzed by three dimensional reconstruction of the synaptonemal complex. *Carlsberg Res. Commun.* 42, 283–323.

Holm, P., S. Rasmussen, D. Zickler, B. Lu and J. Sage 1981. Chromosome pairing, recombination nodules and chiasma formation in the basidiomycete *Coprinus cinereus. Carlsberg Res. Commun.* 46, 305–346.

Loidl, J. 1987. Synaptonemal complex spreading in *Allium ursinum:* pericentric asynapsis and axial thickenings. *J. Cell.Sci.* 87, 439–448.

Maguire, M. 1966. the relationship of crossing over to chromosome synapsis in a short paracentric inversion. *Genetics* 53, 1071–1077.

Maguire, M. 1981. A search for the synaptic adjustment phenomenon in maize. *Chromosoma* 81, 717–725.

Manton, I. 1950. The spiral structure of chromosomes. *Biol. Revs. Cambridge Phil. Soc.* 25, 486–508.

McClintock, B. 1931. Cytological observations of deficiencies involving known genes, translocations, and an inversion in *Zea mays. Missouri Agric. Exp. Stat. Bull.* 163, 1–30.

McClintock, B. 1933. The association of non-homologous parts of chromosomes in mid-prophase of meiosis in *Zea mays. Z. Zellforsch. mikr. Anat.* 19, 191–237.

Moens, P. 1972. Fine structure of chromosome coiling at meiotic prophase in *Rhoeo discolor. Can. J. Genet. Cytol.* 14, 801–808.

Moens, P. 1978. Lateral element cross-connections of the synaptonemal complex and their relationship to chiasmata in rat spermatocytes. *Can. J. Genet. Cytol.* 20, 567-579.

Moens, P. and M. Ashton 1985. Synaptonemal complexes of normal and mutant yeast chromosomes (*Saccharomyces cerevisiae*). *Chromosoma* 91, 113–120.

Moses, M. 1977. Microspreading and the synaptonemal complex in cytogenetic studies. *Chromosomes Today* 6, 71–82.

Moses, M. and P. Poorman 1984. Synapsis, synaptic adjustment and DNA synthesis in mouse oocytes. *Chromosomes Today* 8, 90–103.

Moses, M., P. Poorman, M. Dresser, G. DeWeese and J. Gibson 1985. The synaptonemal complex in meiosis: significance of induced perturbations. In *Aneuploidy: etiology and mechanisms.* V. Dellarco, P. Voytek and A. Hollaender, eds, 337–350. New York: Plenum Press.

Moses, M., P. Poorman, T. Roderick and M. Davisson 1982. Synaptonemal complex analysis of mouse chromosomal rearrangements. *Chromosoma* 84, 457–474.

Moses, M., G. Slatton, T. Gambling and C. Starmer 1977. Synaptonemal complex karyotyping in spermatocytes of the chinese hamster (*Cricetulus griseus*). III. Quantitative evaluation. *Chromosoma* 60, 345–375.

Rasmussen, S. 1976. The meiotic prophase in *Bombyx mori* females analyzed by three dimensional reconstructions of synaptonemal complex. *Chromosoma* 54, 245–293.

Rasmussen, S. 1977. Meiosis in *Bombyx mori* females. *Phil. Trans. R. Soc. London B* 277, 343–350.

Rasmussen, S. and P. Holm 1978. Human meiosis. II. Chromosome pairing and recombination nodules in human spermatocytes. *Carlsberg Res. Commun.* 43: 275–337.

Rasmussen, S. and P. Holm 1979. Chromosome pairing in autotetraploid *Bombyx* females, mechanism for exclusive bivalent formation. *Carlsberg Res. Commun.* 44, 101–125

Stack, S. and L. Anderson 1987. Hypotonic bursting method for spreading synaptonemal complexes of *Zea mays. J. Hered.* 78, 178–182.

Wallace, B. and M. Hultén 1985. Meiotic chromosome pairing in the normal human female. *Ann. Hum. Genet.* 49, 215–226.

Wettstein, R. and J. Sotelo 1967. Electron microscope serial reconstruction of the spermatocyte. I; nuclei at pachytene. *J. Microscopie* 6, 557–576.

Wilson, E. 1928. *The Cell in Development and Heredity.* New York: Macmillan Company.

Zickler, D. 1977. Development of the synaptonemal complex and the "recombination nodules" during meiotic prophase in the seven bivalents of the fungus *Sordaria macrospora* Auersw. *Chromosoma* 61, 289–316.

Synaptonemal complex spreading in human foetal oocytes

R.M. Speed and A.C. Chandley

MRC Clinical and Population Cytogenetics Unit, Western General Hospital, Crewe Road, Edinburgh EH4 2XU, Scotland, UK

The surface spreading technique was introduced over ten years ago for the study of synaptonemal complexes in male insect germ cells (Counce & Meyer 1973). This has since been extended to include a wide variety of male mammals ranging from hamster (Moses 1977a) to human (Solari 1980), and more recently to plants (Gillies 1981). Each new advance has entailed modifications of the original method, especially in the transfer to the more rigid cell-walled plant material. In the case of its application to female mammalian germ cells, modification has not been as great as those problems relating to the difference in timing of the meiotic process in males and females. In female mammals the initial stages of meiosis occur during the period of foetal life. In humans the numbers of oocytes reach a maximum (7 million) at about 5 months of gestation (Baker 1963), declining by a process of atresia to 2 million by birth. The pachytene stage, which is most amenable to analysis by surface spreading (Speed 1984), first appears in humans at approximately 14 weeks of gestation, continuing at least up to 24 weeks, beyond which time, due to termination legislation, ovaries are not available. The technique is equally applicable to the analysis of mouse oocytes (Moses *et al* 1982; Speed 1982). In this case, over the much shorter gestation period, pachytene oocytes are seen from day 14 to birth at day 20. The prime reason for investigating human female pachytene oocytes is that a large number of meiotic non-disjunctional events are thought to arise during female meiosis, giving rise to trisomic conceptuses (see Bond & Chandley 1983 for review). Therefore firstly a study of chromosomally normal foetal ovaries was carried out to determine the variation in synapsis occurring during meiotic prophase, and secondly of chromosomally abnormal foetal ovaries to examine unusual chromosome combinations that might yield novel synaptic forms.

Material and methods

Ovaries ranging from 9–23 weeks of gestation were obtained from normal and abnormal foetuses. The age of the foetus is assessed from both foot length and the date of the mother's last menstrual period. In all normal cases the reason for induction was social, the foetus being described as phenotypically normal.

Foetal blood karyotyping was used to confirm that the genotype was also normal in as many cases as possible. Chromosomally abnormal ovaries were obtained following amniocentesis when either trisomy 18, trisomy 21 or XO chromosome constitutions were recorded. The ovaries were usually collected within 24 hours of extra-amniotic prostaglandin induction. Microspreading was carried out according to standard procedures (Speed 1984) using silver nitrate staining (Howell & Black 1980).

Observations and discussion

A) *Normal foetal ovaries*
The most notable finding from a group of five normal foetuses, all at early to mid gestational ages (17–21 weeks) is the high incidence of synaptic errors and mid-pachytene degeneration (Table 1). From a total of 1200 oocytes examined at

Table 1   The percentage of normal and abnormal oocytes, from five chromosomally normal human foetuses, examined at E.M. level.

Foetus	Age (weeks)	Normal	Z cells	Nonhomologous pairing	Triple pairing	Interchange	Asynaptic regions	Univalents	Duplication
61	17	45.0	28.0	7.0	6.0	3.0	3.0	8.0	–
73	17	62.4	8.3	11.0	3.5	1.0	8.5	5.3	–
75	17	59.5	8.0	8.0	2.0	1.5	9.5	11.5	–
70	20	58.0	14.0	6.0	7.0	5.0	4.0	6.0	–
71	21	50.5	17.5	5.8	3.0	0.5	10.7	8.0	4.0
Mean		55.1	15.2	7.6	4.3	2.2	7.1	7.8	0.8

the electron microscope level, 45% showed abnormal pairing or atresia. This is much higher than seen in human spermatocytes (14% — Speed, unpublished observations). Four main forms of synaptic failure or degeneration were observed:

i) *Partial or complete asynapsis of lateral elements*
This can vary from 5–50% of the total length of an individual SC and was seen in 7.1% of oocytes. Single or multiple asynaptic SCs can be found within an oocyte. Asynaptic regions become thickened and dark-staining with silver. In its extreme form, total failure of pairing is seen and axial elements remain as univalents. This occurred in 7.8% of oocytes. Such disturbances may contribute to oocyte atresia. It was originally proposed by Miklos (1974) that any male germ cell carrying unpaired sex chromosomes would die owing to a failure to saturate pairing sites. This idea was later extended to include autosomes and sex chromosomes of both male and female situations by Burgoyne & Baker (1984).

ii)   Non-homologous pairing
In its simplest form this represents short lengths of axial element or univalents pairing on themselves. More commonly, however, two heterologous axes pair over short distances forming an apparently normal SC (7.6% of oocytes). Included in this grouping was an occasional oocyte where virtually complete non-homologous pairing between different elements had occurred. This resulted in two linked SCs, each with one symmetrical and one asymmetrically paired telomere (Fig. 1a). Oocytes with SCs consistently showing symmetrical sub-telomeric exchanges were seen in 2.2% of oocytes (Fig. 1b). The final form of non-homologous pairing involved the synapsis of three axial elements (4.3% of oocytes, Fig. 1c). This has been observed previously in trisomic foetuses (see below), but is unexpected in karyotypically normal foetuses.

It has been proposed that initial prophase synapsis is strictly homologous (Moses 1977b), non-homologous pairing being a secondary event. However, the majority of the above errors appear to originate at zygotene (Fig. 1d), failure of normal telo- or sub-telomeric pairing allowing almost immediate non-homologous synapsis to occur (Speed 1987).

iii)   Duplication or insertion
In one normal foetus studied, 4% of oocytes exhibited an unpaired region looped out from the central section of one SC (Fig. 1e). The only clear difference between homologues in G and C-banded somatic karyotypes was that one chromosome No. 16 had a smaller block of centric heterochromatin. Localised pairing disturbance due to such differences have been reported previously in red kangaroo by Sharp (1985). Most oocytes appeared normal, possible synaptic adjustment as described by Moses & Poorman (1981) having brought the loop into alignment.

iv)   Mid-pachytene degeneration – "Z" cells
At mid-pachytene the SCs in 15.2% of oocytes appeared to be breaking up into small fragments. These oocytes are termed "Z" cells, as originally used to describe degenerating pachytenes in the rat by Beaumont & Mandl (1962). This breakdown might be associated with the mid-term atresia seen in normal foetal ovaries, but whether it is causal or a consequence to it, remains as yet unresolved.

Why oocytes should be more prone to such synaptic errors is not known, but it might relate to the much greater overall length of the female SC complement. In human oocytes Bojko (1983) has estimated that SC length is about twice that found in spermatocytes within a nucleus of similar dimensions (Fig. 2). The mechanical constraints of such long SCs may thus make location of homologous telomeres, within intranuclear time schedules, more difficult for oocyte chromosomes.

The only evidence as to the fate of the above categories of oocyte, is that such abnormalities were never seen in the few diplotene cells analysed (n = 37), suggesting that they had been selectively removed from the germ line.

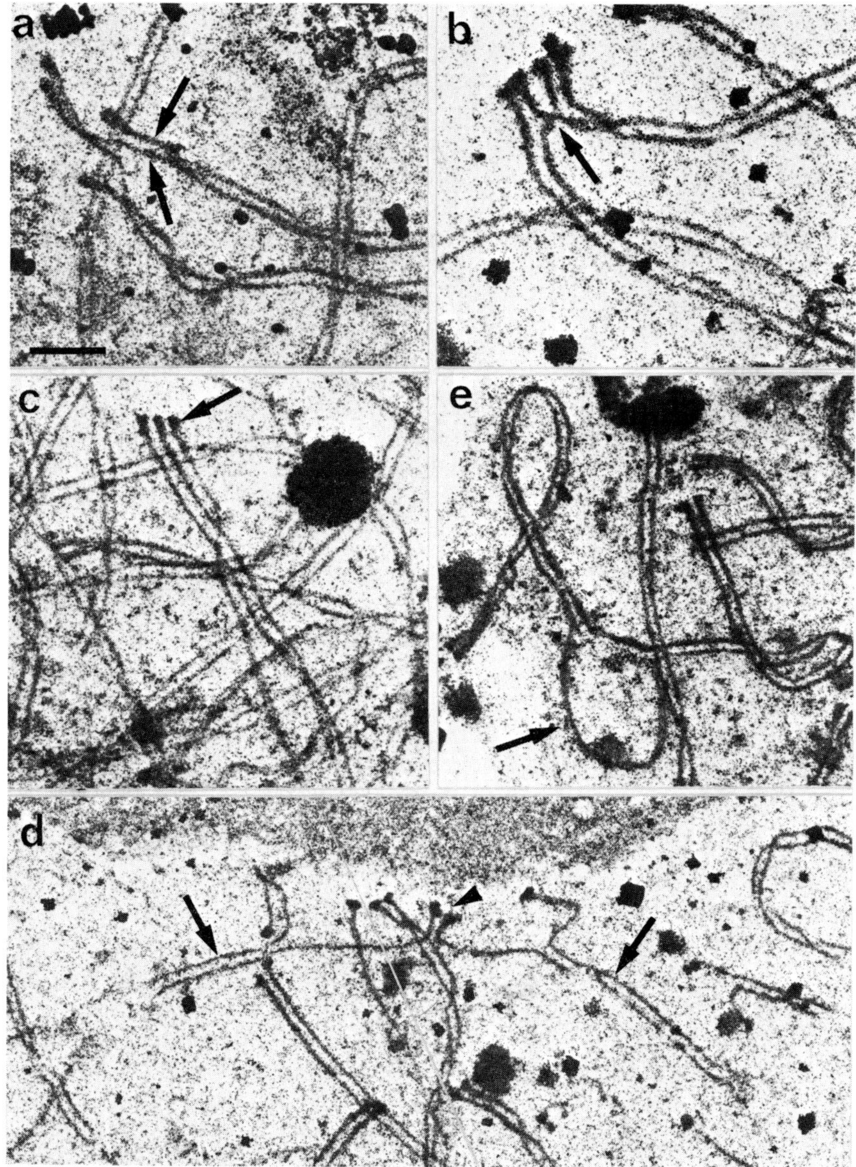

Figure 1. Pairing abnormalities in normal human foetal oocytes, observed at the E.M. level. **a** Virtually complete synapsis between pairs of lateral elements of different lengths (arrows), with homologous terminal synapsis of the larger elements. **b** Subtelomeric exchange. **c** Triple pairing. **d** Possible duplication or insertion. **e** Possible origin of exchange events at zygotene. Two SCs showing homologous subtelomeric pairing (arrows), but with one telomeric axial element from each exhibiting nonhomologous synapsis (arrow head). Bar = 1 μm. (Reproduced by permission of Human Genetics.)

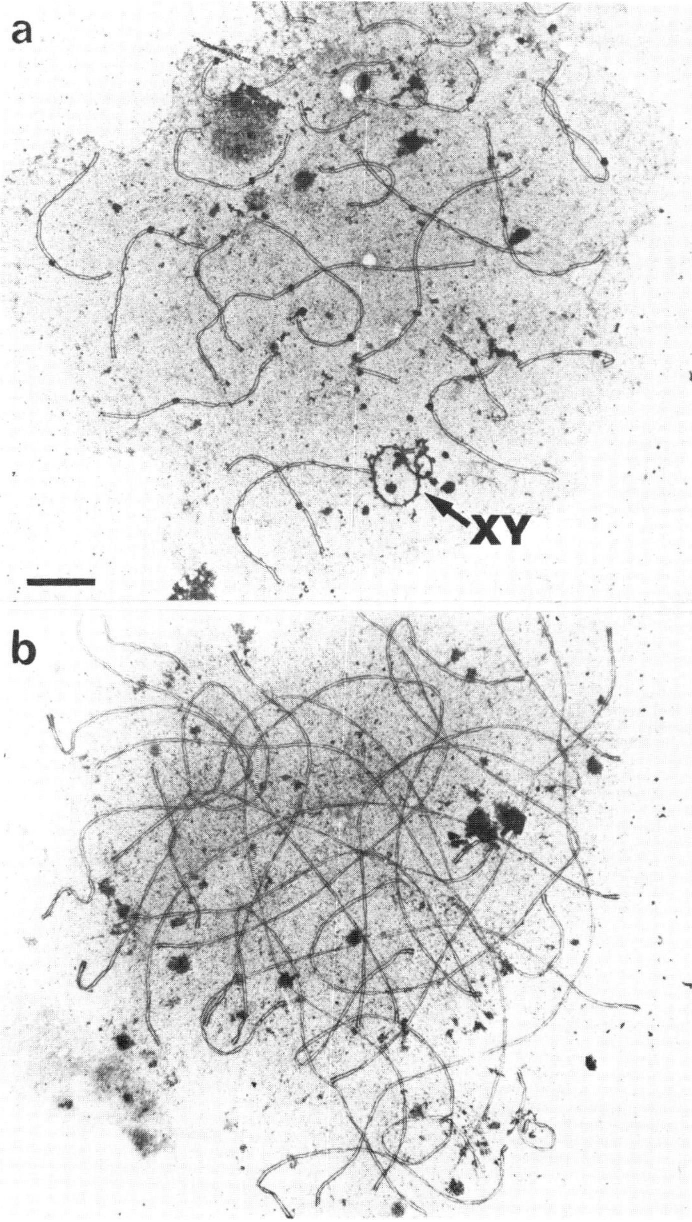

Figure 2. Differing lengths of synaptonemal complexes in male and female germ cells. **a** Human spermatocyte, compared with **b** a human foetal oocyte. Both at mid-pachytene. Bar = 10 μm.

B) Chromosomally abnormal foetal ovaries
i)   Trisomy 21 and 18

The most interesting feature of SC pairing in human foetuses exhibiting trisomy for an autosome, is that the three homologues can simultaneously undergo synapsis. This is contrary to the classical viewpoint that only two-by-two pairing characterises the trivalents of primary trisomics (Sybenga 1975) and triploids (Darlington 1965). In a standard trisomy 21 foetus (Speed 1984) approximately one-third of pachytene oocytes analysed at EM level showed such triple synapsis (Fig. 3a and b). This varied from 5–100% of the length of the three autosomes.

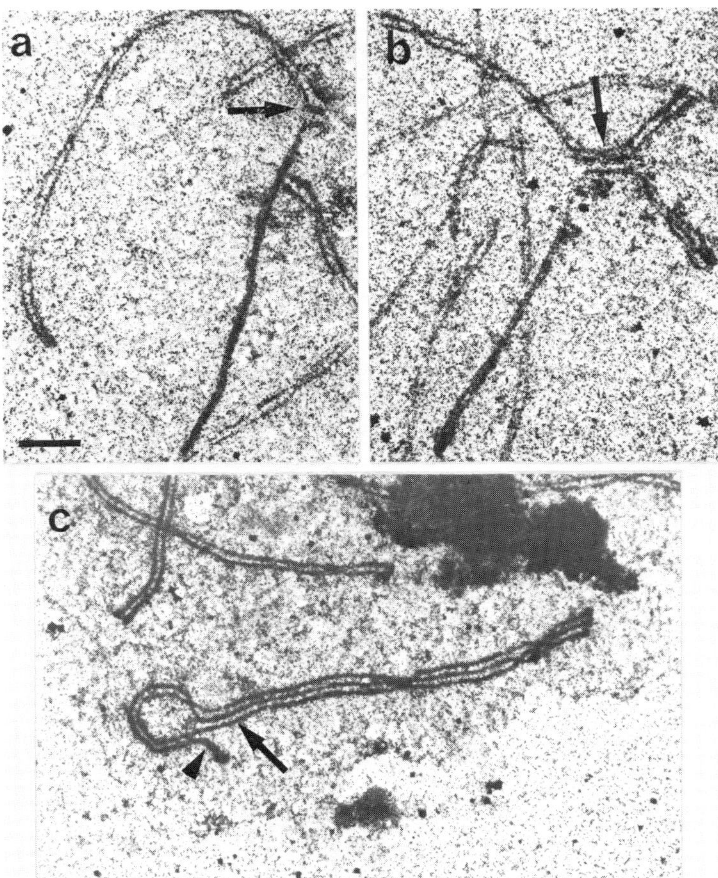

Figure 3. Pairing in chromosomally abnormal human oocytes. **a** Trisomy 21, showing small telomeric region of triple pairing (arrow). **b** Interstitial region of triple pairing (arrow). In both cases, unpaired regions of the third chomosome no. 21, are thickened and dark staining with silver. **c** Trisomy 18, isochromosome 18 (arrow) both self-paired, and synapsed with the normal chromosome 18 (arrow head). Bar = 1 μm. (Reproduced by permission of Human Genetics.)

Any unpaired segments within the trivalent appeared thickened and dark staining with silver. Such thickening was also the case when one of the chromosome 21s remained as a univalent, as occurred in the other two-thirds of the pachytene oocytes. Similarly, in a foetal ovary from a mosaic case of 18p-;iso18q (Speed 1986a), where the latter cell line is effectively trisomic for the long arm of chromosome 18, triple pairing was again frequently observed (Fig. 3c). In two-thirds of the iso18q cell line, the iso-chromosome both paired on itself and with the normal chromosome 18.

ii) Monosomy X (Turner's syndrome)

Our interest in Turner's syndrome stems from the general infertility associated with the human XO chromosome constitution, even though germ cells have been shown to be present during early gestation (Carr *et al.* 1968). In contrast, XO mice were initially thought to have normal fertility (Cattanach 1962), but subsequent studies have shown that they too have fewer oocytes and a shortened reproductive span compared with XX females (Lyon & Hawker 1973). Surface spreading studies undertaken to investigate human and mouse XO foetal oocytes (Speed 1986b) have shown several similar and contrasting features. Firstly, a factor having a bearing on reduced fertility is that the numbers of "Z" cells among XO oocytes are higher in both species than in XX controls. Secondly, in mouse, the survival of a large number of XO germ cells into the pachytene stage contrasts with the progression of very few in humans. A severe germ cell arrest at early prophase in humans appears to be the major cause of oocyte loss, cells from three XO foetuses examined to date showing meiotic arrest at the pre-leptotene stage (Speed 1986b). On examining the SC behaviour, there is again a major difference; in the few human oocytes that reach pachytene the single X axial element remains free-lying and becomes thickened and dark staining with silver (Fig. 4a). In mouse, however, several alternative events can occur. The X chromosome may pair with itself, forming a hairpin or loop (Fig. 4b and c), or the X may synapse with two free telomeric ends of an autosomal SC, forming a triradial structure (Fig. 4d). By day 19 of gestation almost 50% of oocytes showed the mouse X axial element involved in one or other of these synaptic forms.

Our findings therefore indicate that in man at least, deficiences of germ cells arise even before pairing commences. In the XO mouse and possibly in a few oocytes in humans, subsequent germ cell development and survival may depend on the ability of the X axis to form a non-homologous association either with itself or with an autosome.

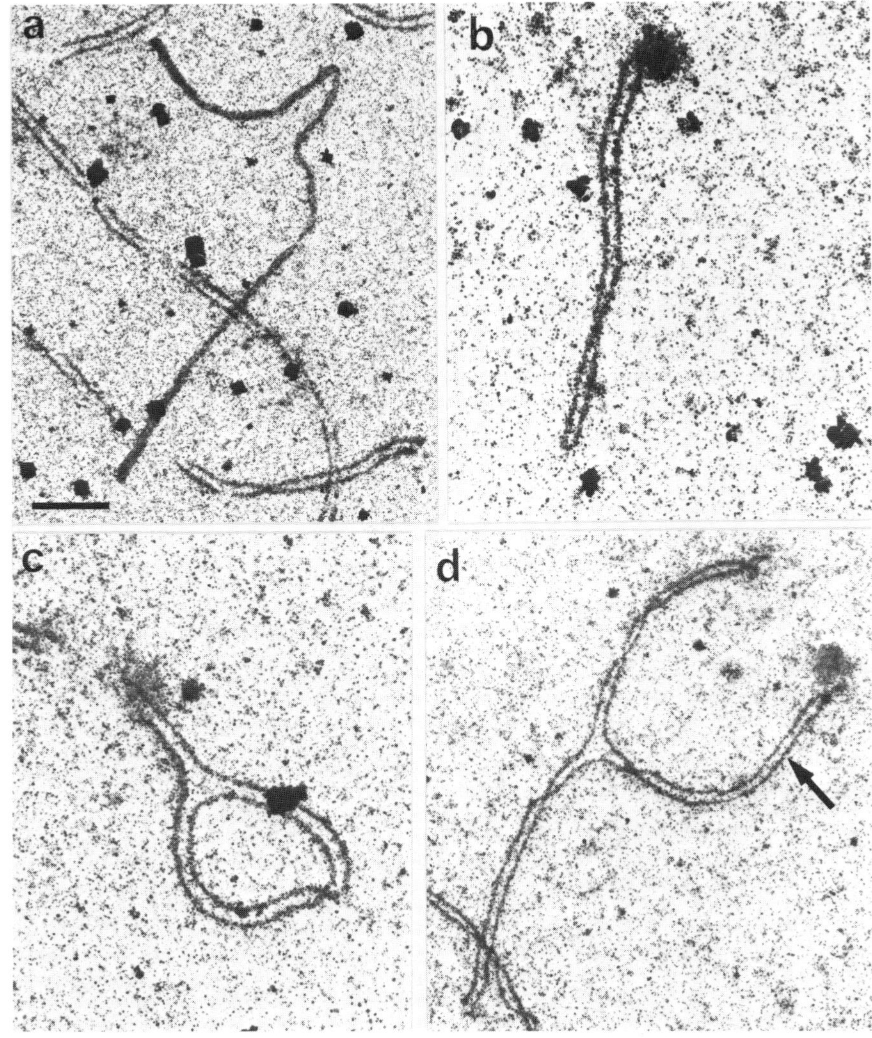

Figure 4. Synapsis in human and mouse XO foetal oocytes. **a** Human X chromosome remains single and thickened at pachytene. **b** Mouse X chromosome showing hairpin (self-synapsed) formation. **c** Mouse X chromosome in loop conformation. **d** Mouse X chromosome (arrow) partially paired with both axial elements of an autosomal SC, to form a triradial structure. Bar = 1 μm. (**b–d** reproduced by permission of Chromosoma.)

References

Baker, T.G. 1963. A quantitative and cytological study of germ cells in human ovaries. *Proc. Roy. Soc. B.* 158, 417–433.

Beaumont, H.M. and A.M. Mandl 1962. A quantitative and cytological study of oogonia and oocytes in the foetal and neonatal rat. *Proc. Roy. Soc. B.* 155, 557–579.

Bojko, M. 1983. Human meiosis VIII. Chromosome pairing and formation of the synaptonemal complex in oocytes. *Carlsberg Res. Commun.* 48, 457–483.

Bond, D.J. and A.C. Chandley 1983. *Aneuploidy.* Oxford Monographs on Medical Genetics No. 11. Oxford University Press, Oxford.

Burgoyne, P.S. and T.G. Baker 1984. Meiotic pairing and gametogenic failure. In *Controlling events in meiosis*, C.W. Evans and H.G. Dickinson, eds, Company of Biologists, Cambridge, 349–362.

Carr, D.H., R.A. Haggar and A.G. Hart 1968. Germ cells in the ovaries of XO female infants. *Am. J. Clin. Pathol.* 49, 521–526.

Cattanach, B.M. 1962. XO Mice. *Genet. Res.* 3, 487–490.

Counce, S.J. and G.F. Meyer 1973. Differentiation of the synaptonemal complex and the kinetochore in *Locusta* spermatocytes studied by whole mount electron microscopy. *Chromosoma* 44, 231–253.

Darlington, C.D. 1965. *Cytology.*, 3rd ed. J. and A. Churchill, London.

Gillies, C.B. 1981. Electron microscopy of spread maize pachytene synaptonemal complexes. *Chromosoma* 83, 575–591.

Howell, W.M. and D.A. Black 1980. Controlled silver-staining of nucleolus organiser regions with a protective colloidal developer: a 1-step method. *Experientia* 36, 1014–1015.

Lyon, M.F. and S.G. Hawker 1973. Reproductive life span in irradiated and unirradiated XO mice. *Genet. Res.* 21, 185–194.

Miklos, G.L.G. 1974. Sex chromosome pairing and male fertility. *Cytogenet. Cell Genet.* 13, 558–577.

Moses, M.J. 1977a. Synaptonemal complex karyotyping in spermatocytes of the Chinese hamster (*Cricetulus griseus*). I. Morphology of the autosomal complement in spread preparations. *Chromosoma* 60, 99–125.

Moses, M.J. 1977b. Microspreading and the synaptonemal complex in cytogenetic studies. In, *Chromosomes Today*, vol. 6, A. de la Chapelle and M. Sorsa, eds, 71–82, Elsevier/North Holland, Biomedical Press, Amsterdam.

Moses, M.J. and P.A. Poorman 1981. Synaptonemal complex analysis of mouse chromosomal rearrangements. II. Synaptic adjustment in tandem duplications. *Chromosoma* 81, 519–535.

Moses, M.J., P.A. Poorman, P.A. Polani, J.A. Crolla and F. Moir 1982. Synapsis, synaptic adjustment and DNA synthesis in meiotic prophase of mouse oocytes. *J. Cell Biol.* 95, 77a.

Sharp, P.J. 1985. Synaptic adjustment at a C-band heterozygosity. *Cytogenet. Cell Genet.* 41, 56–57.

Solari, A.J. 1980. Synaptonemal complexes and associated structures in microspread human spermatocytes. *Chromosoma* 81, 315–337.

Speed, R.M. 1982. Meiosis in the foetal mouse ovary. I. An analysis at the light microscope level using surface spreading. *Chromosoma* 85. 427–437.

Speed, R.M. 1984. Meiotic configurations in female trisomy 21 foetuses. *Hum. Genet.* 66, 176–180.

Speed, R.M. 1986a. Prophase pairing in a mosaic 18p-;iso18q human female foetus studied by surface spreading. *Hum. Genet.* 72, 256–259.

Speed, R.M. 1986b. Oocyte development in XO foetuses of man and mouse: The possible role of heterologous X-chromosome pairing in germ cell survival. *Chromosoma* 94, 115–124.

Speed, R.M. 1987. The possible role of meiotic pairing anomalies in the atresia of human foetal oocytes. *Hum. Genet.* (in press).

Sybenga, J. 1975. *Meiotic configurations.* Springer, Berlin, Heidelberg, New York.

Meiotic roles of nodule structures in zygotene and pachytene nuclei of Angiosperms

G.H. Jones and S.M. Albini

Department of Genetics, University of Birmingham, Birmingham B15 2TT, UK

Nodule structures associated with synaptonemal complexes (SCs) were first thoroughly investigated and described by Carpenter (1975, 1979a) who also proposed a role for them as organelles mediating genetical recombination, based on correlations between their frequencies and distributions and those of genetical crossovers in *Drosophila melanogaster*. These structures have been termed recombination nodules. Equivalent structures, or their close counterparts such as transverse bars have since been reported in a wide variety of species and their involvement in meiotic recombination has been confirmed in many of these cases (reviewed by Carpenter 1979b, 1984; von Wettstein *et al.* 1984). Further investigation of *Drosophila* oocytes led to the recognition of two classes of nodules (termed spherical and ellipsoid) with different chronologies of appearance, frequencies and distributions which suggested that only spherical nodules were associated with reciprocal crossover exchanges while ellipsoid nodules, which occur relatively earlier in pachytene, may be associated with gene conversion events. Investigations of several other species have confirmed that nodules present during early prophase I, chiefly late zygotene to early pachytene, (early nodules) show characteristics differing from the later crossover-associated nodules (late nodules). Early nodules are more numerous than late nodules, although the excess of early nodules as well as their morphology and precise chronology varies from species to species. In most species investigated there is a 1.5 to 2-fold excess of early nodules (von Wettstein *et al.* 1984; Carpenter 1987) but in the tomato the excess of early nodules is much greater, about 15-fold (Stack & Anderson 1986). Early nodules have also been reported to be distributed more randomly than late nodules. Because they show characteristics expected of gene conversion, it has been proposed that early nodules also mediate recombination events, including crossovers and gene conversions, or only gene conversions (Carpenter 1987). In the former case the marked reduction in nodule numbers during prophase I would represent selective loss of the nodules not associated with reciprocal crossovers.

The observation of nodules is apparently independent of the preparative method, provided that an appropriate stain is applied. Nodules have been regularly and consistently reported both from sectioned and surface-spread nuclei. Surface spreading has a considerable advantage for the quantitative analysis of nodules, particularly in terms of the recording of nodule frequencies and distributions from large numbers of bivalents and nuclei and there have been several such analyses of nodules from surface-spread animal meiocytes (see Table 1 in von Wettstein *et al.* 1984). Surface spreading of plant meiocytes offers the same advantages but has taken longer to develop because of the technical obstacle of the greatly thickened pollen mother cell (pmc) walls of plants. However, a number of spreading methods have now been devised for plant pmcs which, in various ways, overcome the cell wall problem (Gillies 1981; Stack 1982; Albini *et al.* 1984). Several recent studies of plant SCs have been based on surface-spreading but relatively few of these have included observations of nodules. The principal reason for this has been the frequent use of silver staining which gives strongly contrasted SCs and axial cores (ACs) and is therefore ideal for producing SC karyotypes and analysing pairing in normal and abnormal situations, but does not reveal nodules. In the present study, surface spreading combined with phosphotungstic acid (PTA) staining to reveal nodules was applied to a number of Angiosperm plant species with the aim of clarifying the relationship between early and late nodules and investigating their relationships to chiasmata (see Albini *et al.* 1984 and Albini & Jones in press for technical details). The main investigation concentrated on a pair of closely related *Allium* species, *A. cepa* and *A. fistulosum* which have very similar chromosome complements but strikingly different chiasma distributions and which therefore constitute a sensitive test system for probing the relationship of nodules to chiasmata. In addition some more fragmentary observations on three other plant species, *Crepis capillaris, Hordeum vulgare* and *Secale cereale,* are presented.

Pachytene nodules

Pachytene nuclei from both *Allium* species typically have eight fully paired SCs, each having a prominent centromere (Fig. 1). Following staining with PTA, spread SCs show low numbers of densely staining ellipsoid or spherical nodules (Figs 2 and 3) which conform in size, shape, staining reaction and location with structures termed recombination nodules (RNs) in other species (Carpenter 1979b). As in other species, these nodules are intimately associated with the central region of the SC, although they may occasionally lie to one side of the SC or overlie one of the lateral elements. A comparison of pachytene nodule numbers and chiasmata (Table 1) shows that both species exhibit fewer nodules than chiasmata, but the deficit is much more pronounced in *A. cepa.* The mean nodule frequency of *A. fistulosum* is 1.34 per SC while the corresponding chiasma frequency is 1.92 per bivalent. In *A. cepa,* on the other hand, the mean bivalent chiasma frequency is higher than in *A. fistulosum* at 2.38, but the mean nodule frequency per SC is only 0.77. While some species show close correspondences between RN and chiasma or crossover frequencies, deficits of pachytene RNs

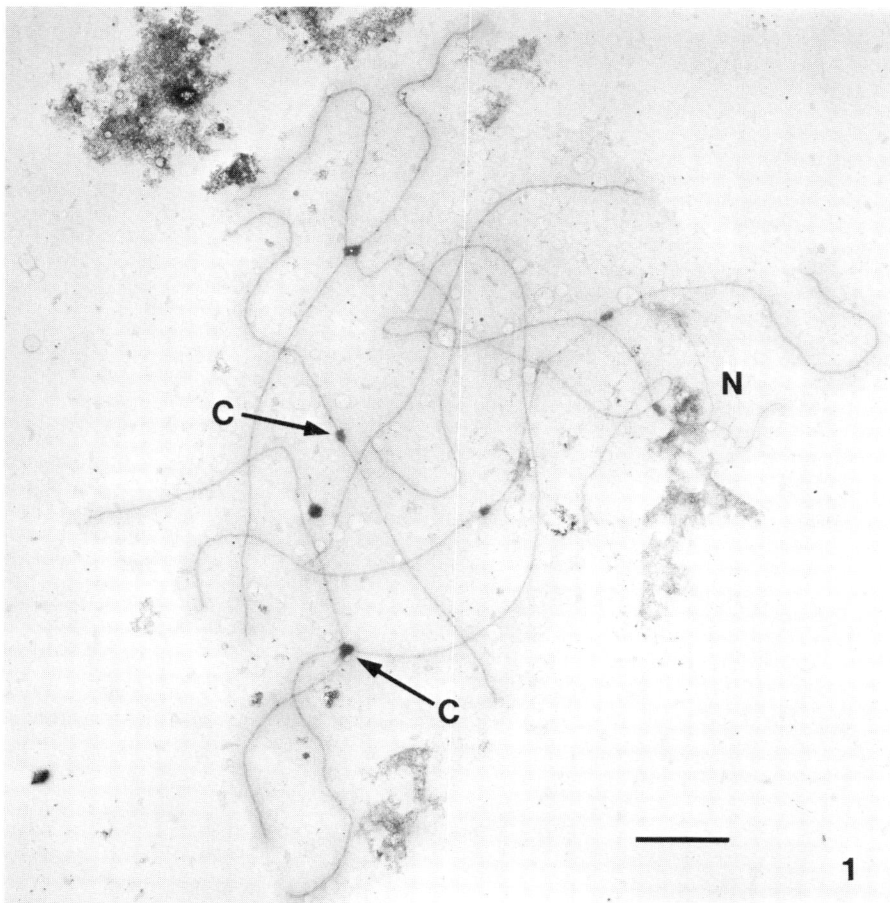

Figure 1. An entire surface-spread PTA-stained pachytene nucleus of *A. fistulosum*. The low magnification prevents the resolution of recombination nodules. C = centromeres, N = nucleolus organising region (unpaired). Bar = 10 μm.

have also been reported (e.g. Gillies 1983; Bernelot-Moens & Moens 1986). These deficits could be attributable to technical losses of nodules during the spreading procedure, but in the *Allium* species investigated the uniformity of the deficits over bivalents, cells, individual plants and accessions suggests a more fundamental cause. Furthermore, similar deficits have been reported from sectioned material in which technical loss is not expected. A plausible and widely quoted alternative explanation for nodule deficits is that not all crossovers occur simultaneously and consequently nodules associated with crossovers are transient or ephemeral structures (Carpenter 1979a; Gillies 1983). The much

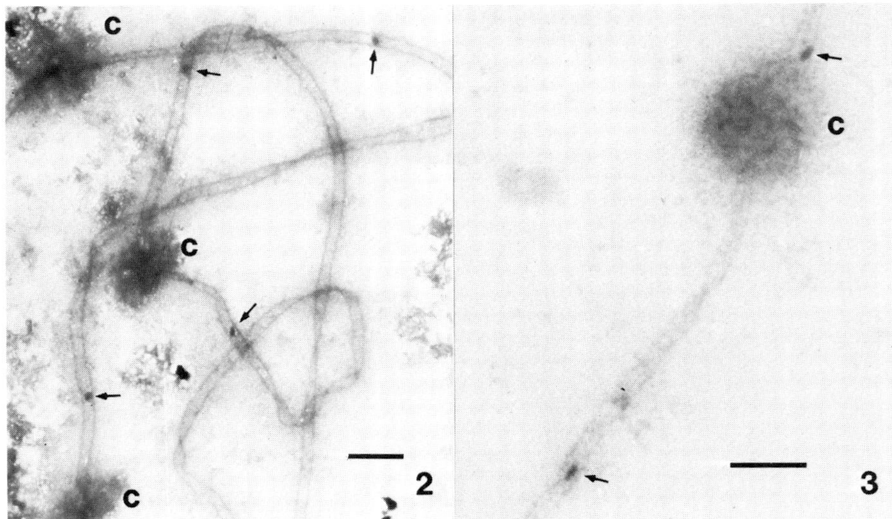

Figures 2 and 3. High power electron micrographs of pachytene SCs from *A. fistulosum* showing the proximal localisation of late nodules. C = centromeres; small arrows indicate nodules. Bars = 1 μm.

greater deficit of pachytene nodules in *A. cepa* argues that nodules are more transient in this species than in *A. fistulosum*.

Table 1. A comparison of mean chiasma frequencies at metaphase I and the mean frequencies of pachytene nodules in A. fistulosum and A. cepa.

	Mean chiasma frequency		Mean nodule frequency	
	per cell	per bivalent	per cell	per SC
A. fistulosum	15.35	1.92	11.3	1.34*
A. cepa	19.0	2.38	6.0	0.77*

*Nodule frequencies per SC combine data from entire nuclei and individual SCs from partially analysable nuclei.

The positional distribution of pachytene nodules along SCs was carefully measured from enlarged electron micrographs using a computerised digitiser, and the results are briefly summarised in Table 2. The two species of *Allium* show a dramatic difference in nodule distribution at pachytene which correlates remarkably closely with their chiasma distributions. In *A. fistulosum* more than

Table 2. A comparison of chiasma and pachytene nodule distributions of <u>A. cepa</u> and <u>A. fistulosum</u>.

		Proximal	Interstitial	Distal
A. cepa	Chiasmata	2.11%	45.00%	52.89%
	Nodules	8.20%	39.55%	52.10%
			Non-Proximal	
A. fistulosum	Chiasmata	92.58%	7.42%	
	Nodules	93.79%	6.21%	

90% of nodules occur in the proximal third of SC arms, that is near the centromeres, in close agreement with the chiasma distribution of this species (also see Figs 2 and 3). Occasional interstitial or distal nodules also occur and similarly some chiasmata are seen in these bivalent regions at metaphase I. *A. cepa*, on the other hand, shows few proximal nodules, most being located interstitially and distally, again in close agreement with chiasma distribution. These comparisons, involving as they do very different chiasma distributions, provide striking evidence that late (in this case pachytene) nodules mediate reciprocal crossover exchanges leading to chiasmata. We have also observed low numbers (0–3 per SC) of relatively localised pachytene nodules in other Angiosperm species including *Secale cereale, Hordeum vulgare* and *Crepis capillaris*. Although these other cases have not been studied quantitatively, the distributions of nodules are in general similar to the chiasma distributions in these species.

Zygotene nodules

A striking feature of PTA-stained zygotene nuclei of both *A. cepa* and *A. fistulosum* is the large numbers of nodules that they possess. These nodules are morphologically similar to pachytene nodules, but they differ markedly in their frequencies and distributions. They are present throughout zygotene and occur in association with unpaired but aligned axial cores (ACs) and with paired SC stretches. In aligned regions, nodules appear between ACs, particularly where ACs converge at intercalary association sites, and also at other places where the ACs have not converged closely although they are closely aligned (see Fig. 4). The nodules persist in association with the SCs of paired regions (Fig. 4) for the

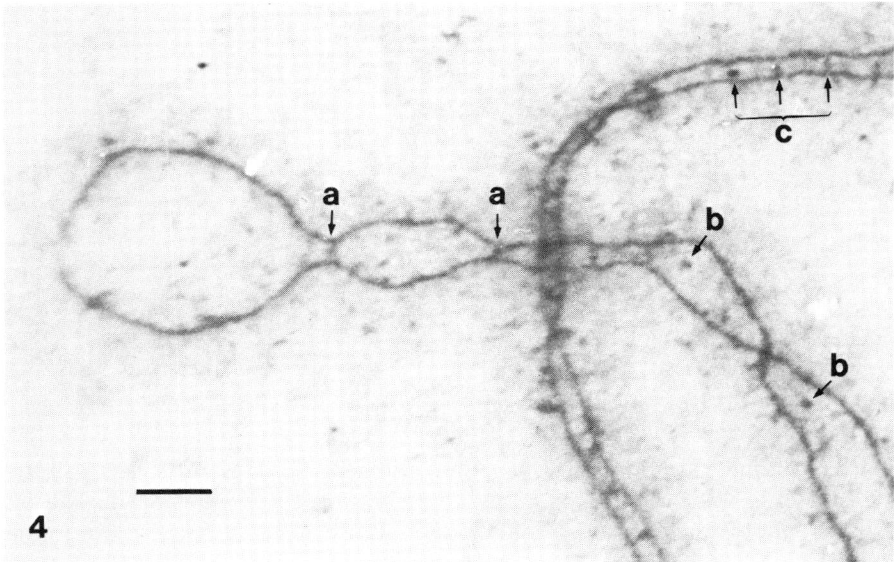

Figure 4. Part of a PTA-stained zygotene spread *A. cepa* showing early nodules at association sites (a), between aligned but not converging axial cores (b) and associated with SC (c). Bar = 1 µm.

duration of zygotene, but they disappear abruptly at or near the end of zygotene. The estimated total numbers of zygotene nodules are very high, about 80 per bivalent in both *Allium* species, and they are evenly distributed, about 1–2 µm apart throughout most SC regions. Relatively large numbers of zygotene nodules have also been observed in *Crepis capillaris* and *Secale cereale*.

Meiotic roles of nodules

There seems to be little doubt from the evidence so far accumulated that late nodules associated with SCs from early or mid-pachytene through late pachytene (chronology varies with species) are directly involved in reciprocal recombination (reviewed by Carpenter 1984). The limited observations hitherto made on Angiosperm plant species in general support this contention. In rye (Abirached-Darmency *et al.* 1983), maize (Gillies 1983) and tomato (Stack & Anderson 1960) late nodules show some degree of distal localisation along SCs which correlates with the localisation of chiasmata in these species. The distributions of pachytene nodules in *A. cepa* and *A. fistulosum*, presented here, closely parallel the contrasting chiasma distributions of these two species and this comparison thus provides further convincing support for the recombinational role of late nodules.

Early prophase I stages of Angiosperms, in common with many animals and fungi (von Wettstein *et al.* 1984), characteristically show high numbers of nodules. In rye (Abirached-Darmency *et al.* 1983), wheat (Hobolth 1981) and tomato (Stack & Anderson 1986) early nodules persist from zygotene to early pachytene, while in potato, a *Tradescantia* species and two species of *Allium* (Stack & Anderson 1986; Albini & Jones 1987) they are apparently confined to zygotene. In comparison with the late nodules discussed above, the meiotic roles of these early nodules are more conjectural. Their postulated roles in gene conversion events and/or in reciprocal recombination (see Introduction) has in general been assumed, and to some extent supported, by previous plant studies (Hobolth 1981; Stack & Anderson 1986). However, the characteristics of early nodules in *Allium* species has prompted a reappraisal of their meiotic role. Their apparent restriction to zygotene, their appearance between closely aligned ACs and in particular their specific locations at association sites where homologous ACs converge to within SC dimensions all suggest strongly, albeit circumstantially, that early nodules play a direct role in mediating homologous pairing.

Although AC convergences (association sites) and zygotene nodules associated with SCs are typically evenly spaced (Albini & Jones 1987) it is at present unknown whether their distributions are "phased", that is they occur repeatedly at specific sites in different nuclei, or whether they are unphased and therefore, in a population of cells, their distribution is random with respect to genes and DNA sequences. The mechanism by which they mediate pairing is also unknown at present, but interestingly Carpenter (1987) has independently suggested that early nodules may mediate searches for DNA homology during meiotic chromosome pairing. According to this hypothesis, homology-checking contacts between chromosomes are checked by comparisons of DNA sequences (heteroduplex DNA) and most simple gene conversion events originate as by-products of this process.

The relationship of early nodules to late nodules remains unresolved. It is possible that early nodules relate to total recombination events, reciprocal and non-reciprocal, and that late nodules are consequently a selected sub-set of early nodules. Alternatively it may be that early and late nodules are different structures with different origins, functions and effects, and their morphological similarities may be superficial and misleading.

References

Abirached-Darmency, M., D. Zickler and Y. Cauderon 1983. Synaptonemal complex and recombination nodules in rye (*Secale cereale*). *Chromosoma* 88, 299–306.

Albini, S.M. and G.H. Jones 1987. Synaptonemal complex spreading in *Allium cepa* and *A. fistulosum* I. The initiation and sequence of pairing. *Chromosoma* 95, 324–338.

Albini, S.M., G.H. Jones and B.M.N. Wallace 1984. A method for preparing two-dimensional surface-spreads of synaptonemal complexes from plant meiocytes for light and electron microscopy. *Exp. Cell Res.* 152, 280–285.

Bernelot-Moens, C. and P.B. Moens 1986. Recombination nodules and chiasma localization in two Orthoptera. *Chromosoma* 93, 220–226.

Carpenter, A.T.C. 1975. Electron microscopy of meiosis in *Drosophila melanogaster* females. II. The recombination nodule — a recombination-associated structure at pachytene? *Proc. Nat. Acad. Sci. USA* 72, 3186–3189.

Carpenter, A.T.C. 1979a. Synaptonemal complex and recombination nodules in wild-type *Drosophila melanogaster* females. *Genetics* 92, 511–541.

Carpenter, A.T.C. 1979b. Recombination nodules and synaptonemal complex in recombination-defective females of *Drosophila melanogaster*. *Chromosoma* 75, 259–292.

Carpenter, A.T.C. 1984. Recombination nodules and the mechanism of crossing-over in *Drosophila*. *Symp. Soc. Exp. Biol.* 38, 233–243.

Carpenter, A.T.C. 1987. Gene conversion, recombination nodules and the initiation of meiotic synapsis. *Bioessays* 6, 232–236.

Gillies, C.B. 1981. Electron microscopy of spread maize pachytene synaptonemal complexes. *Chromosoma* 83, 575–591.

Gillies, C.B. 1983. Ultrastructural studies of the association of homologous and non-homologous parts of chromosomes in the mid-prophase of meiosis in *Zea mays*. *Maydica* 28, 265–287.

Hobolth, P. 1981. Chromosome pairing in allohexaploid wheat var. Chinese Spring. Transformation of multivalents into bivalents, a mechanism for exclusive bivalent formation. *Carlsberg Res. Commun.* 46, 129–173.

Stack, S. 1982. Two-dimensional spreads of synaptonemal complexes from solanaceous plants. I. The technique. *Stain Technol.* 57, 265–272.

Stack, S. and L. Anderson 1986. Two-dimensional spreads of synaptonemal complexes from solanaceous plants. III. Recombination nodules and crossing-over in *Lycopersicon esculentum* (tomato). *Chromosoma* 94, 253–258.

von Wettstein, D., S.W. Rasmussen and P.B. Holm 1984. The synaptonemal complex in genetic segregation. *Ann. Rev. Genet.* 18, 331–413.

Genetic control of synaptonemal complex formation in the Ascomycete *Sordaria macrospora*

D. Zickler, A.D. Huynh, P.J.F. Moreau and G. Leblon

Laboratoire de Génétique, Bât.400, Université de Paris-Sud, 91405 Orsay, France

Crossing-over is fundamental in the meiotic process: mutants that decrease recombination and consequently the number of chiasmata, either directly by reducing the number of exchange events or indirectly by decreasing the amount of pairing, affect proper chromosome disjunction during the first division and lead to production of non-viable meiotic products (Baker *et al.* 1976). This suggested two screeens for the isolation of synaptonemal complex (SC) mutants: a systematic investigation of mutants with increased nondisjunction frequencies and the detection of mutants with defects during the first meiotic division. The analysis of those mutants raised questions about control of meiotic progression and SC assembly.

Isolation of SC mutants in *Sordaria*

The filamentous Ascomycete *Sordaria macrospora* has been chosen since it possesses a number of advantages for studying basic genetic mechanisms and cytology: a) its homothallism (each ascospore is self-fertile) insures the recovery of recessive mutations in the first generation of fructification; b) the four products of meiosis are retained in a single cell and the nuclei resulting from meiosis each undergo one mitosis. Consequently, the linear array of ascospores formed round each of the eight nuclei reflects the order and constitution of the eight DNA strands of the homologous chromosomes; c) with the use of mutants affecting ascospore colour, recombination events (crossing-over and conversion) are thus precisely followed; d) each linkage group has been assigned to its respective chromosome (Zickler *et al.* 1984) and a large number of translocations (Leblon *et al.* 1986) can be used to correlate the physical and genetic maps; e) the seven chromosomes have different lengths and are large enough to allow light microscopic analyses; f) the small size of the nuclei facilitates electron microscopy of serially sectioned nuclei and g) the transformation system is now well-developed.

Meiotic mutants were isolated after ethyl methanesulfonate (EMS) or UV treatment of protoplasts followed by mycelium regeneration and identification of autofructifications that showed reduced or aborted sporulation.

Phenotype of the SC mutants

Among several mutant genes affecting meiosis in *Sordaria*, nine are involved in the pairing and/or recombination processes and show abnormal SC structure. All mutants are recessive and genes mapped on different chromosomes.

Mutant spo 44 exhibits drastic effects on recombination (67% decrease) and irregular segregation of the almost unpaired homologues. Lateral elements (LE) assemble at leptotene, but SCs are rare in spite of the presence of dense material laid down between the chromosomes. Light microscopic observations confirm that the irregular length and number of LEs are due to extensive chromosome breakage (Zickler *et al.* 1985).

Mutant spo 76. In this mutant an early centromere cleavage is followed by a meiotic arrest after anaphase I. The LEs are either normal, especially when engaged in a piece of SC, or are split into two thin elements often widely separated and discontinuous. *spo 76* is also altered in U.V. and X-ray repair and no progeny is formed in the homozygote. In heterozygous crosses, a 50% decrease of the recombination frequencies is observed despite apparently normal SC assembly and pairing (Moreau *et al.* 1985).

Asynaptic mutants. Mutant *spo 77* produces aneuploid ascospores as a consequence of abnormal disjunction. It shows rare SCs and displays modified LEs. Significantly, it is able to suppress the meiotic block of *spo 76* (the double mutant *spo 76*–spo 77 shows a *spo 77* phenotype). Such epistatic effect provided a powerful approach for the screening of other pairing mutants. Among 37 revertants (selected from six U.V. mutageneses on *spo 76)* 34 extragenic suppressors were isolated, which partially restore the sporulation of *spo 76*. They belong to seven genes called *asy 1* to *asy 7* (*asy*naptic). Besides *spo 77*, two other asynaptic mutants, isolated from mutageneses performed on wild-type, are alleles of an *asy* gene (Huynh *et al.* 1986).

All suppressors show a normal centromere cleavage, but only four form bivalents: three (*asy 5–7, asy 1–10, asy 2–19*) with low frequencies and one (*asy 2–17*) with a wild-type frequency. The other suppressor mutants are asynaptic and display modified LEs with thickenings and splittings distributed along all chromosomes. Only *asy 2–17* exhibits normal SCs, but reduced recombination frequencies. One suppressor gene (*asy 4*) is involved in repair functions.

What information about SC assembly can be derived from the study of the SC mutants?

1) Test of epistasis and determination of pathways involved in SC formation
The products of at least nine genes are necessary for wild-type levels of pairing, SC assembly and recombination in *Sordaria*. The isolation of several alleles for most genes indicates that this control system is almost saturated (Huynh *et al.* 1986). Sporulation phenotypes (frequency of viable ascospores) of the *asy* single and double mutant strains have been screened in order to investigate both the

number of independent sequential pathways of events involved during meiotic pairing and to set the order of function of the genes acting on the same pathway. The double-mutants are constructed by the association of an allele of one *asy* gene with an allele of one of the other *asy* genes.

Tests of epistasis have allowed the provisional grouping of the mutations into three pathway structures:

Firstly, in the associations involving mutants of *asy 2, asy 3, asy 5* amd *asy 6* genes, each double-mutant exhibits a sporulation phenotype similar to the phenotype of one parent. For example, if the double-mutant *asy 3–asy 6* sporulates like the single mutant *asy 3*, the *asy 3* mutation is considered as epistatic to *asy 6* and consequently the gene *asy 3* acts in the same pathway but earlier than *asy 6*. It could thus be demonstrated that the four genes function in the same pathway. The method also yields information about the order of gene function when the genes lie on a dependent pathway. i.e. by comparison of all double and single mutants the following order of genes was found: *asy 3, asy 6, asy 5* and *asy 2.*

Secondly, the mutations *asy 1* and *asy 4* associated with any other *asy* mutation express an increase in sporulation deficiency if compared to the phenotype of each single mutant strain. Such failure to show epistasis demonstrates that they cannot be on the previously-defined pathway. Furthermore, the lack of epistasis between mutations from *asy 1* and *asy 4* shows that these two genes also act on different pathways.

Finally the epistatic effect of any *asy* mutation on *spo 76* suggests that all *asy* genes act before *spo 76* in the meiotic process.

2) SC assembly and pairing. The observation that all mutants, and especially all asynaptic mutants, assemble lateral elements during leptotene suggests that such assembly is a characteristic feature of meiotic induced chromosomes.

The 34 asynaptic mutants substantiate the hypothesis that chromosome recognition and SC formation are separated events (see Stern & Hotta 1977 for discussion): some of the *asy* mutants maintain the random distribution of their lateral elements and telomeres (normally observed only during early and mid-leptotene) during the whole prophase I. The other *asy* mutants show a telomere redistribution leading to the characteristic alignment of the homologues seen in the wild-type nuclei just before initiation of pairing by the SC's central elements. Such difference is therefore thought to provide evidence for the existence of a precocious recognition control distinct from the assembly of the central elements and thus not related to the SC functions.

These observations confirm and extend the earlier model proposed for *spo 44* (Zickler *et al.* 1985). In *spo 44* the absence of the roughly homologous alignment at the end of leptotene is followed by an assembly of dense material between the unpaired LEs which can be considered as a precursor of the central elements. The absence of similar material in the wild-type strains reinforces the conclusion that, at least in *Sordaria,* homologous alignment is a prerequisite to SC assembly.

Moreover, the split and/or double LEs of *spo 76* and the asynaptic mutants can be interpreted as being consistent with the suggestion that both sister-chromatids may be involved in the LE assembly, i.e. the two sister-chromatids form two thin elements which cohere together in a thicker LE.

3) SC and meiotic completion. Strikingly, in all the asynaptic mutants, despite an absence of pairing and recombination, meiosis is completed. This demonstrates that although pairing and reciprocal exchanges play essential roles in chromosome disjunction, they are not necessary steps for meiotic completion (see also Baker *et al.* 1976 for review). However, it is worth considering that the meiotic block of *spo 76* indicates that, during meiotic prophase, sister-chromatid association and centromere integrity are required for meiotic completion.

Correlation between split lateral elements and precocious centromeric separation (*spo 76*) supports the proposition (reviewed by Maguire 1982) that the SC contributes to the maintenance of the reinforced sister-chromatid cohesiveness found during the first meiotic division.

Meiotic mutants also provide probes to evaluate the role of SCs in the recombination process

In many species including *Sordaria* (see von Wettstein *et al.* 1984; Carpenter 1987 for review) a good correlation was found between the number and distribution of SC nodules determined by three-dimensional reconstruction of prophase nuclei and the number and distribution of crossover events and/or chiasmata. In *Sordaria,* while most of the meiotic mutants with increased general nondisjunction have their earliest observable defect in pairing, three mutants *spo 44, asy 2-17* and *spo 76* have their presumably primary defect in recombination.

The gene *spo 44* controls recombination over the whole genome. In homozygous mutant crosses 90% of the ascospores abort, but 2% of the asci each contain eight viable ascospores. The frequency of exchange for an interval located on chromosome 4 was estimated to be 10% when random spores were analysed, 21% in the rare asci with eight viable spores and 44% in the wild-type control crosses. Therefore, a decrease a recombination was observed even when correct disjunction occurred. A similar decrease was also found on three other chromosomes. Such a general recombination defect suggests that the *spo 44* wild-type gene has a major function in the exchange events. In this mutant the relationships with recombination nodules are difficult to study because of the low percentage of asci with eight viable spores and as expected almost all reconstructed nuclei showed disturbed chromosome pairing (Zickler *et al.* 1985).

In *asy 2-17* sporulation is almost normal and SCs are steadily assembled during prophase. In this mutant the threefold decrease of crossing-over frequencies corresponds to an almost twofold decrease in the number of recombination nodules. Furthermore, part of the nodules are morphologically abnormal.

No progeny can be analysed in the blocked homozygous *spo 76*, but in heterozygous crosses a 50% decrease of crossing-over frequencies is observed in all chromosomes despite normal SC assembly and regular sporulation. The number and distribution of the SC nodules are indistinguishable from controls. These observations are consistent with the assumption that *asy 2–17* is defective in preconditions for exchange whereas *spo 76* corresponds to a mutant defective in the exchange process itself. *spo 76* probably defines a central event in the meiotic recombination process of *Sordaria*. Besides the recombination defect, *spo 76* is also UV and X-ray sensitive like most *rec* mutants. Taking into account the alterations in LEs discussed above, it is suggested that *spo 76* controls a recombination function related to the spatial organisation of the chromatids.

References

Baker, B.S., A.T.C. Carpenter, M.S. Esposito, R.E. Esposito and L. Sandler 1976. The genetic control of meiosis. *Ann. Rev. Genet.* 10, 53–134.

Carpenter, A.T.C. 1987. Gene conversion, recombination nodules and the initiation of meiotic synapsis. *Bio-Essays* 6, 232–236.

Huynh, A.D., G. Leblon and D. Zickler 1986. Indirect intergenic suppression of a radiosensitive mutant of *Sordaria macrospora* defective in sister-chromatid cohesiveness. *Curr. Genet.* 10, 345–355.

Leblon, G., D. Zickler and Z. Lebilcot 1986. Most UV induced reciprocal translocations in *Sordaria macrospora* occur in or near centromere regions. *Genetics* 112, 183–204.

Maguire, M.P. 1982. Evidence for a role of the synaptonemal complex in provision for normal chromosome disjunction at meiosis II in maize. *Chromosoma* 84, 675–686.

Moreau, P.J.F., D. Zickler and G. Leblon 1985. One class of mutants with disturbed centromere cleavage and chromosome pairing in *Sordaria macrospora*. *Mol. and Gen. Genet.* 198, 189–197.

Stern, H. and Y. Hotta 1977. Biochemistry of meiosis. *Phil. Trans. Roy. Soc. Lond. B. Biol. Sci.* 277, 277–294.

Wettstein von, D., S.W. Rasmussen and P.B. Holm 1984. The synaptonemal complex in genetic segregation. *Ann. Rev. Genet.* 18, 331–413.

Zickler, D., G. Leblon, V. Haedens, A. Collard and P. Thuriaux 1984. Linkage group-chromosome correlations in *Sordaria macrospora*: chromosome identification by three dimensional reconstruction of their synaptonemal complex. *Curr. Genet.* 8, 57–67.

Zickler, D., L. de Lares, P.J.F. Moreau and G. Leblon 1985. Defective pairing and synaptonemal complex formation in a *Sordaria* mutant (*spo 44*) with a translocated segment of the nucleolar organizer. *Chromosoma* 92, 37–47.

Synaptonemal complex immunocytochemistry

P.B. Moens* and C. Heyting†

*Department of Biology, York University, Downsview, Ontario, Canada M3J 1P3
†Institute of Human Genetics, University of Amsterdam, The Netherlands

The immunocytology of synaptonemal complexes has developed only over the last 7 years. Progress has initially been slow because of conceptual and technological difficulties. Recent advances in the isolation of SCs and in the assays for SC purification and for anti-SC antibodies, however, promise rapid progress in the near future. This brief note surveys the results published to date and some of the work in progress.

Naturally-occurring anti-SC antibodies

Synaptonemal complexes (SC) are unexpectedly reactive with naturally-occurring antibodies. Sera from normal humans, from humans with anti-nuclear antibodies and from pre-immunisation mice and rabbits, frequently give immunofluorescence of SCs. It appears that proteins of the SC share epitopes with nuclear and cytoplasmic proteins (Dresser 1987). This common cross-reactivity has caused some experimental and interpretational difficulties. The predicted involvement of actin and myosin in the movement of axial cores and SCs appeared to be borne out by the observed SC immunofluorescence by anti-myosin and anti-actin antibodies (de Martino *et al.* 1980). However, subsequent tests for actin and myosin in the SCs did not provide unambiguous support for those observations (Heyting *et al.* 1984, Spyropoulos & Moens 1984). In retrospect it appears likely that the rabbit serum contained anti-SC antibodies prior to immunisation with gizzard smooth muscle myosin and that the actin-positive serum from a patient with hepatitis and with antinuclear antibodies actually stained centromeres (Dresser 1987).

The serum of humans with the autoimmune CREST syndrome contains anti-centromere antibodies (Fig. 1) and in some cases also stains the SCs (Dresser 1987, Moens *et al.* 1987). An interesting possibility, among several, is that the CREST serum detects suppressed centromeric sites along the SC. A source of SC staining by anti-nuclear antibodies of an autoimmune mouse was analysed by the production of monoclonal antibodies from that mouse (Moses *et al.* 1984). The technique in general is a useful one for the isolation of anti-SC antibodies from among a mixture of antibodies in autoimmune mice or mice immunised with unpurified SCs.

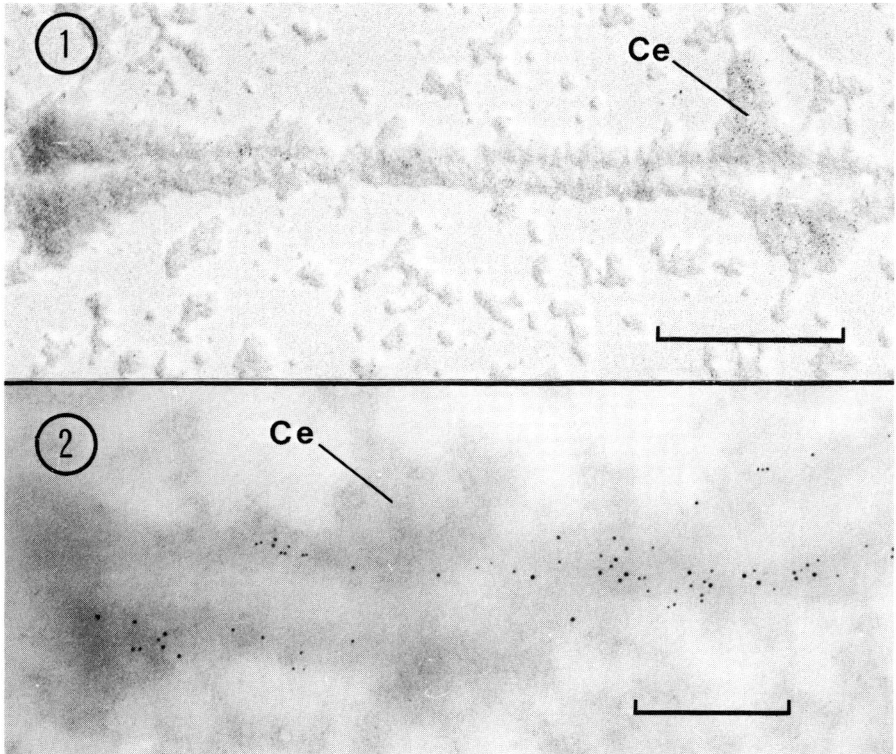

Figure 1. Rat SC treated with human CREST serum that has anti-centromere antibodies, stained with 5 nm gold-conjugated anti-human-IgG antibody, and shadow-cast (gold, platinum, 7 degrees). The centromeric region, Ce, is heavily gold-labelled and a low level of label is present along the SC and in the background. Bar = 0.5 μm. **Figure 2.** Rat SC treated with mouse anti-SC mAb II52F10 and with gold-conjugated goat antimouse IgG antibody. The SC and gold grains (5nm) are enhanced by 1% OsO_4 after fixation. The surface antigens of the lateral elements are gold-labelled. Where one lateral element overlies the other, only the upper one is labelled. The lower one is not accessible with this procedure. The centromere, Ce, has no grains. Bar = 0.25 μm.

Figure 3. Early diplotene rat SC treated with mAb II52F10, immunogold-stained, and shadow-cast. **3a,** Low magnification to illustrate the synapsed and desynapsed portions of the SC. Bar = 1 μm. **3b,** A higher magnification shows the continued presence of the antigen in the desynapsed lateral elements. Bar = 0.5 μm. **Figure 4.** The anti-SC mAb III15B8 mainly recognises an antigen on the inside of the pairing face of the lateral element. Immunogold (5 nm) and shadow-cast. Bar = 0.25 μm.

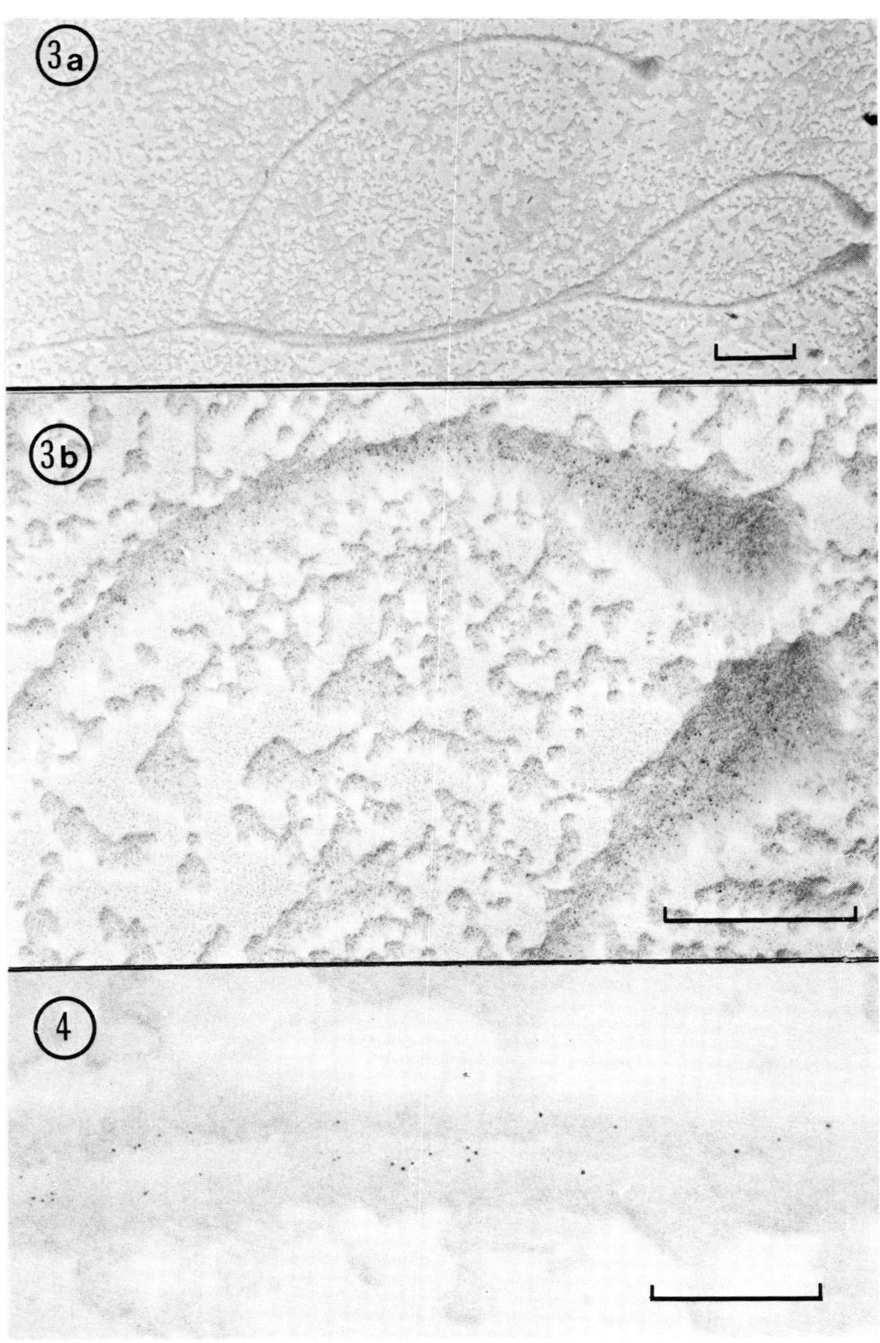

Nonspecific induced anti-SC antibodies

If purified SC antigens are not available, SC-containing material can be used as an inoculum. The monoclonal antibody SC4, induced by whole testis homogenate, labels SCs as well as cytoplasmic filaments (Dresser 1987). Purification of spermatocytes by velocity centrifugation at unit gravity, by centrifugal elutriation, and by equilibrium centrifugation (Meistrich *et al.* 1973, Meistrich 1977), can be used to increase the concentration of SCs. Ierardi, Moss and Bellve (1983) further improved the concentration by the isolation of spermatocyte nuclei and the preparation of nuclear matrices. Although these preparations do not yield free SCs and have complex electrophoretic patterns they have been used for the production of anti-SC mAbs (A.R. Bellve, personal communication). At present the search for anti-yeast-SC antibodies is similarly constrained by the lack of yeast SC purification procedures (M.E. Dresser, N. Kleckner, personal communication). Polyclonal antibodies with anti-yeast-SC activity have been obtained, however, by the use of sporulating yeast cells.

Antibodies generated against isolated SCs

Rat SCs isolated according to the techniques developed by Heyting *et al.* (1985) are morphologically well preserved and contain lateral elements, the central element, and attachment plaques. These isolates produce two major electrophoretic bands at 30 and 33 kD (Heyting *et al.* 1987) and these dominant polypeptides, which are located in the lateral elements of the SC (Heyting *et al.* 1987), generate most of the mAbs against rat SCs (Heyting, unpublished) as determined by immune staining of Western blots. Immunogold staining of spread spermatocytes with mAb II52F10 labels the axial cores and the lateral elements of the SCs throughout meiotic prophase (Figs 2, 3) (Heyting *et al.* 1987, Moens *et al.* 1987). An exceptional mAb, III15B8, does not react with the 30–33 kD bands and it recognises an antigen on the pairing face of the lateral elements when and where they are paired (Fig. 4).

SC antibody production and applications

To promote the detection of stage-specific antigens it may be possible to prepare spermatocyte fractions enriched for specific prophase stages (R.J. Dettmers, personal communication) or to eliminate, in vivo, given developmental stages (A.J.J. Dietrich, personal communication). The production of antibodies against minority proteins may be promoted by inoculation with one or more electrophoretic bands from purified SCs. By selective digestion of isolated SCs the inoculum may be enriched in specific SC structures such as attachment plaques, centromeric regions or recombination nodules.

The SC is associated with a number of functions: DNA binding, synapsis, recombination, and disjunction. It is expected that antibodies will be raised against the proteins responsible for these functions and that their properties will be recognised. Thus, it is feasible to test SC polypeptides on Western blots for their DNA binding properties. The available antibodies can be used to screen cDNA libraries, or in the case of yeast, genomic libraries for clones that produce antigens recognised by anti-SC antibodies. In yeast, with the appropriate genetic manipulation, the function of SC genes and proteins may be assayed more conveniently than in mammals.

References

De Martino, C., E. Capanna, M.R. Nicotra, and P.G. Natali 1980. Immunochemical localization of contractile proteins in mammalian meiotic chromosomes. *Cell Tissue Res.* 213: 159–178.

Dresser, M.E. 1987. The synaptonemal complex and meiosis: An immunocytochemical approach. In *Meiosis.* P.B. Moens, ed. Academic Press, New York. 245–274.

Heyting, C., A.J.J. Dietrich, F. Koperdraad and E.J.W. Redeker 1984. Do synaptonemal complexes contain actin and myosin? In *Chromosomes Today* 8. M.D. Bennett, A. Gropp, and U. Wolf, eds. Allen and Unwin, 316.

Heyting, C., A.J.J. Dietrich, E.J.W. Redeker, and A.C.G. Vink 1985. Structure and composition of synaptonemal complexes, isolated from rat spermatocyes. *Eur. J. Cell Biol.* 36: 307–314.

Heyting, C., P.B. Moens, W. van Raamsdonk, A.J.J. Dietrich, A.C.G. Vink, and E.J.W. Redeker 1987. Identification of two major components of the lateral elements of synaptonemal complexes of the rat. *Eur. J. Cell Biol.* 43: 148–154.

Ierardi, L.A., S.B. Moss, and A.R. Bellve. 1983. Synaptonemal complexes are integral components of the isolated mouse spermatocyte nuclear matrix. *J. Cell Biol.* 43: 148–154.

Meistrich, M.L. 1977. Separation of spermatogenic cells and nuclei from rodent testes. In *Methods in Cell Biology* 15. D.M. Prescott, ed. Academic Press, 15–54.

Meistrich, M.L., W.R. Bruce, and Y. Clermont 1973. Cellular composition of fractions of mouse testis cells following velocity sedimentation separation. *Exp. Cell Res.* 79: 213–227.

Moens, P.B., C. Heyting, A.J.J. Dietrich, W. van Raamsdonk, and Q. Chen 1987. Synaptonemal complex antigen location and conservation. *J. Cell Biol.* 105: 93–103.

Spyropoulos, B., and P.B. Moens 1984. The synaptonemal complex: Does it have contractile proteins? *Can. J. Genet. Cytol.* 26: 776–781.

Index